Environmental Radioactivity

SECOND EDITION

ENVIRONMENTAL SCIENCES

An Interdisciplinary Monograph Series

EDITORS

DOUGLAS H. K. LEE
National Institute of
Environmental Health Sciences
Research Triangle Park
North Carolina

E. WENDELL HEWSON
Department of
Atmospheric Science
Oregon State University
Corvallis, Oregon

DANIEL OKUN
Department of Environmental
Sciences and Engineering
University of North Carolina
Chapel Hill, North Carolina

ENVIRONMENTAL RADIOACTIVITY

SECOND EDITION

Merril Eisenbud

New York University Medical Center
Institute of Environmental Medicine
New York, New York

ACADEMIC PRESS   New York and London   1973

A Subsidiary of Harcourt Brace Jovanovich, Publishers

ACADEMIC PRESS, INC.
111 Fifth Avenue, New York, New York 10003

United Kingdom Edition published by
ACADEMIC PRESS, INC. (LONDON) LTD.
24/28 Oval Road, London NW1

LIBRARY OF CONGRESS CATALOG CARD NUMBER: 72-84367

PRINTED IN THE UNITED STATES OF AMERICA

Contents

Preface

The first edition of this book was published early in 1963 (McGraw-Hill, New York) at a time when worldwide concern existed because of radioactive fallout from testing nuclear weapons. In that year the principal nuclear powers agreed to ban open-air testing, and the environmental levels of radioactivity from that source have accordingly diminished considerably since then despite occasional tests by countries that did not participate in the test-ban agreement. However, interest in the subject of environmental radioactivity has not diminished correspondingly, but has been more than sustained by the rapid growth of the nuclear power industry and the use of radioactive materials in medicine, research, space exploration, and industry.

When the original edition was published, there were only four commercially operated power reactors in the United States, with an installed generating capacity of about 641 megawatts. At this writing, the generating capacity of power plants already in operation in the United States is about 13,400 megawatts and an additional 116,000 megawatts are in the planning stage. Nuclear power will represent an increasing fraction of the electrical generating capacity in many countries, and the potential environmental impact of this new industry is receiving wide attention.

As this second edition goes to press, the United States continues to be embroiled in public and political controversy over the safety of nuclear reactors and the significance of radioactive emissions to the environment. To the best of my ability, I have attempted to provide an objective account of the relevant biological and physical information that has been accumulated in the thirty years since control over the fission process was first achieved. Because this book is intended to be a coherent technical summary, I have dealt only incidentally with the status of the public controversy and the confusion that has been caused by the impact of extreme statements that have been widely publicized in recent years.

The book was written as a reference source for the scientist, engineer, or administrator with a professional interest in the subject, but it may also be of value to the reader who wishes to understand the technical facts behind the public debate. Hopefully, the book will contribute to public understanding of what is fundamentally a complex subject.

The subject of environmental radioactivity is one of vast dimensions, and I cannot claim that this book is an exhaustive treatise on the subject. However, it does represent what I hope is a useful review of what is now known, though perhaps presented with some degree of imbalance due to the fact that my experience is primarily with atomic energy programs in the United States.

I am indebted in many ways to my present and past associates and students for the many thousands of hours of dialogue and research that have made it possible to produce the two editions of this book. My students in particular have proved to be my most effective teachers.

In addition to those to whom I acknowledged assistance in the preparation of the first edition, I feel particularly obliged to mention the help given to me by McDonald E. Wrenn, Norman Cohen, Peter Freudenthal, and Steven Jinks in the preparation of this edition. My work was enormously facilitated also by the assistance of Eleanor Clemm, to whom I am greatly indebted.

MERRIL EISENBUD

Preface to First Edition

It is now more than half a century since the phenomenon of natural radioactivity was first discovered by Becquerel and the Curies, and about a quarter of a century since the discovery of nuclear fission by Hahn and Strassmann. Enormous amounts of energy that have been locked in the nucleus of the atom since the beginning of time can now be released at will, and the greatest challenge in his history confronts man as he decides whether to use this new source of energy to reach ever higher levels of social accomplishment or to destroy what mankind has created.

Whether the fission process is exploited in peace or in war, an unavoidable result will be the production of radionuclides in enormous amounts. If these by-products of the fission process should be produced in a major nuclear war, radioactivity will become one of the dominant features of the environment and one of many major obstacles to survival that will face those who have the good fortune to escape prompt death by blast, radiation, or fire. If, on the other hand, mankind does find a peaceful solution to its political problems, and our future civilian industries are supplied with nuclear power, the radionuclides will be accumulated in such great quantities as to require constant vigilance and great wisdom to prevent the effluents of the nuclear industry from contaminating the environment to a serious extent.

Ever since early in World War II, extensive research has been conducted to understand the physical and chemical properties of radioactive substances, the manner in which they are transported physically through the environment, and the way in which some of them enter into man's food supplies, the water he drinks, and the air he breathes. In 1959, when I accepted an opportunity to develop a graduate teaching program in the general field of radiological hygiene, I was impressed with the need to consolidate the vast amount of information that had been developed on the subject of environmental radioactivity in order that the subject could be presented to students and others in a comprehensive and systematic manner.

The subject of environmental radioactivity has aspects of vast dimensions, and the task of bringing together the pertinent information in so many diverse disciplines proved to be not without its difficulties. There was first the question of what to include, and in this regard I decided that the text should be concerned primarily with the behavior of radioactive substances when they enter the environment. The important and elaborate technology by which passage of radioactive materials to the environment may be prevented and the equally important field of health physics that is concerned with protecting the atomic energy worker were thus placed beyond the bounds of this work, although it has been necessary frequently to deal briefly with both subjects in the present text.

I am greatly indebted to my many associates, past and present, who assisted me in the preparation of this work. It is not possible to acknowledge all the assistance I have received, but certain of my colleagues have been particularly helpful in the review of early drafts of certain chapters. In this regard I am particularly indebted for the help of Norton Nelson, Roy E. Albert, Abraham S. Goldin, Bernard S. Pasternack, Gerard R. Laurer, Harold H. Rossi, and Ben Davidson. The index was prepared by Stephen F. Cleary, and every word of every one of the three to five drafts that ultimately evolved into the seventeen chapters of this book was typed with remarkable proficiency and patience by Patricia S. Richtmann. Finally, as is customary for reasons that can only be understood fully by authors, and authors' families, I wish to express my appreciation for the help and encouragement provided by my wife, Irma, to whom this book is dedicated.

MERRIL EISENBUD

Environmental
Radioactivity

SECOND EDITION

Chapter 1

Introduction

The midtwentieth century discovery of methods by which energy contained within the atomic nucleus can be released is certainly one of the major technological developments in the history of man. The most obvious long-range benefit from exploiting this discovery is that it provides mankind with a source of power that will in time become necessary to replace exhausted reserves of fossil fuels (Hubbert, 1971). Although these reserves may be adequate for the immediate future in most parts of the world, there are some areas in which the cost of fossil fuels is already so high as to have made nuclear power economically feasible since 1963. In time, as the energy requirements of the world increase and as the reserves of fossil fuel become smaller, nuclear energy will play an increasingly important role in our civilian economy. This will be discussed in more detail when we consider the subject of nuclear reactors in Chapter 9.

Copious quantities of radionuclides are a less conspicuous but nevertheless important contribution of the atomic energy industry to mankind. In the fields of medical and biological research, the use of "radioactive isotopes" is now so commonplace that we are no longer conscious of the research that has become feasible because these useful substances have been made available as by-products of the atomic energy industry. The use of isotopes as a research tool has entered rapidly and yet with comparative unobtrusiveness into our research laboratories, and hundreds of discoveries in the biomedical sciences would not have been possible if it were not for the ready availability of radionuclides.

Regrettably, there is a negative side to nuclear energy—the possibility of nuclear war. As with many great technological advances, man possesses the wisdom to use his new knowledge constructively if he wishes to do so, but he is also capable of great folly. Only time will tell if, on balance, nuclear energy has been used to bring blessings to mankind or to hasten his social destruction.

The Early History of Radioactivity

Experience with radioactive materials preceded by many years the discovery of the phenomenon of radioactivity. As will be seen in Chapter 2, the atmospheres of mines in Central Europe that had been exploited for their heavy metals since medieval times were so radioactive that the miners developed a fatal lung disease which was finally diagnosed as lung cancer (Lorenz, 1944) in this century.

Some radioactive substances were used even before it was known that they were radioactive. The Welsbach gas mantle, which was developed in 1885, utilized the incandescent properties of thorium–cerium oxide to greatly increase the luminosity of gaslight in many parts of the world, and uranium oxide has long been used to provide a vivid orange color in ceramic glazes, a practice that continues up to the present time. Other oxides of uranium and thorium have also been used as ceramic glazes and for tinting glass.

Since the turn of the century and continuing up to the present time in some parts of the world (see Chapter 7), natural radioactivity has been exploited for its supposed benefit to health. There is no fully satisfactory explanation as to how this custom originated, but it is known that the popularity of mineral waters in spas around the world led to their establishment as health resorts as long ago as Roman times. It is also known that the laxative properties of spring waters having high mineral content were highly prized, and extensive resorts grew up at places such as Saratoga Springs in New York State and the spas of Europe, Japan, and South America. After the phenomenon of radioactivity was discovered, tests of the mineral waters showed some of them to contain abnormally high concentrations of natural radioelements, and it is this fact that may have given rise to the idea that this new and mysterious property of matter was the reason why the mineral waters had alleged beneficial effects on health. The discovery that radiations from radium could destroy cancerous tissue possibly abetted development of the fad. The spas of the world soon began to advertise the radioactivity of their waters, and to this day the labels on bottled mineral waters in many countries contain measurements of the radioactivity of the spring from which the water was obtained. We will see later that the radioactive sands of Brazilian beaches and the high radon concentrations in the air of old mines in Austria and

Fig. 1-1. See following page for legend.

Fig. 1-1. Early uses of radioactivity in health fads. (a) The "Radiumator": Air was pumped by hand bulb through a small radium source, entraining radon which was bubbled through the glass of drinking water. (b) The Revigator (patented 1912): The cone is a mildly radioactive "ore" which was placed in the drinking water crock. (c) Radioactive compress used for miscellaneous aches and pains. Contained 0.1 mg ^{226}Ra, certified by the Radium Institute of the Faculté des Sciences de Paris.

the United States still attract tourists who believe that exposure to natural radioactivity can cure arthritis, general debility, and a variety of other diseases (Lewis, 1955; Scheminzky, 1961).

In the early 1920's and continuing up to about 1940, radioactive substances, particularly radium and radon, found a place in the medical faddism of the time. Radium was injected intravenously for a variety of ills, but far from being cured, many of the patients later developed bone cancer. Devices sold for home use, bubbled radon into drinking water, and radioactive poultices were prescribed for arthritic wrists and shoulders (Fig. 1-1).

Finally, a word must be said about the early use of radium in luminous paints. Thousands of timepieces were painted with radium-bearing paints during and immediately after World War I, with no precautions to protect the employees. Many cases of aplastic anemia and osteogenic sarcoma developed among the factory workers engaged in applying the paint prior to about 1935, when hygienic practices were developed which proved practical and prevented further injuries from occurring. We will see in Chapter 2 that the information derived from studies of the early radium cases has contributed in a unique and effective way to the excellent safety record of the modern atomic energy program.

The Atomic Energy Industry

In the United States, where the atomic energy program is relatively large compared to the programs of other countries which have made the extent of their efforts in the atomic energy field a matter of public record, the budget for the Atomic Energy Commission in the fiscal year 1971 was about 2 billion dollars, and an additional several hundred million dollars were spent by private industry.

It is estimated that in 1969 the atomic energy program in the United States employed about 150,000 persons (U.S. Atomic Energy Commission, 1970a) distributed throughout a great network of governmental and private industrial mines, laboratories, and industrial plants. This introductory chapter provides a convenient place in which to take a bird's-eye view of this industry, the major components of which will be discussed in greater detail in Chapters 8–12.

One way to visualize the main parts of the atomic energy industry is by means of the flow diagram in Fig. 1-2. After being mined and concentrated, the uranium is shipped to refineries for conversion to a uranium metal or oxide of sufficient purity to be used in reactors (Chapter 8). The refinery products may be shipped directly to fuel-element fabrication plants or the uranium may be converted to UF_4, a green salt, and be transported to gaseous-diffusion plants for conversion to UF_6. The latter is a volatile compound for the gaseous-diffusion process by which uranium is enriched in the isotope ^{235}U. Isotopic enrichment takes place in enormous plants at Oak

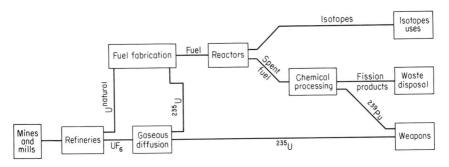

Fig. 1-2. The principal steps in the processing and use of nuclear fuel.

Ridge, Tennessee; Paducah, Kentucky; and Portsmouth, Ohio. Depending on the purpose for which the uranium is intended, the degree of enrichment may vary from a few tenths of 1% to over 90%.

The uranium, either in natural or enriched form or as the metal, oxide, or other compounds or alloys, is then transported to the fuel-element fabrication plants. The exact shape to be taken by the uranium and the manner in which it will be clad with protective cladding of stainless steel, zirconium, or various alloys depend on the needs of the reactor designer.

The fabricated fuel elements are then shipped to reactors (Chapter 9), some of which are used for production of ^{239}Pu. Other fuel elements are fed into power reactors operated by the civilian power industry, the Navy, or various other branches of the government. In addition, some of the fuel will be shipped to research reactors located in laboratories and industrial plants both in the United States and abroad.

The products of these reactors may be heat, plutonium, radioactive isotopes, or radiations for research or industrial purposes. The spent fuel is transported from the reactors to fuel-reprocessing plants in which the fuel is dissolved and the unused uranium and ^{239}Pu are recovered (Chapter 11). The fission products will usually be processed into a form convenient for waste storage and disposal (Chapter 12), but in some cases radioactive isotopes may be separated from these fission products for research, medical, or industrial applications (Chapter 10).

Not all the enriched uranium from the gaseous-diffusion plants goes to nuclear reactors. A portion, together with the plutonium produced by the spent-fuel-reprocessing plants, will be fabricated into components for nuclear weapons. However, the plutonium recovered from the spent fuel of civilian power reactors is not used for weapons but is stockpiled as a future reactor fuel.

In addition to the radioactive by-products produced by the spent-fuel-reprocessing plants, certain isotopes will be produced in research reactors by neutron irradiation. For example, the naturally occurring stable isotope ^{59}Co may be placed in a reactor to produce the widely used radioactive isotope ^{60}Co. Many of the more common radioactive isotopes such as ^{14}C. ^{131}I, and ^{32}P are produced by neutron bombardment of the appropriate parent nuclide.

Some Introductory Comments on the Study of Environmental Radioactivity

In this text we are concerned with the fate of radioactive materials that in one manner or another enter into the general environment. We will be concerned only in an incidental way with the important means by which contamination of the environment with radioactive substances can be reduced or eliminated entirely by proper engineering practices. Our objective in the following chapters will be to achieve an understanding of the manner in which the radioactive materials behave and their ecological effects when they have escaped into the environment, whether by accident or design.

To accomplish this, it will be necessary to deal to an exceptional degree with a variety of scientific disciplines, including knowledge provided by biologists, soil chemists, hydrologists, meteorologists, oceanographers, and others to predict the effects of radioactivity and the manner in which radioactive substances are transported physically and biologically when they enter the environment.

Human beings may be exposed to environmental radioactivity in a number of ways (Table 1-1). When a γ-ray-emitting substance is deposited on surrounding surfaces, the γ-radiation background will be increased and the human body will be irradiated accordingly. If the contaminant is initially airborne, the radioactive particles may be inhaled or may deposit on the skin where, if they are of sufficiently intense β activity, they can produce skin burns. Finally, and perhaps of greatest importance from the point of view of the long-range hazards of the atomic energy industry, the radioactive isotopes may enter food chains and be ingested by man. The effects of exposure to ionizing radiation will be discussed in more detail in Chapter 2.

The manner in which exposures may occur is summarized in Table 1-1 which also enumerates some of the factors which must be considered in any quantitative description of the mechanisms by which human exposure occurs.

These factors are numerous, and Table 1-1 includes only the most important ones. These, together with many of lesser importance, will be discussed in more detail in the appropriate chapters, but it is important to recognize here that man lives in an environment which is not fully understood and which is crisscrossed with physical and chemical pathways linking all living things with the biophysical and biochemical processes on which all organisms depend for their existence.

The study of the interrelationships of organisms or groups of organisms to their environment is known as ecology (Odum, 1963). This fascinating science deals with the biological aspects of the environment in their entirety and with the delicate balance of the numerous factors that comprise the ecosystem.

Throughout man's long history he has been altering his environment in many ways. Some of these changes have been beneficial and have brought

TABLE 1-1

SOME OF THE PRINCIPAL ROUTES OF HUMAN EXPOSURE TO RADIOACTIVE SUBSTANCES

Type of exposure	Steps in passing from environment to man				Factors that influence dose to man
	1	2	3	4	
Inhalation	Release of radioactive particles to atmosphere	Inhalation of airborne particles	—	—	I. Rate of release to atmosphere, duration of release, degree of atmospheric stability, extent of precipitation, wind direction, particle size, chemical and physical form of the radioactive material
Direct γ radiation	Release of γ-emitting solids, liquid, or atmospheric wastes	Direct irradiation from passing cloud	Deposition on surrounding surfaces	—	II. All of above plus characteristics of surfaces (ease of weathering, amount of shielding inherent in configuration of surface structures)
Direct β radiation	Release of airborne β-emitting particles	Deposition on bare skin	—	—	III. All factors in I

Ingestion				
Release of gaseous wastes	Deposition directly on crops	Man eats crops	—	IV. All factors in I plus stage of growing season, dietary habits of human population, agricultural practices, pH of soil
	Uptake by crops from contaminated soil	Animals eat crops	Man eats animals or their dairy products	
Release of liquid wastes	Man drinks water	—	Man eats seafood	V. Chemical properties of the radioactive isotopes, solubility, amount of dissolved minerals and suspended solids, motility of aquatic organisms, dietary habits of local populations, mechanisms of mixing, behavior of bottom sediments
	Concentration by lower aquatic organisms	Lower aquatic organisms eaten by higher forms	—	

about man's mastery of his environment and his ability to attain a mode of existence that is heavily dependent on the environmental changes which he has brought about. However, as is well known, some effects have not been beneficial: Beginning with the kitchen middens of prehistoric time, man has tended to use the environment in which he lives as an unrestricted receptacle for the waste products of his many activities.

The ecological effects of man's activities are of special interest to many physicians, industrial hygienists, sanitary engineers, and others who are concerned with the effects of man's technology on the public health. All these specialties are part of a more general field of specialization that may properly be called anthropecology.

Some of the more simple relationships of ecology to the subject of environmental radioactivity can be given here, and others will be discussed throughout the book. Thus, the hazard to human beings from radioactive iodine suspended in the atmosphere may be due not to inhalation of the dust or vapor but rather to the fact that this radioisotope tends to deposit on the forage consumed by dairy cows, and by this means becomes a contaminant of fresh milk (Chapter 5). Because a dairy cow grazes a large area in a day, it ingests relatively large amounts of radioiodine, and in this way it concentrates the radioiodine into a form in which it can be passed directly to human beings. A somewhat analogous experience has been encountered since early in World War II when atmospheric contamination by a compound of another halogen, fluorine, deposits on forage in the vicinity of aluminum reduction plants and is ingested by cattle in sufficient quantity to cause fluorosis.

Comparison between Radioactive and Chemical Environmental Contamination

Although it is possible to find many analogies, the problem of dealing with radioactive and chemical contaminants in the environment is in fact basically quite different. In order for many elements to be toxic, they must exist in certain molecular forms. Thus, the aromatic hydrocarbon benzol (C_6H_6) is highly toxic when inhaled, manifesting its effect on the blood-forming tissues. If the benzol is fully oxidized, its toxicity disappears and two new compounds are formed (CO_2 and H_2O) which possess relatively innocuous characteristics that are completely different from the toxicological properties of the original compound. Far from being toxic, both carbon and hydrogen (the elements forming benzol) are essential in all living forms.

The ring-shaped molecule C_6H_6 is toxic, not the elements forming the molecule. If, however, the original C_6H_6 were formed from the radioactive isotopes ^{14}C and ^3H, the oxidized products would continue to be radioactive. Changes in the chemical compounds in which radionuclides are incorporated cannot control their basic radiotoxic properties, except as the particular compounds behave differently in the ecologic or metabolic sense.

Many other differences can be identified. Thus, the element zinc may or may not be toxic, depending on not only the chemical form in which it exists but also, in the case of airborne contamination, on the physical state of the compound. The effect of inhaling zinc oxide is different from the effect of inhaling zinc carbonate, and the effects are also different depending on whether the zinc is inhaled as a freshly formed fume or as a dust (Drinker and Hatch, 1954). Moreover, in this case the potential toxicity exists only as long as the zinc is inhaled. Once the contaminant settles from the atmosphere, it becomes part of a pool of inert zinc that already exists in the environment, and although the amount of zinc in this pool may be increased slightly, it is out of the way for all practical purposes. However, if the zinc is in the form of a long-lived radioactive isotope, the pool of environmental zinc becomes contaminated, and the isotope will appear as a radioactive contaminant of living plants from the soil or of marine organisms taken from the sea.

The fact that the radioactive substances are toxic in such small amounts is often given as an additional example of a basic difference between chemically toxic and radiotoxic materials. For example, industrial hygienists permit individuals who are occupationally exposed to lead to breathe air containing as much as 100 μg Pb/cm^3, whereas according to the International Commission on Radiation Protection the maximum permissible concentration of radioactive ^{210}Pb in air is 4×10^{-4} $\mu Ci/m^3$ which is equivalent to only 5×10^{-6} $\mu g/m^3$.

While it is true that atom for atom the radiotoxicity greatly exceeds the chemical toxicity of any given substance, it is likewise true that the amounts of chemically toxic materials with which we are accustomed to deal with are greater by many orders of magnitude than the amounts of radioactive materials. Although the comparison of the maximum permissible concentrations of the radioactive and stable lead would suggest that on an atom-for-atom basis the radioactive isotope is 20 million times more toxic than its nonradioactive cousin, the fact remains that the amounts of ^{210}Pb available for use would rarely be greater than a few micrograms, whereas nonradioactive lead is available in industry in amounts measured in millions of tons. The far greater amounts of chemically toxic materials normally available in some ways offset the higher specific toxicity of the radioactive isotopes.

Attitudes toward Radioactive Contamination of the Environment

Prior to World War II, environmental radioactivity was a natural phenomenon concerning which a limited amount of information was already available, but there was practically no diffusion of this knowledge beyond the relatively few highly specialized laboratories that were then equipped to make measurements of natural radioactivity. The world inventory of radioactive materials was confined to those that occurred in nature, with the insignificant exception of a relatively few millicuries of artificial radioactivity that were produced in cyclotrons during the late 1930's.

During World War II, the construction of large, water-cooled, plutonium-producing reactors at Hanford and the associated operations for extracting the plutonium from the irradiated uranium resulted in the first extensive opportunities for contaminating the environment with radioactive substances. This was also true to a lesser extent at Oak Ridge. The Hanford studies on the behavior of various radionuclides in the environment are classics in the field and have served, on the one hand, to demonstrate the caution one must adopt in discharging radioactive substances to the environment and, on the other hand, the relatively enormous amount of radioactivity that can be discharged safely into the environment if the properties of the environment are well understood (Chapters 6, 9, and 10).

The policies laid down by the Manhattan Engineering District of the Corps of Engineers (Groves, 1962), which was the World War II military organization responsible for the atomic energy program, placed a high priority on the importance of operating in such a way as to keep environmental contamination at a minimum. When the Atomic Energy Commission, a civilian organization, succeeded the Manhattan Engineering District in 1946 (Hewlett and Anderson, 1962), the cautious policies toward release of radioactive materials to the environment were continued; that is, these policies were continued at least insofar as industrial and research types of activities were concerned. In contrast, starting in the late 1940's and continuing at an accelerated rate through the 1950's, there began a series of weapons tests, first in the United States and then in the Soviet Union, the United Kingdom, France, and China, that discharged into the environment the amounts of radioactivity that were large in relation to the prohibitions self-imposed by the AEC in the operation of its industrial plants. The radioisotopes produced in the nuclear explosions in various parts of the world soon permeated the atmosphere, the soils, and the food chains to such an extent that widespread apprehension began to develop, first in certain scientific circles and later among the general public throughout the world. Responding to this concern, the Congress of the United States held a number of hearings on the subject of fallout from weapons testing and also on radioactive waste-disposal practices. At about the same time the National Academy of Sciences in the United States and the Medical Research Council in Great Britain undertook to assess the state of knowledge on the effects of small doses of radioactivity (National Academy of Sciences—National Research Council, 1956; Medical Research Council, 1956), and the United Nations in 1955 appointed a committee, consisting of scientific representatives of 15 nations, to investigate the effects of radiation on man. This United Nations Scientific Committee on the Effects of Atomic Radiation has published a series of authoritative reports that are classics in international scientific collaboration (UNSCEAR, 1958, 1962, 1964, 1966, 1969). The widespread interest in the subject of environmental radioactivity resulted in an enormous acceleration of research in the behavior of trace substances in the environment. Many branches of the biological and physical sciences, including genetics, inorganic chemistry, trace-element metabolism, micrometeorology, upper-atmosphere meteorology, and oceanography, have made great forward strides during the past 2 decades because of the need for a better understanding of environmental radioactivity. Large well-equipped ecological laboratories were established at the major atomic energy production and research centers, and funds and equipment for ecological research were supplied by the government to individual investigators at many universities. The ecological concern that began to pervade the scientific community in the mid-1960's resulted mainly as a result of findings of the complex and subtle interrelationships among the various life forms and their physical environment.

Studies of the behavior of radioactive materials in the environment and their biological effects raised many difficult questions which at first seemed unique to the subject of environmental radioactivity. What are the ecological pathways by which these substances reach man? Do they accumulate in such a way that they can result in unforeseen ecological injury? Are there synergistic effects with other environmental pollutants? By the late 1960's, it was apparent that the same questions could be asked about insecticides, food additives, fossil fuel combustion products, trace metals, and other nonradioactive pollutants of the environment. In many respects, the pioneering studies of the environmental effects of radioactivity provided the tools by which more general problems of environmental pollution could be understood.

Chapter 2

The Biological Basis of Radiation Protection

This chapter is intended to provide sufficient general information about the biological effects of ionizing radiation to permit one to appreciate the relative scale of hazard associated with a given level of exposure and understand the basis on which levels of maximum permissible exposure are established. The reader who wishes to obtain a more comprehensive understanding of the subject is referred to a number of excellent reports which are more detailed in regard to the underlying radiobiological principles and clinical aspects of radiation injuries (Hollaender, 1954, 1955; UNSCEAR, 1966, 1969; National Academy of Sciences—National Research Council, 1956, 1961a–d, International Commission on Radiological Protection, 1969; Behrens, 1959; Saenger, 1963; Van Cleave, 1968; Casarett, 1968; Anderson, 1971).

Any general discussion of radiation effects should begin with certain dichotomies according to the following:

1. whether the source of radiation is external to the body (as in the case of exposure to medical X rays) or is an internally deposited radionuclide (as in the case of radioiodine in the thyroid).

2. whether the dose was from a relatively massive exposure delivered in

a short period of time (less than a few days) or was delivered in small bits over longer periods of time which may extend to many years.

3. whether the effects appear soon after exposure ("acute effects" or are delayed for months or years ("delayed effects").

It will prove helpful if these distinctions are kept in mind when the effects of radiation are being reviewed.

In contrast to many contemporary notions about our ignorance of the effects of ionizing radiation, it is not unfair to say that more is known about this subject than is known about the effects of any other of the many noxious agents which man has introduced into his environment. To a considerable degree, this is the result of the large amount of research that has been performed in this country and abroad since exploitation of nuclear fission began in 1942. The sums of money expended in this field by such agencies as the Atomic Energy Commission, the United States Public Health Service, and the United States Department of Defense have been vast in comparison with expenditures for studies of the effects of the many man-produced chemical pollutants of air, water, and food, whose effects are not yet fully understood.

As a matter of fact, much had been learned about the effects of radiation exposure prior to World War II, partly as a result of excellent research, but primarily as a by-product of tragedies that resulted from the general ignorance of radiation effects that existed early in this century.

Early Knowledge of Radiation Effects

Reports of radiation injury began to appear in the literature with astonishing rapidity after the announcement on November 8, 1895 of Roentgen's discovery of X rays. The very first volume of the American X-ray Journal, published in 1897, included a compilation (Scott, 1897) of 69 cases of X-ray injuries reported from laboratories and clinics in many countries of the world. The reason why so many cases of injury could have been reported in such a short period of time is related to the history of the discovery of X rays and the kind of research that had already been underway for many years. X-ray technology seems to have begun in 1859 when Plücker first recorded the fact that an apple green fluorescence was observable on the inner wall of a vacuum tube through which a current was flowing under high voltage (Grubbé, 1933). Plücker's work was followed by a number of experiments in other laboratories, and in 1875, Sir William Crookes made the first high vacuum tube which thereafter bore his name. Crookes discovered that the apple green fluorescence reported originally by Plücker originated from a discharge at the negative end of the tube and, thus, was inaugurated the term "cathode rays." The fact that cathode rays included a component of radiation that could pass through solid material was apparently discovered by Heinrich Hertz, who is principally remembered for his classic studies of the electromagnetic spectrum and his proof of the electromagnetic nature of light. According to Grubbé, Roentgen's principal contribution was the observation that he could see the shadow created by the bones of his hand when it was placed between a Crookes tube and a screen covered with fluorescent chemical. To this penetrating radiation he gave the name X rays.

The important point is that research with Crookes tubes was underway in many parts of the world, and that unknown to the investigators X rays were being generated by many of these tubes. Grubbé, a physician who was also a manufacturer of Crookes tubes, was studying the fluorescence of chemicals at the same time Roentgen was doing his classic experiment. When he heard of Roentgen's announcement, Grubbé began immediately to experiment with the newly named X rays and by January, 1896 developed an acute dermatitis which was followed by skin desquamation and eventual cancer that many years later necessitated amputation of his hand. The early stages of his injury was presented at a clinical conference at the Hahnemann Medical College in Chicago on January 27, 1896 at which a Dr. Gilman made the interesting observation that "any physical agent capable of doing so much damage to normal cells and tissues might offer possibilities, if used as a therapeutic agent, in the treatment of a pathological condition in which pronounced irritative blistering or even destructive effects might be desirable." Two days later, only 2 or 3 months after the publication of Roentgen's discovery, a patient was referred to Dr. Grubbé with the following note:

January 28, 1896

E. H. Grubbé
12 Pacific Avenue

Dear Sir:
This will introduce Mrs. Rose Lee who has carcinoma of the left breast.
She is willing to have you make X-ray applications.
I hope you can help her.

Very truly yours,
/s/R. Ludlum

The above interesting sequence of events was not reported by Dr. Grubbé until 1933 because he didn't realize the importance of the incident at the time and was under the impression that all of his documentation was lost in a subsequent fire. In his interesting recital of the facts in his 1933 paper, Grubbé made a claim that he was the first exposed person to be injured by X rays, that he was the first person to apply X rays to pathological lesions for therapeutic purposes, and, incidentally, that he was the first to use sheet lead for protection against X-ray effects. More recently, doubt has been cast on the authenticity of the 1933 Grubbé paper by Brecker and Brecker (1969) who found inconsistencies in the chronology as related by Grubbé. Although the question of priorities of radiological discoveries may never be fully satisfied in the historical sense, the 1933 Grubbé paper does, nevertheless, provide unique insight into the early history of radiation injury.

An indirect result of Roentgen's announcement was Bequerel's accidental discovery of radioactivity in the same year. By 1900 the natural radioactivity contained in uranium ore had been sufficiently concentrated so that the first skin burn was produced on the arm of a Mr. Giesel and shortly thereafter confirmed in an experiment by Pierre Curie using a radium extract he reported as having an activity 5000 times that of metallic uranium (Becquerel and Curie, 1901).

By March, 1902, the number of cases reported in the literature totaled at least 147 (Codman, 1902). Interestingly, his analysis of the diminishing numbers of cases being reported in the preceding few years caused Codman to comment, "the main reasons for such a decrease have been the bitter teaching of experience and the fact that the introduction of better apparatus has done away with long exposures and the close approximation of the tube." Codman's observation, though justified in the context of the times, has regrettably not been supported by history. The total number of people who have been injured and killed by the use and misuse of X ray and radium prior to the development of proper standards of radiation hygiene will probably never be known.

During and immediately following World War I, the use of radium in luminous paints was attended by hazards arising out of ignorance of the effects of this radioelement when ingested (Martland, 1951; Evans, 1966; Evans *et al.*, 1969; Finkel *et al.*, 1969). About 40 workers, mostly women, developed bone cancer and aplastic anemia from the practice of pointing the brushes with their lips. In addition, during the 1920's, radium was administered medically as a nostrum for a variety of ailments including arthritis, mental disease, and syphilis; additional deaths were caused by this practice before it was discontinued. More recently, from 1944 to 1951, a compound containing ^{224}Ra (half-life 3.6 days) was injected intravenously into about 2000 German patients with tuberculosis and other diseases

(Spiess and Mays, 1970), and by 1969 bone sarcomas had already appeared in 50 of these patients.

The dial-painting cases have been studied thoroughly by these several investigators, but the main credit belongs to Evans for having worked out the basic biophysical principles of radium injury in sufficient detail so that safe practices could be adopted. These practices proved effective not only for protection against radium but also against many of the later hazards of the atomic energy industry which utilized the information gained with radium to such excellent advantage.

It is frequently noted that in the first 40 years of this century about 2 lb of radium were extracted from the earth's crust and that at least 100 people died from various misuses of this material. In contrast, since 1942 the atomic energy programs have produced the radioactive equivalent of many tons of radium and, up to the present time, except for uranium mining, there have been no deaths that could be attributed to the internal deposition of the wide variety of new radioisotopes, some of which are even more toxic than radium. We shall return to this subject later in this chapter. It had long been known that the men who worked in mines in Eastern Europe (Hartung and Hesse, 1879) were prone to a fatal lung disease, but only in relatively recent times was it learned that the disease was bronchiogenic carcinoma. The mines had been operated for centuries as a source of ores of many metals and, from the beginning of the twentieth century, as a source of pitchblende. When it was realized that cancer could result from chronic irradiation by radionuclides deposited within the body, it was suggested that the high incidence of lung cancer among miners of radioactive ores might be explained by their exposure to radioactive substances in the atmospheres of the mines (Lorenz, 1944). Studies of the mine air revealed the presence of high concentrations of radon, and this radioactive gas came to be regarded by many as the etiological agent in the high incidence of lung cancer. Unfortunately, no additional epidemiological studies in these Eastern European mines have been published since the 1930's, which is regrettable because of the importance of this source of information and its bearing on safety in the uranium mines in the United States.

By 1942, when the wartime atomic energy program was initiated in the United States, the early epidemiological radiation experience had been thoroughly digested, and standards had been developed by 1941 which served well throughout World War II. Ironically, these standards were applied everywhere in the industry except in the uranium mining, where the European mining experience was not sufficiently well heeded. Recent epidemiological studies (Archer and Lundin, 1967; Lundin *et al.*, 1969, 1971) have shown that the pattern of lung cancer among uranium miners seen originally in Eastern Europe is now being repeated in the Southwest United States.

Human experience early in this century also provided the information that ionizing radiations in a sufficient dose could produce sterility, changes in the composition of peripheral blood, and, in the case of acute exposure, a complex of symptoms that came to be known as the acute radiation syndrome. However, not all this information was obtained from the misuse of ionizing radiations. By 1900, it was well known that cancerous tissue was particularly vulnerable to radiation injury, and first X rays and then radium were used in cancer treatment. This gave rise to the need to understand the effects of large doses on healthy as well as cancerous tissue in order to enable the radiologist to limit side effects during radiotherapeutic procedures.

Not all the early knowledge came from studies of effects on people. The early interest of the experimental biologist in the effects of ionizing radiation is illustrated by the fact that the ability of ionizing radiation to produce cataracts in the lenses of exposed animals was discovered in 1897 (Clapp, 1934) only 2 years after Roentgen announced the discovery of X rays. Muller observed in 1926 that mutations could be produced by exposure to X rays and thus opened an era of research on the genetic effects of radiation.

Thus, in one way or another, almost all the major effects of ionizing radiations on man were known prior to World War II. Moreover, the basic techniques for protecting workers were also known and have been utilized ever since with minor basic modifications.

Summary of Present Knowledge of Radiation Effects on Man

A great amount of research conducted since 1942 has been directed toward understanding the mechanisms of radiation injury and the ecological relationships that exist in an environment contaminated with radioactivity.

The effect of exposure to ionizing radiation may differ, depending on whether a given dose is delivered in a short or a long period of time. A sufficiently large acute exposure is capable of provoking a series of effects which are manifest shortly after exposure, but shielding of a relatively small volume of tissue decreases the severity of an otherwise total body exposure (Patt and Brues, 1954). Both acute and chronic exposures are capable of producing a variety of effects which may not appear for many years.

THE EFFECTS OF ACUTE WHOLE-BODY EXPOSURE

When a massive dose of radiation to the whole body is received instantaneously or when the bulk of exposure is received in the first few days, as in the case of the external radiation from fission-product fallout, the effects may be seen as early as the first day and will follow a course dependent on the size of the dose received. Table 2-1, in which the expected effects of acute exposure to external radiation are summarized, shows that relatively minor effects would occur at doses less than 100 rem, but that about 50% fatalities would be expected to occur in the range of 400 to 500 rem. As the whole-body dose approaches 1000 rem, the fatalities would reach 100%.

Significant clinical symptoms are not likely to be seen in individuals who have been exposed to less than 100 rem, but as the dose rises above this figure, vomiting and nausea occur in increasing frequency and will be seen in almost all exposures of about 300 rem.

Nausea and vomiting may be followed by a latent period of as much as 2 weeks when the dose is 100–250 rem, but less than 1 day when the dose is greater than 700 rem. The signs and symptoms which then develop usually include epilation, sore throat, hemorrhage, purpura, petechiae, and diarrhea.

Among the most striking changes are those which occur in the composition of blood. Injury to the blood-forming organs reduces the rate at which the blood elements are produced and has a dramatic effect on the composition of circulating blood (Wald et al., 1962).

At whole-body doses in excess of 1000 rem, the predominant symptoms may be due to injury to the gastrointestinal tract and central nervous system. Disorientation owing to central nervous system injury was a conspicuous feature of at least one case (Shipman, 1961) within a matter of minutes after accidental exposure to several thousand rads.

There has been little experience with acute effects due to inhalation or ingestion of radioactive substances. If highly radioactive substances of low solubility were ingested, acute injury to the gastrointestinal tract lining would occur. Exposure to heavy doses of radioactive iodine can cause thyroid injury, and doses of several thousand rem can result in ablation of the gland. It is possible that for most radionuclides, if generalized fission-product contamination of the environment were to take place, the external radiation dose would be so high as to be overwhelming relative to the dose from absorbed radionuclides. This is evidently not true for ^{131}I, and we will

TABLE 2-1

SUMMARY OF CLINICAL EFFECTS OF ACUTE IONIZING RADIATION DOSES[a]

	Dose (rem)					
	0–100	100–200	200–600	600–1000	1000–5000	Over 5000
Incidence of vomiting	None	100 rem: 5% 200 rem: 50%	300 rem: 100%	100%	100%	100%
Time of onset	—	3 hr	2 hr	1 hr	30 min	30 min
Principal affected organs	None	Hematopoietic tissue			Gastrointestinal tract	Central nervous system
Characteristic signs	None	Moderate leukopenia	Severe leukopenia; purpura; hemorrhage; infection; epilation above 300 rem		Diarrhea; fever; disturbance of electrolyte balance	Convulsions; tremor; ataxia; lethargy
Critical period postexposure	—	—	4 to 6 weeks		5 to 14 days	1 to 48 hr
Prognosis	Excellent	Excellent	Good	Guarded	Hopeless	Hopeless
Convalescent period	None	Several weeks	1 to 12 months	Long	—	—
Incidence of death	None	None	0 to 80% (variable)	80 to 100% (variable)	90 to 100%	90 to 100%
Death occurs within	—	—	2 months		2 weeks	2 days
Cause of death	—	—	Hemorrhage; infection		Circulatory collapse	Respiratory failure; brain edema

[a] From Glasstone (1962).

see in Chapter 16 that ^{131}I doses to the thyroid of 100 to 150 rem, super-imposed on whole-body doses of about 175 rem, during exposure to fallout from weapons testing has resulted in thyroid cancer.

The genetic effects of radiation on human populations are normally of concern only in relation to chronic exposure. Except in nuclear war, the number of people involved in episodes of acute radiation exposure would be so few relative to the total population that the genetic consequences of an acute radiation incident can normally be neglected. This is because the genetic effect is one which ultimately is seen in the population as a whole and is a function of the per capita gonadal dose received by the entire pro-creative population.

DELAYED EFFECTS

In addition to the effects that become apparent immediately following heavy irradiation, some of the consequences may not appear for many years. Some of the delayed effects, such as changes in the texture or pigmentation of hair, will be apparent relatively soon after the initial injury. Other effects such as cataracts and leukemia or other types of cancer may not appear for 5 or more years. Some of the delayed effects result from acute exposure, whereas others are of significance when the dose is delivered in repeated small exposures over a long period of time. The effects which become manifest in the exposed individual are referred to as *somatic effects* to differentiate them from *genetic effects* which are observed in the offspring of the exposed person and are the results of changes transmitted via the genetic mechanisms. Radiation injury to the developing fetus would thus be a somatic rather than genetic effect.

Somatic Effects

The ionizing radiations produce a wide variety of biological effects of which the more important ones are leukemia and other types of cancer, cataracts, and a possible tendency toward reduced life expectancy. The delayed effects of radiation in man and experimental animals have been extensively studied in experimental animals (Casarett, 1965; UNSCEAR, 1966, 1969), but experience with human exposure will be emphasized here.

One of the most important unanswered questions about the delayed somatic effects of radiation is whether a dose threshold exists, below which the effect does not occur, and whether the probability of developing leukemia, or another neoplastic disease, is proportional to dose and independent of dose rate. The arguments pro and con (Brues, 1958) will appear again and again in this and subsequent chapters.

Leukemia. Leukemia is a relatively rare disease which has been observed to occur in increased frequency among Japanese survivors at Hiroshima and Nagasaki (Ishimaru *et al.*, 1971), among children irradiated in infancy for thymic enlargement, among patients irradiated for ankylosing spondylitis, among physicians exposed in the practice of radiology (Cronkite, 1961), and possibly among children who were irradiated *in utero* in the course of pelvic examination during pregnancy (Stewart and Kneale, 1970; UNSCEAR, 1964). Of these groups, only the radiologists can be described as having been exposed to repeated small doses, in contrast to the other groups which were subjected to either single or slightly fractionated exposures.

The ratio of the incidence of leukemia among radiologists and physicians who are not radiologists was found to be 10.3 in the period 1929 to 1943, 6.7 from 1944 to 1948, and 3.6 from 1952 to 1955 (National Academy of Sciences–National Research Council, 1961c). A more recent study by Lewis (1963) in which he analyzed the cause of death of 425 radiologists who died during the 14-year period 1948–1961 showed that the average annual incidence of death from leukemia was 253/million people/year compared with an expected incidence of 85/million people/year, a ratio of about 3.0. Unfortunately, there are no useful estimates of the cumulative exposure of these groups. The downward trend in this ratio may be due in part to the fact that the radiologists have taken greater precautions in more recent years, but perhaps also because nonradiologists have been using ionizing radiations to an increasing extent.

Among the Japanese atom bomb survivors there was a sharp increase in the reported incidence of leukemia, reaching a peak in 1951 at which time the incidence of this disease was 11 times higher than in the nonexposed population. The cases occurred primarily among survivors within 1500 m of the hypocenter, indicating that in the range of dose between 100 and 900 rad the increased incidence of leukemia was proportional to dose at a rate of about 20 cases/million exposed people/rad.

Leukemia data gathered from studies of patients irradiated for ankylosing spondylitis indicate that in the dose range between 300 and 1500 rad the incidence of leukemia is about 0.5/million/year/rad. The lower incidence compared to the Japanese survivors may be due to the fact that in these cases the dose was delivered in several courses of treatment and that perhaps only one-third to one-half of the bone marrow was irradiated (UNSCEAR, 1964).

The incidence of leukemia among children who received X radiation for thymic enlargement has been studied by several investigators and reviewed by Pifer *et al.* (1963). Only in the series reported by Hemplemann was the incidence of leukemia found to be elevated. The thyroid receives a heavy dose during thymus irradiation, and some of the children developed thyroid cancer. Although a few excess cases of leukemia were observed in the Hemplemann series, the data do not permit statistical analysis with respect to the relationship between dose and effect.

There have been several studies of the incidence of leukemia in children who were irradiated *in utero* with conflicting evidence that the fetus may be more sensitive to the leukemogenic effects of radiation.

Stewart and Kneale (1970) and MacMahon (1962) reported an apparent increase in the incidence of leukemia and other neoplasms in the first 10 years of life in children who were irradiated *in utero* in the course of maternal pelvic X-ray examination. This finding is not always supported by other studies. Court Brown, Doll, and Hill (1960) also examined the health records of a large group of children with similar histories but found no increase in leukemia or other cancers.

The fetal dose during pelvic X-ray examination is less than 1 rem. Jablon and Kato (1970) studied the 25-year cancer experience of 1292 children who were irradiated *in utero* at the time of the bombings of Hiroshima and Nagasaki, and who received very much higher doses than the children whose mothers were X rayed during pregnancy. Although some of the Japanese children were estimated to have received as much as 250 rem prenatally, Jablon and Kato could find no increase in the cases of leukemia or other cancers during the first 25 years of life.

Saenger *et al.* (1968) studied the incidence of leukemia in hyperthyroid patients treated by ^{131}I as compared to the incidence in patients treated surgically. In a total population of 36,170 patients treated between 1946 and 1964, among which 96.5% follow-up was achieved, there was no difference in the leukemia incidence of the two groups. However, the hyperthyroid population as a whole had a leukemia incidence about 50% higher than that of the general population. The average dose administered to these patients was 8.9 μCi, which produced a whole-body dose of about 10 rem. Saenger's findings are an outstanding example of how in some instances the medical condition that justified irradiation may also predispose the patient to radiomimetic effects. It has been suggested that the conflicting evidence as to the effects of *in utero* exposure may be due to an association, independent of radiation, between a tendency toward development of leukemia and whatever clinical factors lead to the need for pelvic examination during pregnancy.

From all epidemiological evidence, it can be concluded that for the purposes of estimating the upper limit of risk to an exposed population, a risk coefficient of 20 leukemia cases/rad/million of population can be assumed.

These cases would occur over a 15- to 20-year period during which the normal frequency of leukemia would account for about 1200 cases in a population of 1 million, based on an average incidence of about 6 cases/100,000/year.

Although some experimentalists believe there is a threshold radiation dose below which leukemia will not occur, there is some disagreement. Lewis (1959) reviewed the various available data on the relationship between the incidence of leukemia and the radiation dose in man and concluded that the probability of developing leukemia is $(1-2) \times 10^{-6}$/rad/year. He suggested that this probability was independent of the manner in which the dose was received. Others who reviewed Lewis' assumptions concluded that his estimates should not be extended to total doses received in small bits over a period of many years (National Academy of Sciences–National Research Council, 1961b). The experience to date is limited to the dose range between 100 and 1000 rem, and the number of cases is too few at the lower doses to permit any firm assumption as to the shape of the dose-response curve (Brues, 1959). Thus, whether or not exposures on the order of the natural radioactive background or small multiples of such levels are capable of producing leukemia cannot be answered unequivocally on the basis of existing data. We will see in Chapter 3 that, in setting radiation guidelines, it is assumed that the dose-response curve is linear, independent of dose rate, and without a threshold.

The normal incidence of leukemia is so low and the influence of small doses of ionizing radiation so slight that the shape of the dose-response curve could only be determined from studies of large populations studied over a long period of time. Buck (1959) computed the population sizes

TABLE 2-2

MINIMUM POPULATION SIZES WHICH MUST BE STUDIED AT VARIOUS DOSES TO REVEAL A DETECTABLE INCREASE IN THE INCIDENCE OF LEUKEMIA[a]

Dose from birth to age 34 (R)	Minimum man-years at ages 35 to 44
5	6,000,000
10	1,600,000
15	750,000
20	500,000
50	100,000
100	30,000
200	10,000

[a] Adopted from Buck (1959).

required for investigating the question of whether small doses produce leukemia in man. Her data are given in Table 2-2 and serve to illustrate the difficult practical problems of studying this question.

Neoplasms Other Than Leukemia. Among the survivors of the bombings of Hiroshima and Nagasaki, there is evidence that the incidence of neoplasms other than leukemia has been increased in a population of 20,609 individuals who were less than 10 years of age at the time of bombing and in those who received a dose greater than 100 rem (Jablon *et al.*, 1971; Anderson, 1971).

Several kinds of cancer have been seen in individuals in whom special circumstances resulted in irradiation of a particular organ, such as the skeleton, lung, thyroid, or skin.

Bone cancer. When radium or other radioelements that are chemically similar to calcium are ingested, they can be metabolized into bone from which they are removed very slowly. The continual irradiation of bone into which such substances have been incorporated is capable of producing bone cancer and certain blood diseases. As noted earlier, such cases developed among individuals who were exposed to radium in the course of luminous dial manufacturing and patients who received radium medicinally during the first quarter of this century.

Elements such as strontium and plutonium have similar chemical properties to radium, and their radioactive isotopes have been shown by animal experiments to be capable of producing bone cancer. Fortunately, the early misuse of radium provided an opportunity to learn so much about the cause-and-effect relationships that it became possible to specify protective procedures not only for radium but for other bone-seeking radionuclides as well.

Osteosarcoma among radium dial painters were first observed and diagnosed as "radium jaw" in 1924 by Theodore Blum, a New York dentist. The cases in the New York area originated from a radium dial plant in northern New Jersey, and by the late 1920's it was already understood that the cases of bone cancer being reported among young women who painted radium dials with radium-containing paint were due to the practice of lip pointing of the brush used to paint the numerals. The cases attracted the attention of the forensic pathologist Harrison Martland and the toxicologist A. O. Getler who undertook to study the first several cases (Martland, 1951). It remained for Robley D. Evans, a physicist, who began to collaborate with Martland and other physicians in the mid-1930's, to develop the basic biophysical information in the kind of careful detail that has made it possible to use the information derived from these cases as the basis on which is built much of the art by which people are now protected against the insidious effect of the bone-seeking radionuclides such as radium, strontium, and plutonium. It is estimated that there were about 2000 individuals employed in about 50 dial-painting shops in the United States during and immediately following World War I.

Another series of cases from which valuable epidemiological information was derived were those to whom radium was administered either intravenously or orally for medicinal purposes. The total number of individuals so treated is estimated to have been several thousands (Evans, 1966), and it is known that one clinic alone administered more than 14,000 intravenous injections of radium, usually in doses of about 10 μCi. In addition, a number of popular radium water nostrums were on the market in the early 1920's of which one of the most popular was Radithor, a mixture of 1 μCi of radium and 1 μCi of mesothorium. Among the radium dial painters and those who were administered radium for medical reasons, there have been a total of about 40 deaths. About one-fourth of these were due to carcinomas of the nasal sinus and mastoid air spaces, which Evans believes were due to the accumulation of radon and its daughter products within these cavities. All but a few of the remaining deaths were due to bone cancer.

A very much larger number of individuals have shown lesser bone pathology, ranging from small areas of osteoporosis to extensive necrosis. Advanced osteoporitic or necrotic changes were frequently associated with spontaneous fractures.

Based on his epidemiological work up until about 1942, Evans proposed that 0.1 μCi of radium be adopted as a maximum permissible body burden (Evans, 1943). All the known cases of injury occurred in individuals whose body burden was greater than 0.5 μCi at the time of observation, which was usually many years after cessation of exposure. Based on this finding, Evans has estimated that the maximum permissible body burden has a safety factor of about 15 (Evans, 1967; Evans et al., 1969).

The question of whether a threshold exists for production of bone cancer from internal emitters is of considerable importance in relation to standards of permissible public exposure because traces of bone-seeking nuclides such as ^{90}Sr and [239]Pu are already present in the general environment and can be found in the tissues of the general population. It is also important to establish whether or not the response is linearly related to dose. Based on the study of data from five series of experiments in which bone-seeking β emitters were administered in graded doses to four species of laboratory mammals, Mays and Lloyd (1972a) concluded that the incidence of bone cancer was consistently less than would be predicted by an assumption of a linear dose-response relationship. Mays and Lloyd found that the experimental data fit a sigmoid model rather than a linear model.

The same authors (Mays and Lloyd, 1972b) were more equivocal with respect to the effects of α emitters in bone. Their analyses of data from 12 series of experiments indicated that bone cancers from α emitters seem to increase linearly with dose in some cases and to be best represented by a sigmoid response in others. Evans et al. (1972) analyzed the dose-response relationships among more than 600 people exposed to radium for several decades and concluded that no significant skeletal effects were observed below 1000 rad.

Because of widespread contamination of foods by ^{90}Sr deposited in fallout from open-air nuclear testing of nuclear weapons, and since ^{90}Sr deposits in bone (Chapters 5 and 14), there is considerable interest in the relative risks of ^{90}Sr and ^{226}Ra for production of bone cancer. Marinelli (1969), based on analysis of data from laboratory animals, concluded that the risk ratio ^{90}Sr:^{226}Ra is between 0.05 and 0.10. More data will be required before this important subject can be understood fully.

Lung cancer. Inhalation of air containing a radioactive gas or dust will result in an exposure of the respiratory tract to ionizing radiation and, as noted earlier, the high incidence of lung cancer among miners in eastern Europe has been attributed to the presence of radon gas within the mine atmosphere (Lorenz, 1944; Behounek, 1970). These mines had been exploited for centuries for a number of heavy metals and the miners traditionally developed a lung disease which was diagnosed as lung cancer in the early part of this century. Studies in the 1930's of the presence of radon gas in the mine atmosphere led to the conclusion that atmospheric radioactivity was the responsible factor. By the 1950's, based on work initiated by Harley (1952), it was apparent that the principal dose to the lung from atmospheres containing the gas radon was not due to the radon itself but to the accumulation within the lung of the short-lived daughter products of radon attached to the inert dust normally present in mine atmospheres. More recently, lung cancer has been reported among uranium miners in the United States (Archer and Lundin, 1967; Saccomanno et al., 1964; Lundin et al., 1971; Holaday, 1969; Federal Radiation Council, 1967). An elevated incidence of lung cancer has also been reported among workers in fluorspar mines in which radon enters the mine air via ground water (de Villiers and Windish, 1964).

These experiences as well as experimental evidence that radioactive substances in the lung can cause cancer (National Academy of Sciences—National Research Council, 1961a) have resulted in considerable attention to the methods of protecting employees and the general public from the effects of radioactive atmospheric contamination.

The ability of the lung to concentrate particulates increases the relative risk of inhaling radioactive substances in the form of aerosols compared to the risk of inhaling a radioactive gas. Thus, although it was originally thought that radon was responsible for the lung cancers observed in the

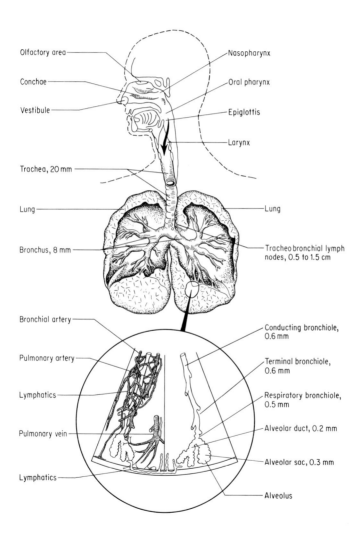

Fig. 2-1. The principal anatomical features of the human respiratory tract (National Academy of Sciences—National Research Council, 1961a).

European miners, it was later observed that the principle dose was not due to exposure to radon per se, but to its daughter products which attach themselves in the dust ordinarily present in the atmosphere and which tend to accumulate in the lung. By this mechanism the daughter products contribute about 20 times as much dose as does the radon (Altshuler et al., 1964).

The respiratory tract (Fig. 2-1) is designed to convey air into the alveoli located deep in the lung where oxygen and carbon dioxide exchange takes place between the blood and air. Much of the dust contained in the inhaled air is removed before the alveoli are reached, the exact fraction depending primarily on the particle size and density of the particles.

Deposition within the respiratory tract may result from inertial impaction, settling, or Brownian motion in the case of the particles less than about 0.1 μm in diameter. In general, the larger particles tend to be removed between the nasal passageways and the lower bronchi, whereas the smaller particles have a high probability of penetrating to the alveoli. If the dust is highly soluble, it will be quickly absorbed from the respiratory tract into the blood by means of which it will pass to other organs of the body. Exposure to soluble dust does not ordinarily constitute a lung hazard.

The manner in which the inhaled dust is deposited in the various parts of the respiratory tract has been studied intensively (National Academy of Sciences—National Research Council, 1961a), as it is known that marked differences in the manner in which the lung rids itself of dust depending on the region of the lung in which the dust is deposited. The respiratory tract above the terminal bronchiole (see Fig. 2-1) is lined with ciliated epithelium that has the ability to move deposits of dust up and out of the respiratory tract, and it has been shown (Albert and Arnett, 1955) that dust deposited on the bronchial epithelium in humans is removed from the lung in a matter of hours, whereas dust that deposits in the alveolar regions of the lung can remain for long periods of time.

Insoluble dust that is deposited in the alveoli is removed by phagocytes which are motile white cells that have the ability to engulf the particles and transport them out of the alveolar spaces. The dust-laden phagocyte may wander onto the ciliated epithelium and be eliminated from the body by this route or it can pass into the lymphatic vessels and hence to the lymph nodes.

The dose received by the different portions of the lungs from inhaled radioactive dust, thus, depends on the concentration of radionuclide in the inhaled air, on the physical properties of the radionuclide, the rate at which the dust is inhaled, the region of the lung in which the dust is deposited, and the rate at which it is removed. The latter factors depend on particle size and density, as well as on physiological factors that probably vary with individuals.

There is considerable uncertainty as to the proper method of computing the dose from radioactive dust in the lung (Parker, 1969). If the total amount of radioactivity in the lung is estimated, one can assume that the energy is absorbed uniformly throughout the total lung mass (customarily 1000 g in the human) and calculate the dose accordingly. It is not known if this is biophysically correct when the total amount of radioactivity originates from relatively few particles emitting α or β radiation. The dose to cells in the vicinity of β particles increases enormously as the particle is approached. This is shown in Table 2–3 in which the dose rates are calculated at various distances around 1-μCi particles containing various β emitters. The implications of this nonuniformity in dose distribution are not known.

The lung cancers in uranium miners are believed to originate in the basal cells of the bronchial epithelium owing to the dose from α radiation emitted by radon daughters that deposit on the ciliated epithelium. The dose delivered in this way will depend on the ratio of the various radon daughters attached to the dust particles and the particle size of the dust (Altshuler *et al.*, 1964).

TABLE 2–3

DOSE RATE IN TISSUE AROUND 1 μCi PARTICLES[a]

Distance (μm)	Dose rate (rem/hr)			
	^{14}C	^{90}Sr[b]	^{32}P	^{90}Y
10	2,000,000	530,000	380,000	270,000
100	1,500	5,000	3,700	2,700
200	40	1,100	930	680
400	0.03	200	230	160
600	—	60	100	70
1,000	—	10	30	26
2,000	—	0.03	7	7
4,000	—	—	0.7	1
6,000	—	—	0.1	0.3
8,000	—	—	0.01	—
10,000	—	—	—	0.02

[a] Adopted from National Academy of Sciences—National Research Council (1961a).
[b] Anomalous spectral distribution; does not include ^{90}Y daughter.

In practice, the term "working level" is used to express the radioactivity of the mine atmosphere and is defined as that concentration of short-lived radon daughter products that emit 1.3×10^5 MeV of α radiation per liter of mine atmosphere. According to the Altshuler model, exposure to one working level for 1 year will result in a bronchial dose of 33 rad. Among the United States uranium miners there is a clear association between exposure to the radon chain and the incidence of lung cancer when the cumulative exposure exceeds 1000 working level months (WLM). Present data do not exclude the possibility of risk at lower cumulative exposures (Federal Radiation Council, 1967).

Thyroid cancer. The thyroid is an organ that must be given special consideration because of its tendency to concentrate radioactive iodine and the possibility of radiation-induced neoplasia (Federal Radiation Council, 1966). When fresh fission products are released to the atmosphere, the dose to the thyroid may be the limiting factor in determining the length of time an individual can be permitted to remain in a given area.

A small percentage of children who were irradiated with X rays for treatment of enlarged thymus developed thyroid cancer (Hemplemann, 1968). The dose in the vicinity of the neck was in excess of 200 rad. No effect has been observed in adults similarly treated (National Academy of Sciences—National Research Council, 1961b). By assuming that a linear relationship exists between thyroid dose and the incidence of thyroid cancer, Lewis (1959) concluded that a dose of 1 rad delivered to the thyroids of 1 million infants would be expected to produce 10 to 100 cases of thyroid cancer over the first 20 years of their lifetime.

In recent years thyroid cancer has been reported among Japanese survivors of the atomic bomb (Wood *et al.*, 1969). The effect was found to be proportionate to radiation dose, and the increase was greater in women than in men.

In 1954, a group of Mashall Islanders on the atoll of Rongelap were irradiated by fallout from a thermonuclear explosion at Bikini (see Chapter 16). Of 67 people irradiated, 21 developed thyroid abnormalities, including three malignant tumors, at the end of about 15 years. The thyroid doses from absorbed radioiodine were estimated to have ranged from 160 to 1400 rem with superimposed whole-body irradiation of about 175 rem (Conard *et al.*, 1970a). The thyroid doses were reconstructed from urine analysis data several weeks postexposure and may not be reliable.

Dolphin (1968) and Dolphin and Marley (1969) have analyzed the various reports of thyroid cancers associated with estimates of radiation exposure. They concluded that in infants the risk from X rays seems to be proportional to dose at a rate of about 100 cancers/million man-rad. The rate in adults exposed to X rays was estimated to be 30 to 40 cases/million man-rad with evidence that the risk per rad is much reduced at doses above 2000 rad. Dolphin and Marley conclude that for purposes of risk estimation it may be assumed that thyroid cancers will develop in a population exposed to radioiodine at a rate of 30 cases/million man-rem in infants and 10 cases/million man-rem in adults.

The practical problems of ascertaining the risk coefficients by epidemiological means are indeed formidable. Patterson and Thomas (1971) have shown that a population containing 10 million man-rem years at risk would be required to demonstrate a risk of cancer induction less than $10^{-4}/$ rad/year.

For purposes of thyroid dose calculation, it is conventionally assumed that the thyroid absorbs 30% of inhaled or ingested radioiodine (International Commission on Radiation Protection, 1959), but according to Dolphin (1968, 1971), thyroidal uptake can vary from 10 to 30% and may actually be a function of the thyroid mass, which is usually taken as 20 g, but has been shown by Mochizuki *et al.* (1963) to vary somewhat around a mean weight of 16.7 $\pm$ 6.9 g in New York City adults. The radioiodine uptake can be almost completely blocked by oral administration of potassium iodide in doses in the range of 50 to 200 mg (Blum and Eisenbud, 1967).

Cataracts. Cataract may develop as a result of exposure of the lens of the eye to heavy doses of X rays, γ rays, β particles, and neutrons. As noted earlier, cataract was reported in experimental animals exposed to X rays within 2 years after Roentgen's discovery.

Cataracts in human beings were observed among survivors of the Japanese bombings (Nefzger *et al.*, 1968), among patients whose eyes were treated by X, γ, or β rays for medical purposes (Merriam and Focht, 1957; National Academy of Sciences—National Research Council, 1961b), and among physicists who were exposed to the radiation from cyclotrons. Although lens changes have been reported from doses as low as 200 rem, the minimum X-ray dose capable of causing clinically significant cataract is thought to lie between 600 to 1000 rad in adults but may be less in children. Neutrons are thought to be five to ten times more effective than X or γ rays in the production of cataracts. Cataract has not been reported as a result of low-level chronic exposure, but further epidemiological data are required before this effect can be excluded unequivocally as a result of repeated low-level exposures.

Effect on Life Span. Whole-body irradiation of experimental animals has been found to result in a shortened life span (Storer, 1965, 1969; Burch,

1969). This is not due to the occurrence of a readily identifiable, radiation-produced, fatal disease such as leukemia but rather to an apparent acceleration of the aging process and an increased susceptibility to all causes of death. Definitive data are as yet available only in rodents and other experimental animals, but the data suggest that the reduction in life span becomes greater as the dose increases (National Academy of Sciences—National Research Council, 1961b). The life expectancies of animals exposed to near-lethal doses have been shown to be 25 to 50% lower than normal. At lower doses (200 to 500 rad) the reduction is from 2 to 4%/100 rad. A single 100 rad dose to mice results in a life-span reduction of 30 days, but is from one-fifth to one-seventh of this figure when the dose is delivered over an extended period of time. Failla and McClement (1957) have suggested that life shortening in man is about 1 day/rad. At lower doses the data are very equivocal, with some suggestion that increased life span may result (Mole, 1957).

A detailed study of the effects of radiation exposure on life span is being conducted in Hiroshima and Nagasaki in a sample of about 100,000 survivors and an equal number of controls. As of 1960, when the last analysis was completed, persons exposed within 1200 m of the hypocenter in the two cities had a mortality rate about 15% higher than the controls (Miller, 1969). Higher mortality has also been observed (Kato, 1971) in a 25-year study of individuals who were *in utero* during the third trimester of pregnancy at the time of the Japanese bombings. However, as reported earlier by Jablon and Kato (1970), there has been no increase in the incidence of neoplasms among the Japanese children exposed prenatally.

There is as yet no reliable information about the effect of chronic exposure on the life span of man, and there are few animal data. It is also not known if there is a basis for extrapolating from the relatively high doses for which data are available to the much smaller doses that are encountered more routinely. One is more frequently concerned with the delayed effects of repeated small doses of radiation which may occur as the normal consequence of day-to-day operation rather than with acute exposure which can result only from a relatively severe accident.

Genetic Effects

As in the case of all higher animal forms, the essential characteristics of man are passed from one generation to the next by means of chromosomes located in the nuclei of reproductive cells. Human cells normally contain 46 chromosomes of which one-half are derived from the mother and the other half from the father. The inheritable characteristics are communicated by means of chromosomal components known as genes which are strung together in beadlike fashion to form tiny filaments that are the chromosomes.

The genes from the mother and father are united within the ovum at conception, and thereafter throughout life the chromosomes are almost always faithfully reproduced at each cell division. The 46 human chromosomes are believed to contain on the order of 10^4 genes, and it is these genes that, when passed on to the next generation, will determine many of the physical and psychological characteristics of the individual.

The genes are large molecules which may undergo structural changes ("mutations") as the result of action by a number of agents including heat, ionizing radiation, and mutagenic chemicals (National Academy of Sciences—National Research Council, 1956; World Health Organization, 1957; UNSCEAR, 1958). When this occurs, the mutated gene may be passed on to offspring in whom the mutant characteristic may ultimately manifest itself. However, many mutations of the type produced by ionizing radiations are recessive and cannot produce a change in the characteristics of the offspring unless a similar mutation is encountered in the chromosomes of the mate. However, the mutated gene will be reproduced and carried by the offspring. As it is highly improbable that two similar mutations would encounter each other in the first postmutation generation, the mutated genes would simply diffuse through the chromosomal material of subsequent generations. The mutated genes would persist from generation to generation and would accumulate in number until they became so numerous that they would encounter each other in the reproductive process. In the offspring produced from such encounters, the mutations would express themselves as inherited changes in the characteristics of individuals.

The genetic injury to a population thus depends on the total number of mutant genes introduced. The measure of potential damage from radiation exposure is the total number of man-roentgens delivered to the gonads or stated another way, the per capita gonadal dose. More precisely, in any given population, allowance must be made for age distribution and the probability of reproduction by the various members of the population. When the per capita gonadal dose is corrected in this way, it is known as the "genetically significant dose."

A small dose delivered to the whole population may thus produce more genetic damage than a much larger dose delivered to a relatively small fraction of the population. Thus, if we assume that 10^5 people received a dose of 5 R/year (the maximum permissible occupational dose), the increment to the per capita dose of a population of 180 million people would be about 2.8 mR/year.

It is thought that in the majority of cases the inherited change will have a deleterious effect on the individual. This may be premature death, inability to produce offspring, susceptibility to disease, or any number of changes of lesser or greater importance.

If a population is subjected to a constant level of ionizing radiation exposure, the total number of mutations produced by the ionizing radiation will continue to increase in the population until the rate at which new mutations are produced is exactly offset by the rate at which carriers of the mutations are dying before reproducing. Such death may result from the tendency of the mutated genes to cause either premature death or infertility. It would take many generations for such an equilibrium to be established.

Based on experiments with the fruit fly, it had been assumed until 1959 that the number of mutations produced is proportionate to total dose and is independent of the rate at which the dose is delivered. It had been further assumed that there is no threshold for radiation-induced mutations. Recently (Russell, 1965, 1968), evidence has developed that indicates the rate at which the dose is delivered does play a role, and that a given dose delivered over an extended period of time produces fewer mutations than if delivered in a short time.

It was found in male mice that a given radiation dose to the spermatogonia delivered at 0.8 R/min produced from one-third to one-fourth as many mutations as the same dose delivered at a rate of 90 R/min. There was no further reduction when the dose rate was reduced to 9 mR/min, nor was there a further increase when the dose rate was raised to 1000 R/min. This dose rate effect is thought to be due to repair mechanisms that operate within the cell but which are overwhelmed at high dose rates. More pronounced dose rate dependence was observed in oocytes to such an extent that at low dose rates the frequency of mutations is indistingusihable from that in controls.

It is estimated that about 4% of live-born infants normally show evidence of hereditary defects. This results from the normal accumulation of mutant genes in the population. To what extent natural radiation contributes to this "load" of mutated genes is not known. It is thought that a radiation dose of about 30 rad delivered to an entire population would cause the spontaneous mutation rate to double. That is, if an average dose of 30 rad was delivered to each generation, a new equilibrium would be reached in time in which mutant characteristics would be seen twice as frequently as in the original population.

Cytological techniques developed during the late 1950's have greatly facilitated studies of chromosomal changes in man. It has been found that aberrations in human chromosome structure occur more frequently in irradiated persons and can be observed in chromosomes from peripheral blood lymphocytes at doses as low as a few rads (UNSCEAR, 1969). The biological significance of such changes, particularly when seen in somatic cells, is not known. Considerable interest in the phenomenon derives from the possibility that the frequency and types of chromosomal aberrations may provide a method of estimating past exposure to ionizing radiation.

Chapter 3

Radiation Protection Standards

One can conclude from Chapter 2 that there may not be an absolutely safe dose of ionizing radiation. It is possible that every bit of exposure to gonadal tissue involves a small probability of genetic injury, and that any amount of whole-body irradiation may result in a higher risk of leukemia or other neoplasms in the exposed individual. We have also seen that the thesis that "all radiation is harmful" cannot be proved, or disproved, because of the low frequency of the events which must be correlated with low levels of radiation and the limitations that exist in the ability of epidemiological and experimental methods to observe low-level effects against the normal "noise level" of morbidity and mortality.

Radiation protection programs are for the most part concerned with relatively small doses delivered over an extended period of time. Actual human experience in which effects have been seen is mainly with doses in excess of 100 rad, and in most cases the dose was administered in single or in a few multiple exposures. Thus, for radiation protection purposes it is frequently necessary to make estimates two orders of magnitude below the lower limit of actual experience. Moreover, we need estimates of the effects produced when the dose is delivered in small bits over a long period of time, sometimes more or less continuously during an entire lifetime.

Figure 3-1 illustrates two types of dose-response curves. Curve A is the usual sigmoid type of response to most forms of stress and it is characterized

38

by a threshold below which no effect is discernible. At doses higher than the threshold, the response follows a sigmoid curve which flattens out at some dose at which all subjects are affected. It is of course possible for a sigmoid type of response to exist with no threshold. There are biophysical reasons, as well as some experimental and epidemiological evidence, to support the hypothesis that the dose-response curve for the genetic effects of ionizing radiation exposure is that shown by B in which there is no threshold, and the effect is linearly proportional to dose over a wide range of doses. Thus, a finite effect can be predicted for any given dose however small. We have seen in Chapter 2 that a no-threshold, linear dose-effect curve has been hypothesized by some investigators for the neoplasms, but in Table 2–2 we saw the difficulties in attempting to design an epidemiological study with sufficient numbers of subjects to answer such questions unequivocally at low doses. The problem becomes even more complicated when we consider the possible effect of dose rate.

In dealing with relatively small populations such as those occupationally exposed, this problem is less important than with respect to the general population because for any given per capita dose the probability of actually causing an injury becomes relatively smaller as the size of the population decreases. This is so despite the fact that the risk to any single individual is a function of the dose to which he is exposed and is independent of population size. When this is so, one can argue that at some point the risk is "acceptable." Thus, Evans (1967) has used the term "practical threshold" to emphasize that in his series of radium cases there seemed to be no effects below some fixed value of skeletal bound radium. This of course does not

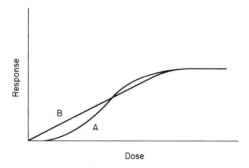

Fig. 3-1. Two possible dose-response relationships: A, classic sigmoid response, with threshold; B, linear response, with no threshold. Other possibilities clearly include curve A with no threshold and curve B with a threshold. It is also possible that over a limited portion of both curves the shapes could be adjusted so that one would be statistically indistinguishable from the other.

exclude the possibility that if the population is sufficiently large or if the life spans could be sufficiently long that an effect might be seen.

Practice in setting maximum permissible radiation limits has for many years been based on the assumption that the dose response is that given by curve B of Fig. 3-1. It is further assumed that the effects are independent of the dose rate, an assumption which is conservative in view of the work of Russell (Chapter 2) that shows the genetic effects of X- and γ-radiation exposure to be dose-rate dependent in mice.

The subject of maximum permissible exposure can be divided into two parts: the levels required to safeguard individuals who are occupationally exposed and the levels required for the general population. The manner in which the maximum permissible doses are established and administered will be reviewed in this chapter.

Evolution of the Maximum Permissible Dose

By 1925 the potential dangers of ionizing radiations were so well recognized and the uses of X ray and radium were becoming so commonplace that there was a clear need to develop uniform standards of radiation protection.

The International Society of Radiology sponsored the formation of the International Commission on Radiological Protection (ICRP) in 1928 (Taylor, 1971). This commission has held a preeminent position in the field of radiation protection ever since and consists of 13 members with 4 committees having a total of some 40 experts from 14 countries. In recent years, the ICRP has received financial support from the World Health Organization, a specialized agency of the United Nations.

Within the United States, 1 year following establishment of ICRP, there was founded the National Council on Radiation Protection and Measurements (NCRP). Until recently, this organization was housed administratively within the Bureau of Standards, but in 1964 NCRP was granted a Congressional charter and has since then operated as an independent organization financed by contributions from government, scientific societies, and manufacturing associations. There are about 60 members on this council and about 175 members on the scientific committees that are given the responsibility for developing its technical reports and recommendations.

The early efforts to develop safe practices were handicapped by the fact that the concept of physical dose had not yet been developed. Until 1928, when the roentgen was adopted, the dose of radiation was measured by the length of time it took for a given radiation flux to produce reddening of the skin, and the maximum permissible dose was expressed as the dose equivalent to 1/1000 of the amount of radiation that would produce erythema in an acute exposure (Failla, 1960). Later, when the roentgen was adopted as the unit of radiation dose, Failla showed the erythema dose of γ radiation from radium was approximately 1800 R, and that the values recommended by various experts between 1925 and 1932 had ranged between 0.04 and 2.0 R/day.

When the first radiation protection standards were proposed for the protection of radiation workers, the total population known to be at risk consisted of a few hundred X-ray and radium workers. Until that time the problem had appeared to be entirely one of external radiation, although, unknown to the scientists then interested in the problem, a number of radium dial painters had already ingested amounts of radium and mesethorium that would soon prove lethal to some of them. The first recommendation of the NCRP for the maximum permissible dose was 0.1 R/day and was proposed by Failla on the basis of his study during the 1930's of three technicians who were exposed to the γ radiation from radium at levels of exposure that did not produce observable damage (Failla, 1960).

Until the early 1930's, the significance of the fact that ionizing radiations could produce genetic mutations was not yet sufficiently appreciated, and the possibility that there might be no threshold for the production of certain types of neoplasms had not yet been suggested. One had reason to be satisfied if no visible damage was being produced among a group of workers by a given level of radiation exposure; at least this was a procedure consistent with the methods successfully employed in establishing maximum permissible levels of exposure to chemically toxic substances, and the procedures proved to be quite adequate from a practical point of view.

The need to extend the concept of the maximum permissible dose to internal emitters arose in the 1930's when it was learned that some of the radium dial painters had ingested radium in harmful amounts. In a group of between 40 and 50 individuals known to have had a measurable radium burden for periods that then ranged up to more than 20 years, it was observed that no apparent injuries occurred if the amount of radium in the body was less than 1.2 μCi. (We saw in Chapter 2 that subsequent work by Evans reduced the lower limit of damage to 0.5 μCi). The NCRP, on the basis of the work of Evans (1943), proposed that the maximum permissible body burden of ^{226}Ra be limited to 0.1 μCi. The number of exposed individuals was not more than a few hundred at that time.

The two benchmarks which had been proposed as maximum permissible radiation levels by NCRP prior to World War II, 0.1 R/day for whole-body external radiation and 0.1 μCi as the maximum body burden of ^{226}Ra, were the standards on which the World War II atomic energy program was conducted with an excellent record of safety. The fact that these important values became available just prior to the start of the atomic energy program in 1942 is a remarkable historical coincidence. If the relatively few cases of radium poisoning had gone unnoticed a decade earlier or if they had not attracted the interest of such perceptive investigators as Martland and Evans, the quantitative basis for the safe handling of bone-seeking internal emitters would not have existed. Similarly, had the medical uses of X ray developed one or two decades later, NCRP would not have had the guidance which the early misadventures provided and the maximum permissible dose might not have been available when needed during World War II. The two all-important benchmarks were developed, as though providentially, just in time to be used by the wartime atomic energy planners.

Much improvising was needed to fill the gap that existed in 1942. The recommendations of the NCRP for protection against external radiation from X ray and radium were quickly adapted to the sudden need to protect people against a wider variety of ionizing radiations, many of which, like neutrons, had not been extensively studied by the biologists. Failla and others devised the concept of relative biological effectiveness, and Parker invented two new units, the rep and the rem* (Cantril and Parker, 1945). This made it possible to extend the use of the recommended limit of 0.1 R/day, which was developed for γ- and X-ray protection, to particulate radiation. Also, the maximum permissible body burden devised for ^{226}Ra could be used as the basis for computing the body burden of a variety of new bone seekers, some of which, such as plutonium, did not exist before the war.

The maximum permissible level of occupational exposure to external ionizing radiations have since been reduced by NCRP as the result of growing conservatism toward ionizing radiation exposure. The original value of 0.1 R/day served as a successful guide, inasmuch as it was the maximum permissible exposure level in effect during the period of great expansion in the medical use of X ray and the early development of the atomic energy program. There was no evidence that individuals exposed

---

* The rep (roentgen equivalent, physical) was the unit of an absorbed dose of radiation having the magnitude of 93 erg/g of absorbing material. It has since been replaced by the rad (D), the unit for an absorbed dose of 100 erg/g. The dose equivalent (H) is the product of D, Q, and N at a point of interest in tissue, where D is the absorbed dose, Q is the quality factor, determined by the type of radiation, and N is the product of any other modifying factors (such as the distribution of absorbed dose in space and time). The special unit of dose equivalent is the rem (International Commission on Radiation Units and Measurements, 1971).

chronically to this level were injured by radiation, but, on the other hand, the negative epidemiological evidence was insufficient to support a conclusion that the level provided adequate safety. In 1949 the value was reduced from 0.1 R to 0.05 R/day not because of evidence that the older figure was unsafe but because the radiations to which people were then being exposed in the atomic energy field were more penetrating than the medium voltage X rays to which people were exposed when the older value was established (Failla, 1960). Noting that the dose to the deeper organs of the body would be somewhat higher in the case of more penetrating radiation, it was decided that the maximum permissible dose should be 0.05 R/day (0.3 R/week) when measured at the depth of bone marrow or other critical organs. The permissible dose to the skin was established at 0.6 R/week which was numerically equivalent to the older figure.

In 1956 a further change was made, this time by the ICRP which limited the dose to the gonads or to the blood-forming organs to 3 rem during any period of 13 consecutive weeks. It was further stipulated that the dose thus accumulated should be limited to $5(N - 18)$ rem in which N is the age in years. This formula reflects the assumption that an individual should not be exposed to ionizing radiation before the age of eighteen. Thus, the average dose over a working lifetime was reduced to about 5 rem/year, one-third of the former weekly rate of 0.3 rem/week. These ICRP values were subsequently adopted by the NCRP (1957), and the new values have since found their way into the regulations of the governmental agencies.

During World War II, there was little time to philosophize about the applicability of the adopted maximum permissible values to the very much larger size of the exposed population caused by the addition of almost 100,000 workers in the wartime atomic energy program. In addition, the successful application of X ray and radium in medical practice had by 1940 swelled the numbers of exposed people to unknown tens of thousands. However, after the war, although there was a general recognition of the excellent job accomplished by the scientists responsible for the wartime program, disquieting questions began to be asked. How should the standards be modified if the population potentially at a risk is the entire population of the world? Worldwide radiation exposures could come about from nuclear war, nuclear weapons testing, the prospective long-term growth of the nuclear power program, or the expected wide use of radioisotopes. Not only would the occupationally exposed population increase enormously, but wastes from the yet-to-be-developed atomic energy program might contaminate the environment, and people might be subject to exposure even though they themselves were not working with radioactive materials.

What about the genetic effects of radiation? Work already completed with lower animals seemed to suggest that production of mutations was linearly proportional to dose and that there might not be a threshold for this effect. Could the incidence of radiation-induced cancer be controlled by a similar dose-effect relation and, if so, could it be that there is no threshold for the production of leukemia and other neoplasms?

These questions resulted in an increasingly conservative attitude toward ionizing radiation exposure. There has been legitimate concern, in particular, about the need to be more cautious in establishing levels of maximum permissible exposure for the population at large. This concern exists for many reasons, of which we cite the following.

1. It seems reasonable to expect an occupationally exposed individual to accept some measure of risk as long as the risk is of a magnitude comparable with other on-the-job risks which are accepted as part of modern living. In most cases the individual accepts occupational risks knowingly. However, these arguments do not hold for the community at large.

2. In the population at large, individuals may be exposed prenatally or in childhood. Since it is believed that children and the fetus may be more susceptible to radiation injury than adults, one must treat the question of total population exposure somewhat more conservatively for this reason alone.

3. The number of exposed individuals must be considered. Whereas it is not likely that the number of occupationally exposed individuals will constitute a sufficiently large fraction of the total population to cause their radiation exposure to be a matter for concern from the genetic point of view, this is obviously not true if the entire population is exposed.

In addition, the task of establishing levels of maximum permissible exposure for nonoccupationally exposed individuals is complicated by the ecological relationships that govern the passage of radioactive materials to man. In occupational exposure, one is normally concerned only with controlling external radiation exposure and radioactive airborne dusts. The opportunities for contamination of food and water in the course of occupational exposure are so readily controlled that problems of this kind rarely develop in actual practice. On the other hand, when one considers the population at large, the principal opportunities for exposure arise from contamination of the environment with radionuclides that thereby become available for ingestion. Except in the case of war or in the event of acute exposure to radioactive substances released from a reactor or chemical-processing-plant catastrophe, the opportunities for public exposure from the use of atomic energy seem to be due mainly to the possibility of contaminated air, food, or water, rather than external radiation.

The National Academy of Sciences Committee on the Biological Effects of Radiation (NAS, 1956) recommended in 1956 that because of genetic considerations the average cumulative exposure to individuals in the population at large should not exceed 10 rem up to age of thirty. It was intended that this exposure should be over and above exposure from natural radioactivity, which amounts to about 3 rem in 30 years. Since it is thought that the medical use of X ray may account for about 5 rem in 30 years, the Committee recommended that the remaining 2 rem be budgeted over the remaining man-made sources, such as occupational exposure, fallout, and other sources of environmental contamination.

The Organizations Involved in Establishing and Implementing Radiation Protection Standards

The radiation protection standards and the procedures by which they are implemented are rooted in the work of many national and international organizations. This discussion will review the manner in which standards are established in the United States and the basic administrative procedures by which radiation protection regulations are promulgated and implemented.

Of primary importance is the need to assemble the published scientific literature on the biological effects of radiation. Since its establishment in 1955, this role has been performed effectively on an international scale by the United Nations Scientific Committee on the Effects of Atomic Radiation, which has from time to time published reports summarizing the state of our basic knowledge with respect to biological effects as well as the behavior of radioactive material in the environment. As noted earlier, the role of establishing numerical standards of permissible exposure has been fulfilled since 1928 by the International Commission on Radiation Protection (ICRP) and by the National Council on Radiation Protection (NCRP) since 1929. The publications of ICRP and NCRP, in addition to providing guidance for the establishment of radiation standards, also include much basic information on the biological effects of radiation. Other major compilations of basic data are to be found in the many records of hearings held by the Joint Committee on Atomic Energy of the United States Congress (JCAE, 1959a, b, c, 1963, 1965, 1969a, b). The Atomic Energy Commission, the Public Health Service, the National Academy of Sciences, the Federal Radiation Council, and other governmental agencies have also produced basic compilations of information which are valuable in developing standards. (See reference list for listings of these organizations).

Having established the necessary numerical values of permissible radiation exposure, there remains the need to translate them into enforceable legal requirements, on the one hand, and recommended safe practices, on the other. This is done by a multitude of organizations, including many government agencies, which sometimes have conflicting and overlapping responsibilities, as well as professional societies (for example, the American Nuclear Society and the American Society of Mechanical Engineers).

An important step in the above processes is to translate the basic numerical value of permissible exposure into standards of permissible concentrations of radioactive substances in foods, water, and air. We will see that the procedures used are not exact, although the methods of computing the maximum permissible concentrations attempt to utilize all existing pertinent knowledge. A basic problem is the wide range in biological variability within and between species. This is not an unusual situation in the public health field since the same deficiencies exist in the procedures for establishing maximum permissible limits of exposure to the chemical pollutants of air and water.

When the maximum permissible concentrations of the various nuclides are adapted for food, air, and water, there remains the need to promulgate codes of operating practice. For example, although the potential hazard from a given isotope may depend on the concentration of the radionuclide in air or water, as a matter of practice it is much more efficient to specify procedures which should be followed in packaging and shipping the isotope than to simply state that the radionuclide may be shipped, provided people along the route are not exposed to such and such concentration in air, food, and water. More and more, as experience is accumulated, the radiation protection programs will be implemented by developing operating practices that preclude the possibility of contamination of the environment above acceptable limits.

A great many organizations play some role either in the establishment of numerical standards or in the development of codes of safe practice.

INTERNATIONAL COMMISSION ON RADIOLOGICAL PROTECTION AND NATIONAL COUNCIL ON RADIATION PROTECTION AND MEASUREMENT

The ICRP was organized in 1928, following establishment in 1925 of the International Commission on Radiological Units and Measurement (ICRU) under the auspices of the First International Congress of Radiology.

From its inception until the present time, the ICRP has been the world's principal standards-setting group in the radiation protection field and has operated harmoniously and effectively by means of a group of subcommittees that now includes about 50 of the world's leaders in the field.

In the United States, the NCRP is the organization that has had the most influence in the establishment of radiation standards (Taylor, 1971). The NCRP was originally formed in 1929 as the Advisory Committee on X-Ray and Radium Protection, and from its inception until 1964 the Council was closely identified with the National Bureau of Standards. In 1964, NCRP received a Congressional charter and became an autonomous, self-supporting organization with financial support from a wide variety of governmental agencies, professional societies, and trade organizations. Many of the NCRP members are also members of one or more of the ICRP subcommittees. Both the ICRP and NCRP issue technical reports in which their recommendations are contained. The ICRP and NCRP have no official status but, in practice, their recommendations serve as the basis for the regulations adopted by official organizations in this country and abroad.

ATOMIC ENERGY COMMISSION

In the Atomic Energy Acts of 1946, 1954, and 1957, Congress gave the Atomic Energy Commission (AEC) prime responsibility for ensuring the safe development of the atomic energy industry. Accordingly, the AEC has evolved a system of licensing and inspection to which users of all but very minimal amounts of radionuclides, fissionable materials, and materials used as sources of fissionable materials manufactured by the AEC are subject. The AEC has neither cognizance over nonfissionable, natural radionuclides such as ^{226}Ra, nor is it responsible for cyclotron-produced isotopes.

The Commission's regulatory program is based on issuances published in the Code of Federal Regulations (CFR) in several parts as follows:

10CFR20: Standards for Protection against Radiation
10CFR30: Licensing of Byproduct Material
10CFR40: Control of Source Material
10CFR50: Licensing of Production and Utilization Facilities
10CFR55: Operator's Licenses
10CFR70: Special Nuclear Material
10CFR71: Regulations to Protect against Accidental Conditions of Criticiality
     in the Shipment of Special Nuclear Material
10CFR100: Reactor Site Criteria

The Commission's basic radiation protection standards are contained in 10CFR20 (Part 20). These regulations are generally based on the NCRP values of maximum permissible exposure, but go somewhat beyond the scope of the NCRP handbooks by prescribing the administrative and technical procedures by which one can determine whether the standards are being met.

A recent change in the Atomic Energy Act permits the AEC to delegate its regulatory responsibility for AEC-produced radionuclides and limited amounts of source materials to states that can demonstrate their technical competence to undertake such a responsibility.

Regulation of Radionuclide Use

Since the Atomic Energy Act limits the responsibility of the AEC for the safety of radionuclide usage to reactor-produced isotopes, radioisotopes that occur naturally or are produced in a cyclotron or other particle accelerators are not covered by AEC regulation.

The section of the CFR applicable to radioisotopes is contained in Title 10, Part 30. The section is limited to by-product material, which is defined as ". . . any radioactive material (except special nuclear material) yielded in or made radioactive in or made radioactive by exposure to the radiation incident to the processing of producing or utilizing special nuclear material." It is because of this wording that only reactor-produced isotopes fall under the cognizance of AEC. The AEC has the authority to transfer to the states its authority to regulate by-product material.

Whereas the CFR, Part 30, determines the conditions under which a license to use by-product materials may be granted, the limitations governing the discharge of radioactive waste by-products to the environment is contained in CFR, Part 20. In general, the maximum permissible amounts that may be discharged to the environment have been derived from the reports of the NCRP.

Reactor Safety

The Atomic Energy Act of 1954 made it possible for private capital to sponsor the development of a civilian nuclear power industry. Prior to that time, the act of 1946 forbade dissemination of much of the information that is necessary to build and operate reactors, and the original act specified that it was not legal for anyone but the government to own a reactor. The 1954 Atomic Energy Act made it possible for the civilian nuclear industry to develop, and the AEC established the procedures necessary to implement the authority and responsibility given to it under the new act.

The rules pertaining specifically to reactors and fuel reprocessing plants are contained in Title 10, Part 50, "Control of Facilities for the Production of Fissionable Material"; and Part 100, "Reactor Site Criteria." These regulations will be discussed in later chapters.

FEDERAL RADIATION COUNCIL

In 1960, in response to criticism that Congress had given AEC responsibility for developing and managing the atomic energy program as well as

TABLE 3–1

FEDERAL RADIATION COUNCIL REPORTS

Date	Staff report number	Staff report title
May, 1960	1	Background Material for the Development of Radiation Protection Standards
Sept., 1961	2	Background Material for the Development of Radiation Protection Standards
May, 1962	3	Health Implications of Fallout from Nuclear Weapons Testing through 1961
May, 1963	4	Estimates and Evaluation of Fallout in the United States from Nuclear Weapons Testing Conducted through 1962
July, 1964	5	Background Material for the Development of Radiation Protection Standards
Oct., 1964	6	Revised Fallout Estimates for 1964–1965 and Verification of the 1963 Predictions
May, 1965	7	Background Material for the Development of Radiation Protection Standards—Protective Action Guides for ^{89}Sr, ^{90}Sr, and ^{137}Cs
Sept., 1967	8	Guidance for the Control of Radiation Hazards in Uranium Mining

the responsibility for its safe conduct, President Eisenhower established a Federal Radiation Council (FRC) to "advise the President with respect to radiation matters directly or indirectly affecting health, including guidance for all federal agencies in the formulation of radiation standards, and in the establishment and execution of programs of cooperation with states." The Council consists of the Secretary of the Department of Health, Education, and Welfare; the Secretary of Defense; the Secretary of Commerce; the Secretary of Labor; and the Chairman of the Atomic Energy Commission. Although the FRC was eliminated when the Environmental Protection Agency (EPA) was established, the original recommendations issued by the FRC in 1960 and 1961 are reproduced in Appendices 3-1 and 3-2 because they contain so much of the thinking that has governed radiation protection in the United States. Additional FRC reports are listed in Table 3-1.

ENVIRONMENTAL PROTECTION AGENCY

In 1970, the President reorganized the environmental functions of the Federal government and created the EPA. Within this agency was established a Radiation Office, which was given responsibility for the development of radiation protection guidelines and environmental radiation standards. This in effect relieved the AEC of the responsibility for the establishment of environmental standards and also took over the functions of the FRC. Surveillance functions that formerly existed in the Public Health Service were also transferred to EPA. Thus, the EPA has the responsibility for establishing Federal standards of radiation protection and also maintaining surveillance over contamination of air, food, and water.

THE STATES

Although the Atomic Energy Act gives the AEC prime responsibility for health and safety relating to its own ramified activities, the act does not apply to radiations that originate from cyclotron-produced isotopes, most natural radionuclides, X-ray machines, etc. Supervision over the public health problems associated with these sources of radiation continues to reside in the states, many of which have issued standards which apply not only to non-AEC regulated sources of ionizing radiation but to the AEC sources as well. Dual jurisdiction thus exists in many areas and, although most states have adopted the basic standards proposed by the AEC, there are some differences in interpretation and procedures which may in the future result in some confusion. The legal question of whether regulations adopted by the states can be held to be applicable in the face of existing AEC regulations has not as yet been decided.

The 1957 act permits the AEC to transfer certain of its responsibilities in the health and safety field to the states, and this has been done insofar as the use of radioisotopes is concerned in a number of states. A basic condition that must be satisfied is that the state should have the technical capability to administer its regulations, which must be compatible with those of AEC. By 1969, 19 states had consummated agreements with the AEC for the transfer of authority for regulation of by-product and source materials.

MUNICIPALITIES

Certain large cities such as New York City have the authority to regulate radiation sources, and, in general, they tend to implement their authority by means of codes very similar to those used by the states.

INTERNATIONAL AGENCIES

A number of international organizations can be expected to issue recommended safe practices from time to time. Among these are the International

Labor Organization, the World Health Organization (WHO), the International Atomic Energy Agency, and the Food and Agriculture Organization. These organizations do not have the authority to require adoption of their recommendations by any of the member countries. The WHO in recent years has been actively supporting the ICRP, and the standards of this group tend to serve as the basis for the codes of practice developed by the international organizations.

The various organizations that contribute toward setting radiation standards can be divided basically into two types. There are the ICRP, NCRP, and other scientific organizations such as the National Academy of Sciences and the British Medical Research Council, which from time to time issue authoritative reports of the basic radiobiological and epidemiological facts with attempts to express quantitative relationships between dose and risk. This information then passes through the second type of organization such as the AEC, the Public Health Service, or the EPA which has the responsibility of taking the basic scientific information and translating it into language suitable for regulatory purposes. The standards of permissible radiation exposure used in most countries of the world originate in the work of these many national and international bodies among whom there is general harmony.

There is apt to be less agreement as to how the basic recommendations that issue from the scientific bodies should be administered in practice; in other words, how does one relate the standards based on epidemiological and experimental evidence to the apparatus of public health regulation. Public health regulation is more an art than a science. The public health administrator as an artisan must start with a mixture of scientific information of various grades of quality which he must evaluate according to contemporary concepts of permissible risks. He must then fabricate a system of regulation that is understandable, that is practical, and that above all protects public health. Whether one is regulating radioactive, chemical, or biological contaminants of the environment, the same approach must be taken. One could cite many examples. Thus, public health officials regulate the biological quality of drinking water by making assays for an innocuous group of coliform organisms because one is basically interested in certain pathogenic organisms that originate in fecal pollution for which the coliform organisms are a useful indicator. This system of control does not provide absolutely safety, but in most cases it provides adequate safety. For many decades this has been a practical system of control which has lent itself to practical systems of enforcement. Other regulatory mechanisms could be designed which could provide more safety, but they might not be practical to administer or enforce. We will see that, in general the present methods of regulating ionizing radiation exposure, for all their defects and inconsistencies, have served well to protect the public health.

The Basic Standards of Maximum Permissible Radiation Exposure

This discussion of the basic radiation standards will draw from three main sources from which the standards used in the United States originate: the ICRP, the NCRP, and the FRC, which was incorporated into the EPA in 1970.

The two most basic documents in the field of radiation protection are Report No. 9 of the ICRP (1966) and "Basic Radiation Protection Criteria," published by NCRP in 1971. These documents provide the main sources from which radiation protection regulations are derived. A third basic document, which deals with the standards for maximum exposure to the internal emitters, is the ICRP report of Committee II, Permissible Dose from Internal Radiation (ICRP, 1960). There is general agreement between the ICRP and NCRP methodology. An important admonition in the documents of both organizations is that there is no such thing as an absolutely safe dose of radiation and that all radiation exposure should be reduced to the extent practicable.

One of the most fundamental criteria for the upper limit of permissible occupational exposure is that an employee should not accumulate more than $5(N - 18)$ rem, where N is the employee's age in years. Stated another way, the employee should not work with ionizing radiation until he is at least 18 years old and then should not be exposed to more than an average of 5 rad/year. The standards then recognize that certain organs of the body should receive no more than is permissible for whole-body exposure, but that other organs or parts of the body may be more resistant to the effects of radiation exposure and, therefore, can be permitted to receive a higher dose. Thus, the skin and portions of the extremities such as the hands and feet can be permitted more than 5 rem/year. The permissible dose equivalents for various types of exposure are given in Table 3–2 (NCRP, 1971).

When internal radiation exposure is involved, the ICRP methodology introduces the concept of the "critical organ" which is the organ in which a given radionuclide tends to accumulate; for example, the critical organ for radioiodine is the thyroid, and for ^{90}Sr it is the skeleton. For occupational exposure, and but for the exceptions noted in Table 3–2, the permissible dose to the individual organs of the body is limited to 15 rem/year.

For occupational external exposure the maximum permissible dose is administered over a 13-week period, and although the permissible annual

TABLE 3–2

DOSE-LIMITING RECOMMENDATIONS PROPOSED BY NCRP[a]

Combined whole-body occupational exposure	
Prospective annual limit	5 rem in any one year
Retrospective annual limit	10–15 rem in any one year
Long term accumulation to age N years	(N-18) $\times$ 5 rem
Skin	15 rem in any one year
Hands	75 rem in any one year (25/qtr)
Forearms	30 rem in any one year (10/qtr)
Other organs, tissues, and organ systems	15 rem in any one year (5/qtr)
Fertile women (with respect to fetus)	0.5 rem in gestation period
Dose limits for the public or occasionally exposed individuals	0.5 rem in any one year
Population dose limits	
Genetic	0.17 rem average per year
Somatic	0.17 rem average per year
Emergency dose limits—life saving	
Individual (older than 45 years if possible)	100 rem
Hands and forearms	200 rem, additional (300 rem, total)
Emergency dose limits—less urgent	
Individual	25 rem
Hands and Forearms	100 rem, total

[a] Adopted from National Council on Radiation Protection and Measurements (1971).

dose is limited to 5 rem, the quarterly (13 week) dose is limited to 3 rem. For occupational exposure, it is seen from Table 3–2 that the NCRP has recommended a limit of 100 rem for persons older than 45 years who become involved in life-saving emergencies such as might be necessary in the event of an industrial accident. The hands and forearms can receive an additional 200 rem.

The NCRP recommendations recognize that although one may plan to limit the exposure to 3 rem/quarter or 5 rem per year, that this may be exceeded for various unanticipated reasons. The NCRP position is that no deviation from sound protection is implied if the retrospective dose does not exceed 10 to 12 rem for dose increments well distributed over time or even 15 rem for exceptionally well-distributed increments. Of course, the age-dependent cumulative dose limitation, $5(N - 18)$, is overriding.

The ICRP, NCRP, and FRC are for the most part consistent insofar as permissible exposure of the general public is concerned. The recommended

TABLE 3–3

INTAKE AND EXCRETION OF THE STANDARD MAN[a]

Water balance			
Intake (cm³/day)		Excretion (cm³/day)	
Food	1000	Urine	1400
Fluids	1200	Sweat	600
Oxidation	300	From lungs	300
		Feces	200
Total	2500	Total	2500

	O_2 (vol. %)	CO_2 (vol. %)	N_2 + others (vol. %)
	Air balance		
Inspired air	20.94	0.03	79.03
Expired air	16	4.0	80
Alveolar air (inspired)	15	5.6	—
Alveolar air (expired)	14	6.0	—

Vital capacity of lungs	3–4 liters (men)
	2–3 liters (women)
Air inhaled during 8-hr workday	10^7 cm³/day
Air inhaled during 16 hr not at work	10^7 cm³/day
Total air inhaled	2×10^7 cm³/day
Interchange area of lungs	50 m²
Area of upper respiratory tract, trachea, bronchi	20 m²
Total surface area of respiratory tract	70 m²
Total water in body	4.3×10^4 g
Average life span of man	70 years
Occupational exposure time of man	8 hr/day;
	40 hr/week;
	50 weeks/year;
	50 years total time

[a] Adopted from International Commission on Radiological Protection (1960). The data of Tables 3-3 to 3-6 are reproduced in this text for purposes of illustration and discussion and should not be used for dose calculations without first ascertaining that they have not been revised.

TABLE 3-4

Organs of Standard Man[a] (Mass and Effective Radius of Organs of the Adult Human Body)

Organs	Mass (m) (g)	Percent or total body[b]	Effective radius (X) (cm)
Total body[b]	70,000	100	30
Muscle	30,000	43	30
Skin and subcutaneous tissue[c]	6,100	8.7	0.1
Fat	10,000	14	20
Skeleton			
Without bone marrow	7,000	10	5
Red marrow	1,500	2.1	—
Yellow marrow	1,500	2.1	—
Blood	5,400	7.7	—
Gastrointestinal tract[b]	2,000	2.9	30
Contents of GI tract			
Lower large intestine	150	—	5
Stomach	250	—	10
Small intestine	1,100	—	30
Upper large intestine	135	—	5
Liver	1,700	2.4	10
Brain	1,500	2.1	15
Lungs (2)	1,000	1.4	10
Lymphoid tissue	700	1.0	—
Kidneys (2)	300	0.43	7
Heart	300	0.43	7
Spleen	150	0.21	7
Urinary bladder	150	0.21	—
Pancreas	70	0.10	5
Salivary glands (6)	50	0.071	—
Testes (2)	40	0.057	3
Spinal cord	30	0.043	1
Eyes (2)	30	0.043	0.25
Thyroid gland	20	0.029	3
Teeth	20	0.029	—
Prostate gland	20	0.029	3
Adrenal glands or suprarenal (2)	20	0.029	3
Thymus	10	0.014	—
Ovaries (2)	8	0.011	3
Hypophysis (pituitary)	0.6	8.6×10^{-6}	0.5
Pineal gland	0.2	2.9×10^{-6}	0.04
Parathyroids (4)	0.15	2.1×10^{-6}	0.06
Miscellaneous (blood vessels, cartilage, nerves, etc.)	390	0.56	—

[a] See footnote a of Table 3-3.

[b] Does not include contents of the gastrointestinal tract.

[c] The mass of the skin alone is taken to be 2000 g.

limit for the maximum exposed individual is 0.5 rem/year with 0.17 rem as the maximum average dose to the general population. The maximum dose to members of the public is thus limited to one-tenth of the occupational dose, and the average dose is limited to one-thirtieth.

The standards for permissible human exposure to ingested or inhaled radioactive substances are based on the assumption that the permissible amount of any radioactive material in the body or in the critical organ should not exceed the annual permissible dose. These figures are then translated into maximum permissible concentrations of each radionuclide in air or water using a set of physiological parameters that describe the movement of each element in and out of the critical organ, the mass of the organ, and the daily rate at which the contaminents are inhaled or ingested. For purposes of uniformity, the ICRP has defined the characteristics of the "standard man," and from these calculations the ICRP and NCRP have prepared tabulations of the maximum permissible concentrations of each radionuclide in air or water which if inhaled or ingested continuously

TABLE 3–5

PARTICULATES IN RESPIRATORY TRACT OF THE STANDARD MAN[a,b]

Distribution	Readily soluble compounds (%)	Other compounds (%)
Exhaled	25	25
Deposited in upper respiratory passages and subsequently swallowed	50	50
Deposited in the lungs (lower respiratory passages)	25[c]	25[d]

[a] See footnote a of Table 3–3.

[b] Retention of particulate matter in the lungs depends on many factors, such as the size, shape, and density of the particles, the chemical form, and whether or not the person is a mouth breather; however, when specific data are lacking it is assumed the distribution is as shown.

[c] This is taken up into the body.

[d] Of this amount, half is eliminated from the lungs and swallowed in the first 24 hr, making a total of $62\frac{1}{2}\%$ swallowed. The remaining $12\frac{1}{2}\%$ is retained in the lungs with a half-life of 120 days, it being assumed that this portion is taken up into body fluids.

TABLE 3–6

GASTROINTESTINAL TRACT OF THE STANDARD MAN[a]

Portion of GI tract that is the critical tissue	Mass of contents (g)	Time food remains (day)	Fraction from lung to GI tract (fa)	
			Sol.	Insol.
Stomach	250	1/24	0.50	0.625
Small intestine	1100	4/24	0.50	0.625
Upper large intestine	135	8/24	0.50	0.625
Lower large intestine	150	18/24	0.50	0.625

[a] See footnote *a* of Table 3–3.

would, in a lifetime exposure of 50 years, result in a body burden that would deliver the maximum permissible dose to one or more organs of the body (ICRP, 1960; NCRP, 1959). The values for the principal parameters of the ICRP standard man are given in Tables 3–3 to 3–6.

PERMISSIBLE EXPOSURE TO INTERNAL EMITTERS

The NCRP and ICRP from time to time publish tabulations of maximum permissible values for more than 130 radionuclides (NCRP, 1959; ICRP, 1960). The two sets of values are in essential agreement, as indeed they should be, inasmuch as the NCRP is well represented in the membership of the ICRP, and there is good communication between the two groups. The American values are now essentially the basis for the AEC regulations as expressed in 10CFR20.

The permissible dose of internal emitters is calculated on the assumption that there is no exposure to external radiation. Where such exposure exists, the permissible concentration of radionuclides in air, water, or food must be reduced so that the total dose to the organ from both internal and external sources does not exceed the basic guide. To accomplish this, the MPC values must be reduced by the factor $(D - E)/D$, where D is the maximum permissible dose permitted to an organ and E is the dose received from external radiation.

The published tabulations of the permissible concentrations in air and drinking water for exposure of the general public are a convenient method of presenting the individual calculated values but are an oversimplification, since human exposure may result from contamination of foods, in which case biological concentration may require that the permissible concentration in air or water should in some situations be lower than the concentrations that are acceptable for direct inhalation or ingestion. The FRC recognized this in their proposed guidelines, which give the maximum permissible daily intake from all sources. This possibility is also recognized by the AEC in 10CFR20. Although the limits allowed for the various nuclides are basically those given the NCRP and ICRP for air and drinking water, AEC regulations require the licensee to monitor the environment in such a way that the permissible limits would be revised downward if it is shown that biological concentration is taking place in the environs.

USE OF DERIVED WORKING LIMITS

In the United Kingdom, considerable reliance is placed on the use of the derived working limit (DWL), which is the maximum rate of release to the environment that will not exceed the maximum permissible dose to nearby inhabitants (Bryant, 1970a). DWL's have been developed for atmospheric releases of [131]I (Bryant, 1964), ^{90}Sr and [137]Cs (Bryant, 1966, 1969), and [129]I (Bryant, 1970b). The DWL's are secondary standards that have no legal status, but their numerical values are established for each facility by agreement between the facility operator and the governmental authorities. A similar technique is used in this country in which the operating licenses for nuclear facilities usually contain limits of discharge rates based on an analysis of the local environmental factors in relation to the maximum permissible levels of public exposure. A method of deriving working limits for discharges to the marine environment has been described by Freke (1967).

METHODS OF COMPUTING MAXIMUM PERMISSIBLE CONCENTRATIONS FOR OCCUPATIONAL EXPOSURE

The maximum permissible concentrations of radionuclides in air and water must satisfy the requirement that the dose rate to the critical organ after 50 years of occupational exposure should not exceed the maximum permissible dose rates in the basic guides given above. For most isotopes, equilibrium in the critical organ is reached long before the 50-year exposure period has passed. This, however, is not so for the relatively long-lived bone-seeking radionuclides such as ^{90}Sr, [226]Ra, and [239]Pu, or for other rare-earth elements.

The basic procedure in setting the maximum permissible concentrations is to calculate that amount of each radionuclide in air and water which after 50 years of either continuous exposure or exposure during a 40-hr work week would result in a dose to the critical organ equal to the maximum permissible dose. In some cases the radionuclides will decay to form radioactive daughters, and the calculations must then take into consideration the dose delivered by the daughters.

The mathematical relationships that govern the accumulation of a radionuclide in an organ during a period of exposure are not known exactly. The ICRP and NCRP have assumed that for each isotope there is one or more critical organ having a characteristic rate of accumulation and excretion. The radionuclides are assumed to accumulate in the critical organ at a rate determined by the amount present in food or water, the rate of inhalation or ingestion, the rate of metabolic transfer to the critical organ, and the rate at which the body eliminates the radionuclide. It is assumed that elimination is exponential, a constant fraction of the organ burden being eliminated in unit time.

For certain substances and organs, notably bone, the rate of elimination may be better described by a power function of the following form:

$$R_t - At^{-n} \qquad (3\text{-}1)$$

where R_t is the fraction of radioactivity remaining in an organ at time t, A is the fraction of radioactivity remaining at end of unit time, and n is a constant. Use of the power function yields values of permissible daily intake that are higher in most cases than the values calculated with the exponential model.

Use of the exponential model permits the concept of the "biological half-life" Tb, a factor which can be used conveniently in conjunction with the radioactive half-life Tr. The overall rate at which the radioactivity owing to a radionuclide diminishes in the critical organ can be expressed as the effective half-life.

$$T = TrTb/(Tr + Tb) \qquad (3\text{-}2)$$

The exponential model is undoubtedly an oversimplification but is used because of its convenience as well as its conservatism. Rather than a constant rate of clearance, it is known that the rate frequently varies as some inverse power of time and is best expressed by the power function. Radium freshly deposited in bone is eliminated more rapidly than radium that has been fixed in the skeleton for a period of years. Moreover, many organs have several "compartments," each of which has its characteristic rate of clearance. Thus, dust deposited in the lungs is eliminated at different rates, depending on whether it is deposited on the bronchial epithelium or in the alveoli.

In order to ensure uniformity in the basic computations, it has become customary to compute the maximum permissible concentrations (MPC's) for a "standard man," whose characteristics are given in Table 3–3 to 3–6. The actual dose which an individual will receive from the maximum permissible body burden will thus depend on whether he is larger or smaller than the standard man and whether his relevant physiological processes are on the high or low side of normal.

The values for bone-seeking radionuclides such as ^{90}Sr and ^{239}Pu are not established by this type of calculation but are computed from a comparison with ^{226}Ra, the maximum permissible body burden of which is based on epidemiological evidence. In one additional exception, the soluble long-lived uranium compounds, the limiting body burden is based on the chemical rather than the radiological toxicity.

BODY BURDEN BASED ON COMPARISON WITH RADIUM

The maximum permissible body burden of ^{226}Ra and its daughter products corresponds to an average dose rate to the bone of 0.56 rem/week, or about 0.06 rad/week.

The maximum permissible body burdens of other bone-seeking radionuclides are calculated from this basic value by calculating the amount of the radionuclide that would deliver the dose equivalent to the critical organ divided by N, the relative damage factor for radionuclides deposited in the bone. This factor takes into consideration the fact that rad for rad some radionuclides produce greater damage than others to the bones of experimental animals. This relative damage factor is taken as 5 for bone-seeking radionuclides other than radium.

It should be noted that based on the epidemiological studies of radium poisoning 0.1 μCi of ^{226}Ra is a somewhat larger amount of radium than would be permissible if the maximum permissible body burden (MPB) were calculated on the basis of a maximum permissible dose of 0.3 rem/week, as is true for most other organs of the body. However, the RBE dose from 0.1 μCi Ra includes the RBE for α particles, which is taken as 10, a figure for which there is as yet little experimental basis. If, for example, the RBE for neoplastic changes in bone proved to be only 1 for α radiation, the MPB for radium would deliver a dose of only 0.06 rem/week.

COMPUTATION OF BODY BURDEN BASED ON PERMISSIBLE DOSE RATE TO
    THE CRITICAL ORGAN

If the maximum permissible weekly dose rate is R rem, then the maximum body burden q in microcuries may be calculated as follows:

$$q = \frac{100mR}{3.7 \times 10^4 \times 1.6 \times 10^{-6} \times 6.05 \times 10^5 f(\Sigma E)(QF)} \tag{3-3}$$

$$q = \frac{2.8 \times 10^{-3}mR}{f(\Sigma E)(QF)}$$

and where $R = 0.3$ rem/week

$$q = \frac{8.4 \times 10^{-4} m}{f(\Sigma E)\,(QF)} \tag{3-4}$$

where

q = maximum permissible body burden (μCi)
3.7×10^4 = disintegrations/sec/μCi
1.6×10^{-6} = erg/MeV
6.05×10^5 = sec/week
100 = erg/g/rad
m = mass of organ (g)
f = fraction of isotope in organ, relative to amount in whole body
ΣE = energy absorbed in organ (MeV)
QF = 1 for β rays and γ radiation and 10 for α particles*

The above equation is a slightly simplified form of the one used by the ICRP. It is applicable only to cases in which equilibrium exists between a parent and its daughters. For cases in which this is not so, the ICRP report (1960) should be consulted.

COMPUTATION OF MAXIMUM PERMISSIBLE CONCENTRATIONS IN AIR AND WATER

As noted earlier, the ICRP method of computating the MPC's in air and water is based on an exponential model for the accumulation of a radionuclide in the critical organ. It is assumed that the daily intake of the critical organ can be expressed as

$$\frac{d(qf)}{dt} = P - \lambda(qf) \tag{3-5}$$

where

P = rate of uptake of radionuclide by the critical organ (μCi/day)
qf = burden of radionuclide in organ
λ = 0.693/T
T = effective half-life
t = period of exposure (set at 50 years)

---

* There is only one measured value of the QF for α radiation, and it was determined by Burns, Albert, and Heimbach (1968) to be 2.9 for tumor production in rat skin.

If $qf = 0$ when $t = 0$, then

$$qf = \frac{P(1 - e^{-\lambda t})}{\lambda} \tag{3-6}$$

When qf is equal to the MPB as calculated from Eq. (3-6), P is the maximum permissible daily intake. One may then proceed to a computation of the MPC in air, $(MPC)_a$ and in water, $(MPC)_w$, using the physiological parameters provided by the definition of the standard man. We see from Table 3–3 that it is assumed that one consumes 2200 ml of water and breathes 2×10^7 cm³ of air per 24-hr day. Half of these daily rates are assumed to be applicable to exposure in an 8-hr workday because of the individual's greater activity during that period. Two factors f_a and f_w enable one to estimate the fraction of the inhaled or ingested element that reaches the critical organ.

Assuming that the work schedule for the standard man is 8 hr/day, 5 days/week, and 50 weeks/year,

$$P = 1100 \times \tfrac{5}{7} \times \tfrac{50}{52}(MPC)_w f_w \tag{3-7}$$

$$P = 750(MPC)_w f_w$$

and similarly

$$P = 6.9 \times 10^6 (MPC)_a f_a \tag{3-8}$$

If the value given for P in Eqs. (3-7) and (3-8) is substituted in Eq. (3-6), one can obtain

$$(MPC)_a = \frac{10^{-7} qf}{Tf_a(1 - e^{-693t/T})} \ \mu Ci/cm^3 \tag{3-9}$$

and

$$(MPC)_w = \frac{9.2 \times 10^{-4} qf}{Tf_w(1 - e^{-693t/T})} \ \mu Ci/ml \tag{3-10}$$

The values for occupational exposure can then be established by setting $t = 50 \times 365$ days and by using the values of f, f_a, and f_w as given in the ICRP report.

When the radionuclide in question disintegrates into a series of daughters, the mathematics becomes more complicated. The various equations which may be used and their derivations will be found in the 1959 report of the ICRP (ICRP, 1960).

When insoluble substances are ingested, the critical organ may be any one of the several segments of the gastrointestinal tract. Here the computation will be somewhat different, inasmuch as the radionuclides are moving through the gastrointestinal tract with the rate of movement varying in each section of the tract. In addition, allowance must be made for absorption of a portion of the energy within the contents of the gastrointestinal tract, and the dose to the wall of the tract must then be computed. This method of computation is given in detail in the ICRP report.

The foregoing discussion is intended only to acquaint the reader with the basic approach to the methods used by the ICRP in computing MPC's of radionuclides in air and water. The calculated values, together with the basic data from which the values are derived, are published in full in the reports of the ICRP. The various reports are so comprehensive and so basic to the subject of environmental radioactivity that they should be considered to be basic references on the subject of MPC's of internal emitters.

The MPC's of some of the more important nuclides in air and water, as adopted by AEC from the ICRP, are given in Table 3–7.

Maximum Permissible Concentration of Unidentified Radionuclides

Unidentified β emitters may be detected in air or water, and the question may arise as to what the MPC of such unidentified radioactivity should be. The ICRP has considered this question and has issued the recommendations contained in Tables 3–8 and 3–9. These values in turn have been adopted by the NCRP and the AEC.

The adopted approach in the absence of quantitative radiochemical analyses of a mixture of radionuclides is to regard the unidentified radioactivity as consisting of the most radiotoxic radionuclides with provision for step-by-step increases in the permissible concentration, depending on whether the presence of certain of the more toxic radionuclides can be ruled out. For example, ^{226}Ra and ^{228}Ra and other radionuclides of heavy elements can certainly be ruled out in the event of a release of fission products from a reactor. As a matter of fact, these isotopes can be eliminated from consideration in almost all atomic energy installations. Similarly, the presence of ^{90}Sr in significant quantities can frequently be ruled out as well since this isotope is only an insignificant fraction of fresh fission products and because of fractionation is usually a minute fraction of the radioactivity contained in reactor effluents.

Maximum Permissible Concentrations for Exposure of the General
  Public

It has become general practice to use one-tenth of the ICRP–NCRP values as the MPC's for individuals living in the vicinity of plants or

TABLE 3-7

AEC PERMISSIBLE LEVELS FOR CERTAIN RADIONUCLIDES[a]

Element	Isotope	Type of exposure	Occupational exposure (restricted areas)		Nonoccupational exposure (unrestricted areas)	
			Air (μCi/ml) (1)	Water (μCi/ml) (2)	Air (μCi/ml) (1)	Water (μCi/ml) (2)
Carbon	^{14}C	S[b]	4×10^{-6}	2×10^{-2}	1×10^{-7}	8×10^{-4}
	(CO_2)	Sub[c]	5×10^{-5}	—	1×10^{-6}	—
Cesium	^{137}Cs	S	6×10^{-8}	4×10^{-4}	2×10^{-9}	2×10^{-5}
		I[d]	1×10^{-8}	1×10^{-3}	5×10^{-10}	4×10^{-5}
Cobalt	^{60}Co	S	3×10^{-7}	1×10^{-3}	1×10^{-8}	5×10^{-5}
		I	9×10^{-9}	1×10^{-3}	3×10^{-10}	3×10^{-5}
Hydrogen	^{3}H	S	5×10^{-6}	1×10^{-1}	2×20^{-7}	3×10^{-3}
		Sub	2×10^{-3}	—	4×10^{-5}	—
Iodine	^{131}I	S	9×10^{-9}	6×10^{-5}	1×10^{-10}	3×10^{-7}
		I	3×10^{-7}	2×10^{-3}	1×10^{-8}	6×10^{-5}
Iron	^{59}Fe	S	1×10^{-7}	2×10^{-3}	5×10^{-9}	6×10^{-5}
		I	5×10^{-8}	2×10^{-3}	2×10^{-9}	5×10^{-5}
Krypton	^{85}Kr	Sub	1×10^{-5}	—	3×10^{-7}	—
Manganese	^{54}Mn	S	4×10^{-7}	4×10^{-3}	1×10^{-8}	1×10^{-4}
		I	4×10^{-8}	3×10^{-3}	1×10^{-9}	1×10^{-4}
Plutonium	^{239}Pu	S	2×10^{-12}	1×10^{-4}	6×10^{-14}	5×10^{-6}
		I	4×10^{-11}	8×10^{-4}	1×10^{-12}	3×10^{-5}
Polonium	^{210}Po	S	5×10^{-10}	2×10^{-5}	2×10^{-11}	7×10^{-7}
		I	2×10^{-10}	8×10^{-4}	7×10^{-12}	3×10^{-5}
Radium	^{226}Ra	S	3×10^{-11}	4×10^{-7}	3×10^{-12}	3×10^{-8}
		I	5×10^{-11}	9×10^{-4}	2×10^{-12}	3×10^{-5}
Radon	^{222}Rn	S	1×10^{-7}	—	3×10^{-9}	—
Strontium	^{89}Sr	S	3×10^{-8}	3×10^{-4}	3×10^{-10}	3×10^{-6}
		I	4×10^{-8}	8×10^{-4}	1×10^{-9}	3×10^{-5}
	^{90}Sr	S	1×10^{-9}	1×10^{-5}	3×10^{-11}	3×10^{-7}
		I	5×10^{-9}	1×10^{-3}	2×10^{-10}	4×10^{-5}
Sulfur	^{35}S	S	3×10^{-7}	2×10^{-3}	9×10^{-9}	6×10^{-5}
		I	3×10^{-7}	8×10^{-3}	9×10^{-9}	3×10^{-4}
Thorium	^{nat}Th	S	3×10^{-11}	3×10^{-5}	1×10^{-12}	1×10^{-6}
		I	3×10^{-11}	3×10^{-4}	1×10^{-12}	1×10^{-5}
Uranium	^{235}U	S	5×10^{-10}	8×10^{-4}	2×10^{-11}	3×10^{-5}
		I	1×10^{-10}	8×10^{-4}	4×10^{-12}	3×10^{-5}
	^{nat}U	S	7×10^{-11}	5×10^{-4}	3×10^{-12}	2×10^{-5}
		I	6×10^{-11}	5×10^{-4}	2×10^{-12}	2×10^{-5}
Zinc	^{65}Zn	S	1×10^{-7}	3×10^{-3}	4×10^{-9}	1×10^{-4}
		I	6×10^{-8}	5×10^{-3}	2×10^{-9}	2×10^{-4}

laboratories that discharge wastes into the environment. The AEC in regulation 10CFR20 has given these values official status by making them generally applicable to installations under licence to the AEC.

However, it should be noted that the AEC and other regulatory agencies that take their guidance from the ICRP, NCRP, and FRC require that exposure of the public be maintained at the lowest practicable value, the so-called "permissible dose" notwithstanding.

The AEC uses the MPC in drinking water as the MPC of radionuclides in liquids being discharged into the uncontrolled environment. In so doing, it assumes that if the concentrations being discharged are below the maximum permissible concentrations for the radionuclides in drinking water the concentration beyond the point of discharge would also be safe. In many instances, this is a very conservative assumption, as where the total volume of liquids being discharged is relatively small in relation to the body of water into which the discharge is taking place. Thus, for example, a few microcuries per day being discharged at the MPC into a relatively large body of water such as the ocean would certainly result in concentrations in the ocean that are undetectable, even a short distance from where the discharge is taking place. On the other hand, where relatively large total amounts of radioactivity are discharged, the possibility exists that biological reconcentration may result in hazardous amounts of a radionuclide being present in food. As noted earlier, the AEC requires its licensees to demonstrate by actual measurement that this does not take place.

---

[a] USAEC, 10CFR20 (1969a).

[b] Soluble.

[c] "Sub" means that values given are for submersion in an infinite cloud of gaseous material.

[d] Insoluble.

Note: In any case where there is a mixture in air or water of more than one radionuclide, the limiting values for purposes of this table should be determined as follows.

If the identity and concentration of each radionuclide in the mixture are known, the limiting values should be derived as follows: Determine, for each radionuclide in the mixture, the ratio between the quantity present in the mixture and limit otherwise established in this table for the specific radionuclide when not in a mixture. The sum of such ratios for all the radionuclides in the mixture may not exceed "1" (i.e., "unity").

Example. If radionuclides *A*, *B*, and *C* are present in concentrations C_A, C_B, and C_C, and if the applicable MPC's are MPC_A, MPC_B, and MPC_C, respectively, then the concentrations shall be limited so that the following relationship exists:

$$\frac{C_A}{MPC_A} + \frac{C_B}{MPC_B} + \frac{C_C}{MPC_C} \leq 1.0$$

TABLE 3-8

MAXIMUM PERMISSIBLE CONCENTRATION OF UNIDENTIFIED RADIONUCLIDES IN WATER, (MPCU)$_W$ VALUES, FOR CONTINUOUS OCCUPATIONAL EXPOSURE[a]

Limitations	μCi/cm^3 of water[b]
If none of the radionuclides ^{90}Sr, ^{126}I, ^{129}I, ^{131}I, ^{210}Pb, ^{210}Po, ^{211}At, ^{223}Ra, ^{224}Ra, ^{226}Ra, ^{228}Ra, ^{227}Ac, ^{230}Th, ^{231}Pa, ^{232}Th, and ^{nat}Th is present, then the (MPCU)$_W$ is	3×10^{-5}
If none of the radionuclides ^{90}Sr, ^{129}I, ^{210}Pb, ^{210}Po, ^{223}Ra, ^{226}Ra, ^{228}Ra, ^{231}Pa, and ^{nat}Th is present, then the (MPCU)$_W$ is	2×10^{-5}
If none of the radionuclides ^{90}Sr, ^{129}I, ^{210}Pb, ^{226}Ra, and ^{228}Ra is present, then the (MPCU)$_W$ is	7×10^{-6}
If neither ^{226}Ra nor ^{228}Ra is present, then the (MPCU)$_W$ is	10^{-6}
If no analysis of the water is made, then the (MPCU)$_W$ is	10^{-7}

[a] See footnote a of Table 3-3.
[b] Use one-tenth of these values for interim application in the neighborhood of an atomic energy plant.

TABLE 3-9

MAXIMUM PERMISSIBLE CONCENTRATION OF UNIDENTIFIED RADIONUCLIDES IN AIR, (MPCU)$_A$ VALUES, FOR CONTINUOUS OCCUPATIONAL EXPOSURE[a]

Limitations	μCi/cm^3 of air[b]
If there are no α-emitting radionuclides and if none of the β-emitting radionuclides ^{90}Sr, ^{129}I, ^{210}Pb, ^{227}Ac, ^{228}Ra, ^{230}Pa, ^{241}Pu, and ^{249}Bk is present, then the (MPCU)$_A$ is	10^{-9}
If there are no α-emitting radionuclides and if none of the β-emitting radionuclides ^{210}Pb, ^{237}Ac, ^{228}Ra, and ^{241}Pu is present, then the (MPCU)$_A$ is	10^{-10}
If there are no α-emitting radionuclides and if the α-emitting radionuclide ^{227}Ac is not present, then the (MPCU)$_A$ is	10^{-11}
If none of the radionuclides ^{227}Ac, ^{230}Th, ^{231}Pa, ^{232}Th, ^{nat}Th, ^{238}Pu, ^{239}Pu, ^{240}Pu, ^{242}Pu, and ^{249}Cf is present, then the (MPCU)$_A$ is	10^{-12}
If none of the radionuclides ^{231}Pa, ^{nat}Th, ^{239}Pu, ^{240}Pu, ^{242}Pu, and ^{249}Cf is present, then the (MPCU)$_A$ is	7×10^{-13}
If no analysis of the air is made, then the (MPCU)$_A$ is	4×10^{-13}

[a] See footnote a of Table 3-3.
[b] Use one-tenth of these values for interim application in the neighborhood of an atomic energy plant.

We have seen that the ICRP–NCRP method of computing MPC's assumes that an individual drinks 2200 ml/day and that his total daily intake of radioactivity is derived from water rather than food. However, it is the maximum permissible daily intake from all sources that determines if a hazard exists, and it is proper to apportion this maximum permissible daily intake among the various items of the diet including water. For example, Pritchard (1958) has shown what the effect would be on MPC's if a person received 50% of his protein requirements from seafood harvested from waters containing the MPC's of a number of isotopes. Table 3–10, which has been adopted from Pritchard (1959), lists some of the more important corrosion and fission products which are likely to be discharged from a light-water reactor and illustrates the way in which the MPC's would be altered by concentrations of the biological concentration that might take place. In this hypothetical marine environment, which is assumed to produce 50% of the protein consumed by a population, the MPC of the individual isotopes in river water proves to be very much less than in drinking water. If these isotopes were present in the water at a concentration safe for drinking, the permissible concentration in seafood would be exceeded by a considerable factor (Table 3–10).

TABLE 3–10

Maximum Permissible Concentrations for Selected Radioisotopes in Seafood and in Coastal Waters for a Selected Population That Obtained One-Half its Protein Requirement from Seafood Harvested from the Coastal Waters[a,b]

Isotope	Conc. factors for			Weighted mean conc. factor	MPC drinking water (μCi/ml)	MPC in seafood (μCi/g)	MPC in coastal waters (μCi/ml)
	Marine invertebrates	Fish flesh	Fish bone				
^{60}Co	10^4	(10^3)	(10^3)	5×10^3	3×10^{-3}	1×10^{-4}	2×10^{-8}
^{55}Fe	10^4	10^3	5×10^3	6×10^3	8×10^{-4}	3×10^{-3}	6×10^{-7}
^{59}Fe	10^4	10^3	5×10^3	6×10^3	5×10^{-5}	2×10^{-4}	3×10^{-8}
^{182}Ta	?	?	?	?	4×10^{-5}	2×10^{-4}	?
^{65}Zn	5×10^3	10^3	10^3	3×10^3	1×10^{-4}	4×10^{-4}	1×10^{-7}
^{137}Cs	50	10	(10)	30	2×10^{-5}	8×10^{-5}	3×10^{-6}
^{90}Sr	10	1	2×10^2	15	1×10^{-7}	4×10^{-7}	3×10^{-8}
^{95}Zr	2×10^3	(10^2)	(10^2)	10^3	6×10^{-6}	2×10^{-4}	2×10^{-7}
^{95}Nb	2×10^2	10^2	(10^2)	150	1×10^{-4}	4×10^{-4}	3×10^{-6}
^{106}Ru	3×10^3	(10^3)	(10^3)	2×10^3	1×10^{-5}	4×10^{-5}	2×10^{-8}
^{144}Ce	8×10^3	12	(10)	4×10^3	1×10^{-5}	4×10^{-5}	1×10^{-8}

[a] Adopted from Pritchard (1958).
[b] For other assumptions, see text.

The above philosophy was adopted by the FRC in recommendations they submitted to the President for radiation protection guides applicable to exposure of population groups (FRC, 1962a,b). These recommendations, which were approved by the President and were issued to the Federal agencies for their guidance, are reprinted in Appendix 3-1. The recommendations are limited to four radioisotopes: ^{226}Ra, ^{131}I, ^{90}Sr, and ^{89}Sr. The guides recommended the establishment of three ranges of daily intake with three grades of required action as given in Appendix 3-1. The upper limit of Range II is considered an acceptable risk for a lifetime, but it was recommended that for administrative purposes the period of evaluation be taken as 1 year. The actual levels recommended for the upper limit of daily intake for Range II are based on the criterion of Table A3-1-1 of Appendix 3-1, which lists the recommended radiation-protection guide for thyroid, bone marrow, and bone for population groups. Consistent with the earlier recommendations of the FRC, it was recommended that since it is not feasible to sample every individual of the population the average dose to the individuals comprising a sample be limited to one-third of the value recommended in the radiation-protection guide. It is assumed that if the average value is one-third of the maximum level suggested for individuals, no individual will be exposed to a level greater than that given in the radiation-protection guide.

The ranges of transient rates of intake in microcuries per day to be used in conjunction with the graded scale of actions indicated in Table A3-2-2 of Appendix 3-2 are given in Table A3-2-3 of Appendix 3-2.

DEFINITION OF LOWEST PRACTICABLE DOSE

Reports from the ICRP and NCRP since the mid-1950's have consistently noted that because it is prudently assumed that no threshold exists, that the recommended permissible levels of exposure notwithstanding, all radiation exposure should be reduced to the lowest practicable dose. The AEC has administratively required all licensees to take all reasonable steps to reduce radiation exposure to both employees and the public, but no regulations were promulgated to provide a numerical definition of the meaning of "lowest practicable" until 1971 when the AEC responded to the general trend to interpret "lowest practicable" as meaning as low as is feasible with existing technology. Recognizing the performance of the light-water reactors, definitions were promulgated that required designers of this class of reactors to limit the offsite exposures to about 1% of those recommended by the NCRP and listed in Part 20 of the Commission's regulations.

MAXIMUM PERMISSIBLE CONCENTRATION FOR BRIEF EXPOSURE

Although there is a need for permissible standards of short-term exposure, none has yet been published officially. To some extent the presently used MPC's for continuous exposure may be adopted to short-term conditions, since the basic radiation guides simply define the amounts of radiation which may be received over a relatively long unit of time. Thus, if a given concentration is based on an assumed exposure of 40 hr/week, but if the individual is only exposed for 1 hr/week, the permissible concentration for 1 hr may be increased fortyfold. One need only avoid exceeding the maximum permissible dose given as quarterly or annual limits. For example, an individual exposed to radioiodine in the course of his work is permitted to receive a thyroid dose of up to 4 rem in a 13-week period. If he normally

TABLE 3–11

SINGLE EXPOSURE VALUES FOR INHALATION OF SELECTED SOLUBLE RADIOACTIVE MATERIAL WHEN BODY ORGANS OTHER THAN THE GASTROINTESTINAL TRACT OR LUNGS ARE THE CRITICAL BODY ORGANS[a]

		Microcuries inhaled in 8 hr that will present dose of		
Isotope	Critical organ	0.3 rem in 1 week	15.7 rem in 1 year	150 rem in 70 year
^{3}H(HTO or ^{3}H$_2$O)	T. body	4.6×10^4	5.2×10^5	5.0×10^6
^{14}C (CO$_2$)	Fat	1.4×10^3	9.5×10^3	9.1×10^4
^{24}Na	T. body	8.5×10^2	4.4×10^4	4.3×10^5
^{32}P	Bone	1.6×10^2	2.4×10^3	2.3×10^4
^{35}S	Skin	1.4×10^3	1.7×10^4	1.7×10^5
^{45}Ca	Bone	10^2	2.0×10^2	1.5×10^3
^{56}Mn	Kidneys	3.5×10^3	1.8×10^5	1.8×10^6
	Liver	4.8×10^3	2.5×10^5	2.4×10^6
^{59}Fe	Blood	42	3.6×10^2	3.5×10^3
^{60}Co	Liver	1.1×10^3	2.6×10^4	2.5×10^5
^{65}Zn	Bone	4.5×10^3	4.9×10^4	4.7×10^5
^{89}Sr	Bone	30	1.4×10^2	1.3×10^3
^{90}Sr + ^{90}Y	Bone	16	17	15
^{131}I	Thyroid	2.1	52	5.0×10^2
^{137}Cs + ^{137}Ba	Muscle	4.3×10^2	5.5×10^3	5.2×10^4
^{144}Ce + ^{144}Pr	Bone	28	51	3.7×10^2
^{nat}Th	Bone	4.8×10^{-2}	4.6×10^{-2}	7.9×10^{-3}
^{nat}U	Kidneys	0.11	0.85	8.1
^{239}Pu	Bone	0.38	0.36	6.1×10^{-2}

[a] Morgan *et al.* (1956).

TABLE 3-12

MAXIMUM PERMISSIBLE CONCENTRATION (μCi/cm³) IN WATER AND AIR OF RADIOACTIVE FALLOUT FOR 7 EXPOSURE TIMES (n) AT 11 DIFFERENT TIMES AFTER BURST TO DELIVER A DOSE TO THE CRITICAL ORGAN OF 15 rem IN 90 DAYS (t)[a]

n(days)	Exposure	Time after fission										
		3.5 hr	12 hr	1 day	2 days	4 days	7 days	14 days	28 days	105 days	210 days	365 days
1	Water	2.5	0.81	0.48	0.29	0.18	0.13	6.9×10^{-2}	7.1×10^{-2}	0.10	0.10	8.7×10^{-2}
	Air	3.7×10^{-4}	1.2×10^{-4}	6.9×10^{-5}	4.0×10^{-5}	2.4×10^{-5}	1.7×10^{-5}	8.2×10^{-6}	6.9×10^{-6}	6.4×10^{-6}	6.0×10^{-6}	5.7×10^{-6}
7	Water	0.54	0.18	0.10	5.8×10^{-2}	3.6×10^{-2}	2.7×10^{-2}	1.2×10^{-2}	1.2×10^{-2}	1.5×10^{-2}	1.4×10^{-2}	1.3×10^{-2}
	Air	7.7×10^{-5}	2.5×10^{-5}	1.4×10^{-5}	7.4×10^{-6}	4.5×10^{-6}	2.4×10^{-6}	1.2×10^{-6}	1.2×10^{-6}	1.5×10^{-6}	1.5×10^{-6}	1.3×10^{-6}
14	Water	0.37	0.12	6.9×10^{-2}	3.9×10^{-2}	3.0×10^{-2}	1.8×10^{-2}	7.6×10^{-3}	7.1×10^{-3}	7.0×10^{-3}	6.3×10^{-3}	6.5×10^{-3}
	Air	5.0×10^{-5}	1.8×10^{-5}	9.0×10^{-6}	5.0×10^{-6}	3.0×10^{-6}	2.0×10^{-6}	8.1×10^{-7}	6.1×10^{-7}	8.3×10^{-7}	7.7×10^{-7}	6.4×10^{-7}
21	Water	0.32	0.11	5.9×10^{-2}	3.3×10^{-2}	2.4×10^{-2}	1.5×10^{-2}	5.8×10^{-3}	5.4×10^{-3}	4.6×10^{-3}	4.3×10^{-3}	4.0×10^{-3}
	Air	4.4×10^{-5}	1.4×10^{-5}	7.7×10^{-6}	4.2×10^{-6}	2.4×10^{-6}	1.8×10^{-6}	6.2×10^{-7}	5.4×10^{-7}	5.7×10^{-7}	5.5×10^{-7}	4.5×10^{-7}
30	Water	0.29	9.7×10^{-2}	5.4×10^{-2}	3.1×10^{-2}	2.0×10^{-2}	1.2×10^{-2}	5.3×10^{-3}	4.7×10^{-3}	3.4×10^{-3}	3.1×10^{-3}	2.8×10^{-3}
	Air	3.7×10^{-5}	1.2×10^{-5}	6.7×10^{-6}	3.1×10^{-6}	2.1×10^{-6}	1.3×10^{-6}	5.3×10^{-7}	4.5×10^{-7}	4.4×10^{-7}	4.1×10^{-7}	3.3×10^{-7}
60	Water	0.25	8.3×10^{-2}	4.6×10^{-2}	2.6×10^{-2}	1.5×10^{-2}	1.1×10^{-2}	4.9×10^{-3}	3.6×10^{-3}	2.5×10^{-3}	2.3×10^{-3}	2.0×10^{-3}
	Air	3.2×10^{-5}	1.0×10^{-5}	5.4×10^{-6}	2.9×10^{-6}	1.7×10^{-6}	1.1×10^{-6}	4.3×10^{-7}	3.3×10^{-7}	2.7×10^{-7}	2.4×10^{-7}	1.8×10^{-7}
90	Water	0.24	8.0×10^{-2}	4.4×10^{-2}	2.4×10^{-2}	1.4×10^{-2}	9.8×10^{-3}	3.7×10^{-3}	2.6×10^{-3}	2.1×10^{-3}	1.7×10^{-3}	1.4×10^{-3}
	Air	3.1×10^{-5}	1.0×10^{-5}	5.4×10^{-6}	2.9×10^{-6}	1.7×10^{-6}	9.7×10^{-7}	3.6×10^{-7}	2.4×10^{-7}	1.5×10^{-7}	1.3×10^{-7}	1.2×10^{-7}

[a] Adopted from Teresi and Newcombe (1961).

TABLE 3-13

MAXIMUM PERMISSIBLE CONCENTRATION ($\mu Ci/cm^3$) IN WATER AND AIR OF RADIOACTIVE FALLOUT FOR 5 EXPOSURE TIMES (n) AT 11 DIFFERENT TIMES AFTER BURST TO DELIVER A DOSE TO THE CRITICAL ORGAN OF 150 rem IN 30 DAYS (t)[a]

n(days)	Exposure	Time after fission										
		3.5 hr	12 hr	1 day	2 days	4 days	7 days	14 days	28 days	105 days	210 days	365 days
1	Water	27	8.5	5.1	3.0	2.0	1.4	0.78	0.83	1.2	1.3	1.1
	Air	4.1×10^{-3}	1.3×10^{-3}	7.7×10^{-4}	4.5×10^{-4}	2.8×10^{-4}	1.9×10^{-4}	1.0×10^{-4}	9.8×10^{-5}	1.2×10^{-4}	1.2×10^{-4}	1.2×10^{-4}
7	Water	5.9	1.9	1.1	0.63	0.40	0.26	0.14	0.14	0.18	0.19	0.15
	Air	8.6×10^{-4}	2.8×10^{-4}	1.3×10^{-4}	8.0×10^{-5}	5.6×10^{-5}	3.7×10^{-5}	1.8×10^{-5}	1.7×10^{-5}	2.0×10^{-5}	2.0×10^{-5}	1.9×10^{-5}
14	Water	4.3	1.4	0.78	0.43	0.27	0.18	8.8×10^{-2}	8.7×10^{-2}	0.12	0.10	7.9×10^{-2}
	Air	7.2×10^{-4}	2.3×10^{-4}	1.3×10^{-4}	6.3×10^{-5}	3.8×10^{-5}	2.5×10^{-5}	1.1×10^{-5}	1.0×10^{-5}	1.1×10^{-5}	1.1×10^{-5}	1.0×10^{-5}
21	Water	3.7	1.2	0.67	0.38	0.23	0.15	7.3×10^{-2}	6.8×10^{-2}	8.1×10^{-2}	6.6×10^{-2}	5.2×10^{-2}
	Air	5.4×10^{-4}	1.8×10^{-4}	9.7×10^{-5}	5.5×10^{-5}	3.3×10^{-5}	2.0×10^{-5}	9.8×10^{-6}	6.8×10^{-6}	8.9×10^{-6}	8.6×10^{-6}	8.7×10^{-6}
30	Water	3.6	1.1	0.62	0.34	0.21	0.14	6.8×10^{-2}	6.0×10^{-2}	6.3×10^{-2}	5.1×10^{-2}	3.8×10^{-2}
	Air	5.4×10^{-4}	1.7×10^{-4}	8.4×10^{-5}	5.3×10^{-5}	2.8×10^{-5}	2.0×10^{-5}	9.2×10^{-6}	7.9×10^{-6}	7.7×10^{-6}	7.1×10^{-6}	6.4×10^{-6}

[a] Adopted from Teresi and Newcombe (1961).

works 40 hr/week for 13 weeks, his exposure would be received over a 520-hr period. The maximum permissible dose for a 1-hr exposure could thus be about 500 times the maximum permissible value given on the basis of a 40-hr week.

In addition to this rationale for permitting exposure to higher concentrations over relatively short periods of time, it is recognized that from time to time an individual may be exposed in an accident. The AEC in part 10CFR20 permits individuals to receive up to 25 rem in the event of accidents, assuming that such accidents would be rare and would not likely occur more than once in the lifetime of any one individual.

Morgan, Snyder, and Ford (1956) have extended the methods of calculation used by the ICRP and NCRP to estimate the MPC values for single exposures to radioactive materials. The maximum permissible intake (MPI) for 8 hr of inhalation was calculated for three conditions, assuming that the permissible dose is 0.3 rem in 1 week, 15.7 rem in 1 year, or 150 rem in 70 years. The latter limit was imposed arbitrarily by the postulate of Morgan and his associates that the lifetime dose should not exceed 150 rem. The calculated MPI's for the more important radionuclides are given in Table 3–11. Although these values do not yet have official status, they are useful guides in the planning of emergency procedures.

The problem of establishing maximum permissible emergency values becomes more complicated when one deals with mixed fission products, as in the case of fallout from a nuclear weapon. In such cases the fission product distribution is complex and is changing with time. The exact composition in the case of weapons debris will depend on the time since detonation, whereas reactor fission products vary in composition depending on the radiation history of the reactor prior to the accident as well as the interval of time since the accident.

Teresi and Newcombe (1961) have undertaken a complex set of computations in which they have estimated the MPC's of radioactive fallout in water and air for consumption during selected periods at various times after a weapons burst. The values have been computed for two different criteria of maximum permissible exposure, namely, 15 rem in 90 days and 150 rem in 30 days. The values proposed by these investigators are given in Tables 3–12 and 3–13.

Appendix 3-1

Federal Radiation Council

RADIATION PROTECTION AND GUIDENCE FOR FEDERAL AGENCIES

Memorandum for the President

Pursuant to Executive Order 10831 and Public Law 86-373, the Federal Radiation Council has made a study of the hazards and use of radiation. We herewith transmit our first report to you concerning our findings and our recommendations for the guidance of Federal agencies in the conduct of their radiation protection activities.

It is the statutory responsibility of the Council to "***advise the President with respect to radiation matters, directly or indirectly affecting health, including guidance for all Federal agencies in the formulation of radiation standards and in the establishment and execution of programs of cooperation with States ***".

Fundamentally, setting basic radiation protection standards involves passing judgment on the extent of the possible health hazard society is willing to accept in order to realize the known benefits of radiation. It involves inevitably a balancing between total health protection, which might require foregoing any activities increasing exposure to radiation, and the vigorous promotion of the use of radiation and atomic energy in order to achieve optimum benefits.

The Federal Radiation Council has reviewed available knowledge on radiation effects and consulted with scientists within and outside the Government. Each member has also examined the guidance recommended in

73

this memorandum in light of his statutory responsibilities. Although the guidance does not cover all phases of radiation protection, such as internal emitters, we find that the guidance which we recommend that you provide for the use of Federal agencies gives appropriate consideration to the requirements of health protection and the beneficial uses of radiation and atomic energy. Our further findings and recommendations follow.

Discussion

The fundamental problem in establishing radiation protection guides is to allow as much of the beneficial uses of ionizing radiation as possible while assuring that man is not exposed to undue hazard. To get a true insight into the scope of the problem and the impact of the decisions involved, a review of the benefits and the hazards is necessary.

It is important in considering both the benefits and hazards of radiation to appreciate that man has existed throughout his history in a bath of natural radiation. This background radiation, which varies over the earth, provides a partial basis for understanding the effects of raidation on man and serves as an indicator of the ranges of radiation exposures within which the human population has developed and increased.

The Benefits of Ionizing Radiation

Radiation properly controlled is a boon to mankind. It has been of inestimable value in the diagnosis and treatment of diseases. It can provide sources of energy greater than any the world has yet had available. In industry, it is used as a tool to measure thickness, quantity or quality, to discover hidden flaws, to trace liquid flow, and for other purposes. So many research uses for ionizing radiation have been found that scientists in many diverse fields now rank radiation with the microscope in value as a working tool.

The Hazards of Ionizing Radiation

Ionizing radiation involves health hazards just as do many other useful tools. Scientific findings concerning the biological effects of radiation of most immediate interest to the establishment of radiation protection standards are the following.

1. Acute doses of radiation may produce immediate or delayed effects or both.

2. As acute whole body doses increase above approximately 25 rems (units of radiation dose), immediately observable effects increase in severity with dose, beginning from barely detectable changes, to biological signs clearly indicating damage, to death at levels of a few hundred rems.

3. Delayed effects produced either by acute irradiation or by chronic irradiation are similar in kind, but the ability of the body to repair radiation damage is usually more effective in the' case of chronic than acute irradiation.

4. The delayed effects from radiation are in general indistinguishable from familiar pathological conditions usually present in the population.

5. Delayed effects include genetic effects (effects transmitted to succeeding generations), increased incidence of tumors, lifespan shortening, and growth and developmental changes.

6. The child, the infant, and the unborn infant appear to be more sensitive to radiation than the adult.

7. The various organs of the body differ in their sensitivity to radiation.

8. Although ionizing radiation can induce genetic and somatic effects (effects on the individual during his lifetime other than genetic effects), the evidence at the present time is insufficient to justify precise conclusions on the nature of the dose-effect relationship at low doses and dose rates. Moreover, the evidence is insufficient to prove either the hypothesis of a "damaged threshold" (a point below which no damage occurs) or the hypothesis of "no threshold" in man at low doses.

9. If one assumes a direct linear relation between biological effect and the amount of dose, it then becomes possible to relate very low dose to an assumed biological effect even though it is not detectable. It is generally agreed that the effect that may actually occur will not exceed the amount predicted by this assumption.

Basic Biological Assumptions

There are insufficient data to provide a firm basis for evaluating radiation effects for all types and levels of irradiation. There is particularly uncertainty with respect to the biological effects at very low doses and low-dose rates. It is not prudent therefore to assume that there ls a level of radiation exposure below which there is absolute certainty that no effect may occur. This consideration, in addition to the adoption of the conservative hypothesis of a linear relation between biological effect and the amount of dose, determines our basic approach to the formulation of radiation protection guides.

The lack of adequate scientific information makes it urgent that additional research be undertaken and new data developed to provide a firmer basis for evaluating biological risk. Appropriate member agencies of the Federal Radiation Council are sponsoring and encouraging research in these areas.

Recommendations

In view of the findings summarized above the following recommendations are made:

It is recommended that:

1. There should not be any man-made radiation exposure without the expectation of benefit resulting from such exposure. Activities resulting in man-made radiation exposure should be authorized for useful applications provided in recommendations set forth herein are followed.

It is recommended that:

2. The term "Radiation Protection Guide" be adopted for Federal use. This term is defined as the radiation dose which should not be exceeded without careful consideration of the reasons for doing so; every effort should be made to encourage the maintenance of radiation doses as far below this guide as practicable.

It is recommended that:

3. The following Radiation Protection Guides be adopted for normal peacetime operations.

The following points are made in relation to the Radiation Protection Guides herein provided.

(1) For the individual in the population, the basic Guide for annual whole body dose is 0.5 rem. This Guide applies when the individual whole

TABLE A3-1-1

RADIATION PROTECTION GUIDES

Type of exposure	Condition	Dose (rem)
Radiation worker		
(a) Whole body, head and trunk, active blood forming organs, gonads, or lens of eye	Accumulated dose 13 weeks	5 times the number of years beyond age 18 3
(b) Skin of whole body and thyroid	Year 13 weeks	30 10
(c) Hands and forearms, feet and ankles	Year 13 weeks	75 25
(d) Bone	Body burden	0.1 µg ^{226}Ra or its biological equivalent
(c) Other organs	Year 13 weeks	15 5
Population		
(a) Individual	Year	0.5 (whole body)
(b) Average	30 year	5 (gonads)

body doses are known. As an operational technique, where the individual whole body doses are not known, a suitable sample of the exposed population should be developed whose protection guide for annual whole body dose will be 0.17 rem per capita per year. It is emphasized that this is an operational technique which should be modified to meet special situations.

(2) Considerations of population genetics impose a per capita dose limitation for the gonads of 5 rems in 30 years. The operational mechanism described above for the annual individual whole body dose of 0.5 rem is likely in the immediate future to assure that the gonadal exposure Guide (5 rem in 30 years) is not exceeded.

(3) These Guides do not differ substantially from certain other recommendations such as those made by the National Committee on Radiation Protection and Measurements, the National Academy of Sciences, and the International Commission on Radiological Protection.

(4) The term "maximum permissible dose" is used by the National Committee on Radiation Protection (NCRP) and the International Commission on Radiation Protection (NCRP) and the International Commission on Radiological Protection (CRP). However, this term is often misunderstood. The words "maximum" and "permissible" both have unfortunate connotations not intended by either the NCRP or the ICRP.

(5) There can be no single permissible or acceptable level of exposure without regard to the reason for permitting the exposure. It should be general practice to reduce exposure to radiation, and positive effort should be carried out to fulfill the sense of these recommendations. It is basic that exposure to radiation should result from a real determination of its necessity.

(6) There can be different Radiation Protection Guides with different numerical values, depending upon the circumstances. The Guides herein recommended are appropriate for normal peacetime operations.

(7) These Guides are not intended to apply to radiation exposure resulting from natural background or the purposeful exposure of patients by practitioners of the healing arts.

(8) It is recognized that our present scientific knowledge does not provide a firm foundation within a factor of two or three for selection of any particular numerical value in preference to another value. It should be recognized that the Radiation Protection Guides recommended in this paper are well below the level where biological damage has been observed in humans.

It is recommended that:

4. Current protection guides used by the agencies be continued on an interim basis for organ doses to the population.

Recommendations are not made concerning the Radiation Protection Guides for individual organ doses to the population, other than the gonads. Unfortunately, the complexities of establishing guides applicable to radiation exposure of all body organs preclude the Council from making recommendations concerning them at this time. However, current protection guides used by the agencies appear appropriate on an interim basis.

It is recommended that:

5. The term "Radioactivity Concentration Guide" be adopted for Federal use. This term is defined as the concentration of radioactivity in the environment which is determined to result in whole body or organ doses equal to the Radiation Protection Guide.

Within this definition, Radioactivity Concentration Guides can be determined after the Radiation Protection Guides are decided on. Any given Radioactivity Concentration Guide is applicable only for the circumstances under which the use of its corresponding Radiation Protection Guide is appropriate.

It is recommended that:

6. The Federal agencies, as an interim measure, use radioactivity concentration guides which are consistent with the recommended Radiation Protection Guides. Where no Radiation Protection Guides are provided, Federal agencies continue present practices.

No specific numerical recommendations for Radioactivity Concentration Guides are provided at this time. However, concentration guides now used by the agencies appear appropriate on an interim basis. Where appropriate radioactivity concentration guides are not available, and where Radiation Protection Guides for specific organs are provided herein, the latter Guides can be used by the Federal agencies as a starting point for the derivation of radioactivity concentration guided applicable to their particular problems. The Federal Radiation Council has also initiated action directed towards the development of additional Guides for radiation protection.

It is recommended that:

7. The Federal agencies apply these Radiation Protection Guides with judgment and discretion, to assure that reasonable probability is achieved in the attainment of the desired goal of protecting man from the undesirable effects of radiation. The Guides may be exceeded only after the Federal agency having jurisdiction over the matter has carefully considered the reason for doing so in light of the recommendations in this paper.

The Radiation Protection Guides provide a general framework for the radiation protection requirements. It is expected that each Federal agency, by virtue of its immediate knowledge of its operating problems, will use these Guides as a basis upon which to develop detailed standards tailored to meet its particular requirements. The council will follow the activities of the Federal agencies in this area and will promote the necessary coordination to achieve an effective Federal program.

If the foregoing recommendations are approved by you for the guidance of Federal agencies in the conduct of their radiation protection activities, it is further recommended that this memorandum be published in the FEDERAL REGISTER.

ARTHUR S. FLEMMING
Chairman
Federal Radiation Council

The "recommendations numbered "1" through "7" contained in the above memorandum are approved for the guidance of Federal agencies, and the memorandum shall be published in the FEDERAL REGISTER.

DWIGHT D. EISENHOWER
MAY 13, 1960

(F.R. Doc. 60-4539; Filed, May 17, 1960; 8:51 a.m.)

Appendix 3-2

Federal Radiation Council

RADIATION PROTECTION GUIDANCE FOR FEDERAL AGENCIES

Memorandum for the President

SEPTEMBER 13, 1961

Pursuant to Executive Order 10831 and Public Law 86-373, the Federal Radiation Council herewith transmits its second report to you concerning findings and recommendations for guidance for Federal Agencies in the conduct of their radiation protection activities.

Background

On May 13, 1960, the first recommendations of the Council were approved by the President and the memorandum containing these recommendations was published in the FEDERAL REGISTER on May 18, 1960. There was also released at the same time, Staff Report No. 1 of the Federal Radiation Council, entitled, "Background Material for the Development of Radiation Protection Standards", dated May 13, 1960. The first report of the Council provided a general philosophy of radiation protection to be used by Federal agencies in the conduct of their specific programs and responsibilities. It introduced and defined the term "Radiation Protection Guide" (RPG). It provided numerical values for Radiation Protection Guides for the whole body and certain organs of radiation workers and for the whole body of individuals in the general population, as well as an average population gonodal dose. It introduced as an operational technique, where individual whole body doses are not known, the use of a "suitable sample" of the exposed population in which the guide for the average exposure of the sample should be one-third the RPG for the individual members of the group. It emphasized that this operational technique should be modified to meet special situations. In selecting a suitable sample particular care should be taken to assure that a disproportionate fraction of the average dose is not received by the most sensitive population elements. The observations, assumptions, and comments set out in the memorandum published in the FEDERAL REGISTER, May 18, 1960, are equally applicable to this memorandum.

This memorandum contains recommendations for the guidance of Federal agencies in activities designed to limit exposure of members of population groups to radiation from radioactive materials deposited in the body as a result of their occurrence in the environment. These recommendations include: (1) Radiation Protection Guides for certain organs of individuals in the general population, as well as averages over suitable samples of exposed groups; (2) guidance on general principles of control applicable to all radionuclides occurring in the environment; and (3) specific guidance in connection with exposure of population groups to radium-226, iodine-131, strontium-90, and strontium-89. It is the intention of the Council to release the background material leading to these recommendations as Staff Report No. 2 when the recommendations contained herein are approved.

Specific attention was directed to problems associated with radium-226, iodine-131, strontium-90, and strontium-89. Radium-226 is an important naturally occurring radioactive material. The other three were present in fallout from nuclear weapons testing. They could, under certain circumstances, also be major constituents of radioactive materials, released to the environment from large scale atomic energy installations used for peaceful purposes. Available data suggest that effective control of these nuclides, in cases of mixed fission product contamination of the environment, would provide reasonable assurance of at least comparable limitation of hazard from other fission products in the body.

Establishment of the Federal Radiation Council followed a period of public concern incident to discussions of fallout. While strontium-90 received the greatest popular attention, exposures to cesium-137, iodine-131, strontium-89 and, in still lesser degrees to other radionuclides, are involved in the evaluation of over-all effects. The characteristics of cesium-137 lead to direct comparison with whole body exposures for which recommendations by the Council have already been made.

Studies by the staff of the Council indicate that observed concentrations of radioactive strontium in food and water do not result in concentrations in the skeleton (and consequently in radiation doses) as large as have been assumed in the past. However, concentrations of iodine-131 in the diets of small children, particularly in milk, equal to those permitted under current standards would lead to radiation doses to the child's thyroid which, in comparison with the general structure of current radiation protection standards, would be too high. This is because current concentration guides for exposure of population groups to radioactive materials in air, food, and water have been derived by application of a single fraction to corresponding occupational guides. In the case of iodine-131 in milk, consumption of milk and retention of iodine by the child may be at least as great as by the adult, while the relatively small size of the thyroid makes the radiation dose to the thyroid much larger than in the case of the adult. In addition, there is evidence that irradiation of the thyroid involves greater risk to children than to adults.

Recommendations as to Radiation Protection Guides

The Federal Radiation Council has previously emphasized that establishment of radiation protection standards involves a balancing of the benefits to be derived from the controlled use of radiation and atomic energy against the risk of radiation exposure. In the development of the Radiation Protection Guides contained herein, the Council has considered both sides of this balance. The Council has reviewed available knowledge, consulted with scientists within and outside the Government, and solicited views of interested individuals and groups from the general public. In particular, the Council has not only drawn heavily upon reports published by the National Committee on Radiation Protection and Measurements (NCRP), and the National Academy of Sciences (NAS), but has had during the development of the report the benefit of consultation with, and comments and suggestions by, individuals from NCRP and NAS and of their subcommittees. The Radiation Protection Guides recommended below are considered by the Council to represent an appropriate balance between the requirements of health protection and of the beneficial uses of radiation and atomic energy.

It is recommended that:

1. The following Radiation Protection Guides be adopted for normal peacetime operations.

It will be noted that the preceding table provides Radiation Protection Guides to be applied to the average of a suitable sample of an exposed

TABLE A3-2-1

RADIATION PROTECTION GUIDES (RPG) FOR CERTAIN BODY ORGANS IN
RELATION TO EXPOSURE OF POPULATION GROUPS

Organ	RPG for individuals	RPG for average of suitable sample of exposed population group
Thyroid	1.5 rem/year	0.5 rem/year
Bone marrow	0.5 rem/year	0.17 rem/year
Bone	1.5 rem/year	0.5 rem/year
Bone (alternate guide)	0.003 μg ^{226}Ra in the adult skeleton or the biological equivalent of this amount of ^{226}Ra	0.001 μg ^{226}Ra in the adult skeleton or the biological equivalent of t

population group which are one-third of these applying to individuals. This is in accordance with the recommendations in the first report of the Council concerning operational techniques for controlling population exposure. Since in the case of exposure of a population group to radionuclides the radiation doses to individuals are not usually known, the organ dose to be used as a guide for the average of suitable samples of an exposed population group is also given as an RPG.

Recommendations as to general principles. Control of population exposure from radionuclides occurring in the environment is accomplished in general either by restriction on the entry of such materials into the environment or through measures designed to limit the intake by members of the population of radionuclides already in the environment. Both approaches involve the consideration of actual or potential concentrations of radioactive material in air, water, or food. Controls should be based upon an evaluation of population exposure with respect to the RPG. For this purpose, the total daily intake of such materials, averaged over periods of the order of a year, constitutes an appropriate criterion.

The control of the intake by members of the general populations of radioactive materials from the environment can appropriately involve many different kinds of actions. The character and import of these actions may vary widely, from those which entail little interference with usual activities, such as monitoring and surveillance, to those which involve a major disruption, such as condemnation of food supplies. Some control actions may require prolonged lead times before becoming effective, e.g., major changes in processing facilities or water supplies. The magnitude of control measures should be related to the degree of likelihood that the RPG may be exceeded. The use of a single numerical intake value, which in part has been the practive until now, does not in many instances provide adequate guidance for taking actions appropriate to the risk involved. For planning purposes, it is desirable that insofar as possible control actions to meet contingencies be known in advance.

It is recommended that:

2. The radiological health activities of Federal agencies in connection with environmental contamination with radioactive materials be based, within the limits of the agency's statutory responsibilities, on a graded series of appropriate actions related to ranges of intake of radioactive materials by exposed population groups.

In order to provide guidance to the agencies in adapting the graded approach to their own programs, the recommendations pertaining to the specific radionuclides in this memorandum consider three transient daily rates of intake by suitable samples of exposed population groups. For the other radionuclides, the agencies can use the same general approach, the details of which are considered in Staff Report No. 2. The general types of action appropriate when these transient rates of intake fall into the different ranges are also discussed in Staff Report No. 2. The purpose of these actions is to provide reasonable assurance that average rates of intake by a suitable sample of an exposed population group, averaged over the sample and averaged over periods of time of the order of one year, do not exceed the upper value of Range II. The general character of these actions is suggested in the following table.

Recommendations on Ra-226, I-131, Sr-90, and Sr-89

The Council has given specific consideration to the effects on man of rates of intake of radium-226, iodine-131, strontium-90, and strontium-89 resulting in radiation doses equal to those specified in the appropriate RPG's. The Council has also reviewed past and current activities resulting

TABLE A3-2-2

GRADED SCALES OF ACTION

Ranges of transient rates of daily intake	Graded scale of action
I	Periodic confirmatory surveillance as necessary
II	Quantitative surveillance and routine control
III	Evaluation and application of additional control measures as necessary

in the release of these radionuclides to the environment and has given consideration to future developments. For each of the nuclides three ranges of transient daily intake are given which correspond to the guidance contained in Recommendation 2, above. Routine control of useful applications of radiation and atomic energy should be such that expected average exposures of suitable samples of an exposed population group will not exceed the upper value of Range II. For iodine-131 and radium-226, this value corresponds to the RPG for the average of a suitable sample of an exposed population group. In the cases of strontium-90 and strontium-89, the Council's study indicated that there is currently no known operational requirement for an intake value as high as the one corresponding to the RPG. Hence, a value estimated to correspond to doses to the critical organ not greater than one-third of the RPG has been used.

The guidance recommended below is given in terms of transient rates of (radioactivity) intake in micromicrocuries per day. The upper limit of Range II is based on an annual RPG (or lower, in case of radioactive strontium) considered as an acceptable risk for a lifetime. However, it is necessary to use averages over periods much shorter than a lifetime for both radiation dose rates and rates of intake for administrative and regulatory purposes. It is recommended that such periods should be of the order of one year. It is to be noted that values listed in the tables are much smaller than any single intake from which an invidual might be expected to sustain injury.

It is recommended that:

3. (a) The guidance on daily intake be adopted for normal peacetime operations to be applied to the average of suitable samples of an exposed population group as shown in Table III.

   (b) Federal agencies determine concentrations of these radionuclides in air, water, or items of food applicable to their particular programs which are consistent with the guidance contained herein on average daily intake for the radionuclides radium-226, iodine-131, strontium-90, and strontium-89. Some of the general considerations involved in the derivation of concentration values from intake values are given in Staff Report No. 2.

It is recommended that:

4. For radionuclides not considered in this report, agencies use concentration values in air, water, or items of food which are consistent with recommended Radiation Protection Guides and the general guidance on intake.

In the future, the Council will direct attention to the development of appropriate radiation protection guidance for those radionuclides for

TABLE A3-2-3

RANGES OF TRANSIENT RATES OF INTAKE (μCi/day)
FOR USE IN GRADED SCALE OF ACTIONS SUMMARIZED
IN TABLE II

Radionuclides	Range I	Range II	Range III
^{226}Ra	0–2	2–20	20–200
^{131}I[a]	0–10	10–100	100–1000
^{90}Sr	0–20	20–200	200–2000
^{89}Sr	0–200	200–2000	2000–20000

[a] In the case of ^{131}I, the suitable sample would include only small children. For adults, the RPG for the thyroid would not be exceeded by rates of intake higher by a factor of 10 than those applicable to small children.

which such consideration appears appropriate or necessary. In particular, the Council will study any radionuclides for which useful applications of radiation or atomic energy require release to the environment of significant amounts of these nuclides. Federal agencies are urged to inform the Council of such situations.

ABRAHAM RIBICOFF
Chairman
Federal Radiation Council

The recommendations numbered "1" through "4" contained in the above memorandum are approved for the guidance of Federal agencies, and the memorandum shall be published in the FEDERAL REGISTER.

JOHN F. KENNEDY

SEPTEMBER 20, 1961

Chapter 4

Mechanisms of Transport in the Atmosphere

Release of radioactive contaminants to the atmosphere may result in inhalation of the contaminated air or, under extreme conditions, in γ-radiation exposure from a passing cloud or from radionuclides deposited on surfaces. In addition, direct deposition of radioactive particles of sufficiently high activity can produce skin burns, and deposition on the surfaces of plant parts may result in contamination of man's food supplies. The atmosphere may also serve as the medium of transport from the source of contamination to soil and water.

In order to evaluate the consequences of a release of radioactivity to the atmosphere, one must forecast the fate of the effluent in space and time. The potential hazard of a release depends on the way in which it is diluted and transported by the atmosphere and on the mechanisms by which the contaminants deposit on surfaces.

This chapter is intended to acquaint the nonmeteorologist with the basic principles by which calculations are made of diffusion and deposition of pollutants released to the atmosphere. The treatment in this chapter will be of necessity superficial but hopefully adequate for an appreciation of the basic methodology that can be employed.

87

Properties of the Atmosphere

Until World War II, man-made atmospheric contaminants were injected into the atmosphere at relatively low heights from the ground, and if one wished to study the properties of the atmosphere that governed the transport of such contaminants it was necessary only to study the physical behavior of the lowest levels of the atmosphere. Practically all man-made effluents were injected into the surface friction layer which extends to about 100 m from the ground and which has properties somewhat different from the rest of the atmosphere because of the influences of surface features on atmospheric flow. The study of this boundary layer has been called micrometeorology. Transport and mixing within this layer has been studied primarily by meteorologists concerned with dilution of industrial effluents and chemical warfare agents (Sutton, 1953).

The advent of atomic energy, and more particularly the testing of nuclear weapons, extended the problem of forecasting the fate of atmospheric contaminants considerably above the friction layer to altitudes of 100,000 ft or greater. In more recent years the movement of the upper atmosphere, even to the fringes of outer space, has become important. The properties of the upper atmosphere must be thoroughly understood to permit the precise calculations of trajectories of rockets and satellites. In the future, high performance aircraft may operate at altitudes above 60,000 ft, and studies of the fate of pollutants introduced by them will require additional knowledge of the physics and chemistry of the upper atmosphere.

In this chapter we shall devote most of our attention to the properties of the friction layer where most radioactive air-pollution problems arise and where our knowledge of diffusion phenomena is most complete. To the extent permitted by the somewhat scanty state of our knowledge, we shall also consider the upper region of the atmosphere.

Although the atmosphere contains many gases, as shown in Table 4-1, more than 99.9% of its weight is due to nitrogen, oxygen, and argon (Mason, 1960). The relative proportions of these gases remain constant to great heights, but separation due to differences in molecular weight does occur above 60 kilometers (km). The total mass of the dry atmosphere is thought to be about 50×10^{17} kilograms (kg), to which may be added about 1.5×10^{17} kg of water vapor, the most variable constituent of the atmosphere and that which governs many of its thermodynamic characteristics. Dry air has a density of 0.0013 g/cm³ at the surface of the earth, where pressure due to the weight of the atmosphere is 760 mm of mercury. At 50 km above sea level, the atmospheric pressure has dropped to 10^{-3} of the sea-level pressure and has attenuated by a factor of 10^{-6} at an alti-

TABLE 4–1

THE AVERAGE COMPOSITION OF THE ATMOSPHERE[a]

Gas	Composition by volume (ppm)	Composition by Weight (ppm)	Total mass ($\times 10^{22}$ g)
N_2	780,900	755,100	38.648
O_2	209,500	231,500	11.841
A	9,300	12,800	0.655
CO_2	300	460	0.0233
Ne	18	12.5	0.000636
He	5.2	0.72	0.000037
CH_4	1.5	0.9	0.000043
Kr	1	2.9	0.000146
N_2O	0.5	0.8	0.000040
H_2	0.5	0.03	0.000002
O_3[b]	0.4	0.6	0.000031
Xe	0.08	0.36	0.000018

[a] Adopted from Mason (1960).
[b] Variable, increases with height.

tude of 100 km. The thinness of the atmosphere at these altitudes is illustrated by the length of the mean free path between molecules, about 2.5 cm at 100 km and about 25 m at 300 km (Petterssen, 1958). Above an altitude of about 600 km, the molecules are thought to behave as satellites describing free eliptical orbits about the earth (Mason, 1960).

The atmosphere contains natural and man-made aerosols that originate from many sources. In addition to the air pollutants introduced as the result of human activities, meteorites, volcanic activity, dust storms, forest fires, and ocean spray all combine to contribute great quantities of suspended solids.

When a gas vapor or aerosol is introduced into the atmosphere, it dilutes by either molecular or turbulent diffusion. In practice, one can usually neglect the contribution of molecular diffusion in which the coefficients of diffusivity are several orders of magnitudes smaller than those due to turbulent diffusion. The total range of values of the diffusivity coefficients that control the rates of atmospheric dilution is enormous, ranging from 0.2 cm²/sec for molecular diffusion to 10^{11} cm²/sec for the diffusion due to large-scale cyclonic storms in the atmosphere (U.S. Weather Bureau, 1955). The atmospheric motions that contribute to the mixing processes thus vary in scale from almost microscopic eddies to much larger motions in which the scales of distances are measured in hundreds of kilometers.

The motions of turbulent diffusion are so complicated mathematically that exact theories are not available to describe the manner in which a contaminant introduced into the atmosphere behaves in space and time. However, we shall see later that methods have evolved which are basically statistical in nature and which make it possible for one to predict the manner in which a contaminant will diffuse in the atmosphere under a given set of meteorological conditions. These techniques are of enormous value for the purposes of estimating the consequences of releasing a noxious emission to the atmosphere.

The mixing characteristics of the atmosphere are governed in a major way by its vertical temperature gradient. A typical temperature profile in the Temperate Zone is illustrated in Fig. 4-1. As one rises in height above the ground, the temperature normally decreases at a rate called the "lapse rate of temperature," about $-3.5°F$ per 1000 ft ($-6.5°C/km$). It is seen from Fig. 4-1 that the temperature decreases with height to about 11 km, where one encounters an isothermal region of the atmosphere which, in the example shown, extends to a height of about 32 km. The lower region of the atmosphere in which the temperature normally diminishes with height is called the troposphere, which contains about 75% of the mass of the atmosphere and almost all its moisture and dust (Petterssen, 1958). Above the troposphere, separated by an imaginary boundary called the tropopause, is the stratosphere. The height of the tropopause varies with latitude and with the season of the year. In contrast to the stratosphere, the troposphere is a relatively unstable (turbulent) region of the atmos-

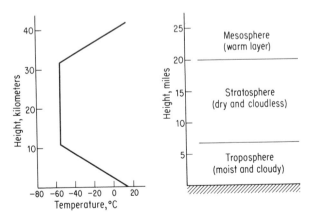

Fig. 4-1. Idealized profile of the atmospheric temperature gradients. From "Introduction to Meteorology" by S. Pettersen. Copyright 1958, McGraw-Hill. Used with permission of McGraw-Hill Book Company.

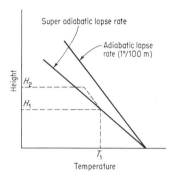

Fig. 4-2. The instability of the superadiabatic atmosphere. A parcel of air raised in height from H_1 to H_2 cools adiabatically, and its rate of rise is accelerated because it becomes warmer and, therefore, less dense than the ambient atmosphere.

phere. The isothermal stratosphere continues to a height of about 100,000 ft, where the temperature begins to rise. The nomenclature for this portion of the atmosphere is not always consistent in the literature; in this presentation we shall call the warm layer the mesosphere. It extends upward to the ionosphere, a densely ionized region that begins at altitudes greater than 200,000 ft.

Under normal daytime conditions, the earth's surface absorbs solar radiation and becomes warmer than the air above. The lapse rate then becomes superadiabatic; that is, the temperature decreases at a rate exceeding that which would occur if a parcel of air was raised and, with the drop in pressure, was permitted to expand adiabatically. The dry adiabatic lapse rate is normally about $-1°C/100$ m. The influence of the vertical temperature gradient on the stability of the atmosphere is shown in Fig. 4-2. If a parcel of air having a temperature T_1 at altitude H_1 is raised to altitude H_2, it will cool adiabatically at a rate of $-1°C/100$ m. Since the superadiabatic conditions shown in Fig. 4-2 is assumed to exist, the parcel of air initially at H_1 will be warmer than the ambient atmosphere when it reaches H_2. It will thus be of lower density than the surrounding air and will continue to rise. Similarly, if a parcel of air is lowered in altitude in a surrounding atmosphere which is in the superadiabatic condition, the parcel will be denser than the ambient atmosphere and will continue to descend. Thus, in the superadiabatic condition, all vertical motions tend to be accelerated, and the atmosphere is said to be unstable. Figure 4-3 shows that the reverse situation exists when the lapse rate is less than adiabatic.

In fact, as shown in Fig. 4-3, the temperature gradient may become

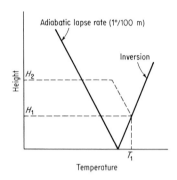

Fig. 4-3. The inherent stability of the inverted temperature gradient. A parcel of air raised in height from H_1 to H_2 cools adiabatically and sinks to its original position because it becomes more dense than the ambient atmosphere.

"inverted" (i.e., the temperature may increase with height). This results in the very stable condition called an inversion. Inversions may be caused by the overrunning of warm air over cold (as along a front between air masses), by advection of cool air at low level (as during a sea breeze), or by diurnal cooling of the atmosphere. This last condition usually develops after sunset when the surface of the earth, having been warmed during the day, begins to emit blackbody radiation at a rate faster than its gaseous atmosphere. As the night proceeds and the surface of the earth becomes cooler relative to the atmosphere, the temperature gradient may in time become positive.

The manner in which this happens is shown in Fig. 4-4 (Holland, 1953), which gives an example of how the vertical temperature gradient may change in a 24-hr period. A sharp inversion to a height of about 500 ft is evident at 0600 and has begun to weaken slightly at 0700. By 0800 the morning sun has heated the ground, and the inversion begins to disappear with a superadiabatic gradient developing between 1100 and 1300 hr. By late afternoon, as the ground cools, an inversion develops again and persists throughout the night. If the sun rises on a clear day, the superadiabatic condition will again develop by late morning, and the cycle will be repeated. However, should cloud cover limit solar heating of the earth's surface, the inversion may persist through the day.

The change in temperature profiles greatly affects the characteristics of plumes of stack gases, and it is seen immediately from Fig. 4-5 that the concentration of a contaminant at ground level is greatly affected by the effect of the vertical temperature gradient on atmosphere stability.

The difference between turbulent and laminar flow is the subject of a familiar lecture-hall demonstration in which thin filaments of a dye are introduced in a direction parallel with the motion of water in a long, straight glass tube. When the velocity of the water in the tube is sufficiently slow, the filaments retain their form, and there is little or no visible mixing between the dye and clear water. As the velocity increases, the filaments break up and mix rapidly across the entire section of the tube. In this experiment the motion starts as laminar flow, and as the velocity increases the flow becomes turbulent. The physics of laminar and turbulent flow was first understood by Reynolds in classic hydraulic investigations which he conducted in the latter part of the nineteenth century. The basic mathematical relationships which he derived are also applicable in aerodynamics (Sutton, 1953).

Turbulence is characterized by highly nonregular motions which have

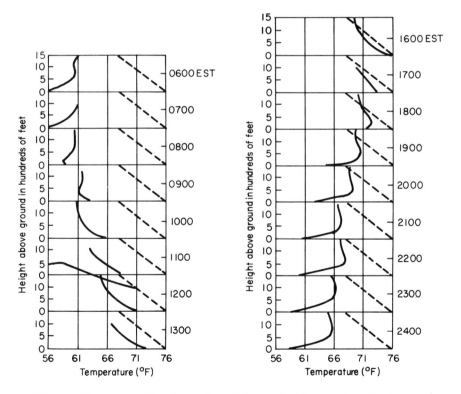

Fig. 4-4. The average diurnal variation of the vertical temperature structure at the Oak Ridge National Laboratory during the period September–October, 1950 (solid lines). The dashed lines represent the adiabatic lapse rate (Holland, 1953).

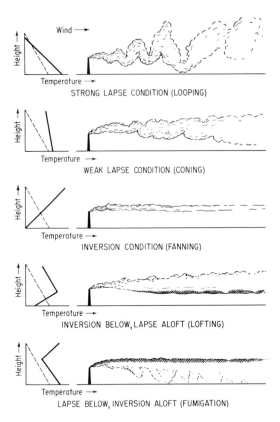

Fig. 4-5. Schematic representation of stack-gas behavior under various conditions of vertical stability. Actual temperature (solid line) and dry adiabatic lapse rate (dashed line) are shown (U.S. Weather Bureau, 1955).

thus far escaped exact mathematical definition. In true laminar flow, mixing between adjacent layers is the result of random, Brownian molecular motions that are insignificant compared to the effects of the much larger random motions of turbulent flow. Atmospheric motions in nature are turbulent, and atmospheric mixing is the result of turbulent diffusion which, for all practical purposes, completely masks the effects of molecular diffusion.

The apparent randomness of turbulent motion has resulted in the evolution of equations of atmospheric mixing that are based on the mathematical analogy of turbulent and molecular diffusion. The original equations for the molecular diffusion of gases were developed by the physiologist Fick and depended on the analogy with Fourier's laws of heat conduction.

Fick's law of molecular diffusion states that diffusion of material is in the direction of decreasing concentration and is proportional to the concentration gradient. From this principle the following differential equation has been derived (U.S. Weather Bureau, 1955):

$$\frac{d\chi}{dt} = \frac{\partial}{\partial x}\left(K_x \frac{\partial \chi}{\partial x}\right) + \frac{\partial}{\partial y}\left(K_y \frac{\partial \chi}{\partial y}\right) + \frac{\partial}{\partial z}\left(K_z \frac{\partial \chi}{\partial z}\right) \tag{4-1}$$

where χ is the concentration and x, y, and z are distances in three-coordinate system. K_x, K_y, and K_z, are coefficients of diffusion.

It has been noted that the values of these exchange coefficients vary with the scale of atmospheric motion, from 0.2 cm²/sec for molecular diffusion to 10^{11} cm²/sec for the effect of large-scale cyclonic storms. The rate of diffusion thus depends on the scale of motions involved in any given stage of dispersion. In practice, one cannot readily estimate the magnitude of the K's as a function of time.

Because the values of K are not known, an exact solution to Eq. (4-1) has not been formulated. In order to approximate the solution, most formulas assume that the distribution of pollutants in a plume is Gaussian distributed in both the horizontal and vertical directions.

DIFFUSION IN THE FRICTION LAYER FROM A CONTINUOUS-POINT SOURCE

In order to determine ground-level concentrations downwind from a continuous-point source, the following equation is commonly used (Slade, 1968):

$$\chi_{(x,y)} = \frac{Q}{\pi \sigma_y \sigma_z \bar{u}} \exp -\left(\frac{h^2}{2\sigma_y^2} + \frac{y^2}{2\sigma_z^2}\right) \tag{4-2}$$

where

χ = concentration (Ci/m³) at downwind point (x, y, o)
Q = source strength (Ci/sec)
σ_y, σ_z = crosswind and vertical plume standard deviations (m). Both are functions of x
$\bar{u}$ = mean wind speed (m/sec) at the stack elevation, h_s (m)
h = effective stack height ($h_s + \Delta h$, the plume rise) (m)
x, y = downwind and crosswind distances (m)

Numerical values for σ_y and σ_z will vary according to stability conditions, wind shear, and roughness of the terrain. Therefore, if extremely high accuracy is required, diffusion coefficients must be determined experimentally at site locations. However, for practical application, Meade

TABLE 4-2

Relation of Turbulence Types to Weather Conditions[a,b]

Surface wind speed (m/sec)	Daytime insolation			Night-time conditions	
	Strong	Moderate	Slight	Thin overcast or $\geq \frac{4}{8}$ cloudiness[c]	$\geq \frac{3}{8}$ cloudiness
<2	A	A–B	B	—	—
2	A–B	B	C	E	F
4	B	B–C	C	D	E
6	C	C–D	D	D	D
>2	C	D	D	C	D

[a] Slade (1968).
[b] Conditions: A, extremely unstable; B, moderately unstable; C, slightly unstable; D, neutral (applicable to heavy overcast, day or night); E, slightly stable; F, moderately stable.
[c] The degree of cloudiness is defined as that fraction of the sky above the local apparent horizon which is covered by clouds.

(1960) and Pasquill (1961, 1962) defined values for σ_y and σ_z by classifying stability conditions according to prevailing conditions of average wind speed and estimated radiation balance, as shown in Table 4-2. Figure 4-6 and 4-7 present values of σ_y and σ_z as a function of distance for each stability class.

A great many field tests of the diffusion equation have been made, aimed at fitting values of the turbulence and diffusion parameters and determining the applicability of the equation for a variety of terrain types. Ratios of average to peak values have been determined, and the manner in which the concentration varies along the y and z axis across sections of the plume (Slade, 1968). In general, the observed hourly average concentrations given by Eq. (4-2) are shown to be about fourfold higher than actually observed under lapse conditions (Smith, 1951).

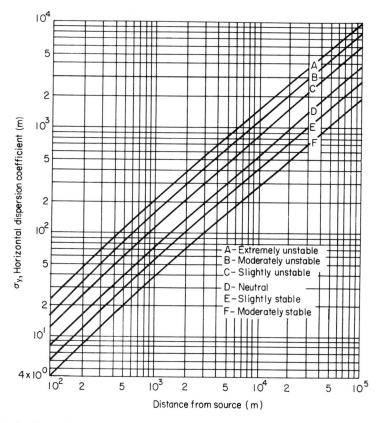

Fig. 4-6. Lateral diffusion (σ_y) *versus* downwind distance from source for Pasquill's turbulence types (U.S. Atomic Energy Commission, 1968b).

This is partly because the diffusion equation predicts the concentration at the center line of the cloud, whereas the cloud during the sampling period meanders on either side of its mean center line. The average value of χ will, therefore, be lower than predicted. Another reason is that the equation assumes the surface of the earth to be aerodynamically smooth, whereas in actual practice the surface is almost invariably aerodynamically rough (Gosline et al., 1956). In general, it may be concluded that the values predicted by the diffusion equations will be higher than the observed values by a factor that will vary depending on the length of the sampling period and on topographical factors.

The predictions are least reliable during extremely stable inversion conditions. Under such conditions, plumes have been observed to remain aloft

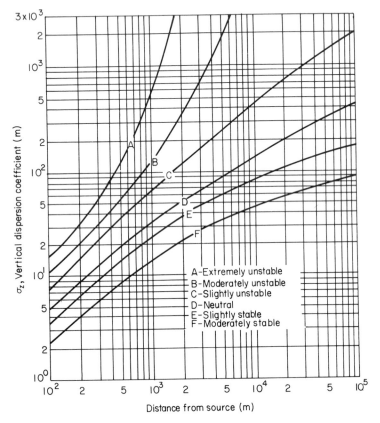

A-Extremely unstable
B-Moderately unstable
C-Slightly unstable
D-Neutral
E-Slightly stable
F-Moderately stable

Fig. 4-7. Vertical diffusion (σ_z) *versus* downwind distance from source for Pasquill's turbulence types (A–F) (U.S. Atomic Energy Commission, 1968b).

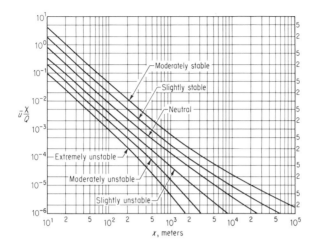

Fig. 4-8. Values of $\bar{u}\chi/Q$ as a function of downwind distance for a source located at the surface. Pasquill's diffusion categories are represented by solid curves (Hilsmeier and Gifford, 1962).

indefinitely, with no evidence of ground-level contamination for more than 25 miles downwind from a 400 ft stack (Smith, 1951).

Gifford (1961) and Hilsmeier and Gifford (1962) have developed curves of concentration normalized to wind speed and emission rate $(\bar{u}\chi/Q)$ for various stack heights. These curves, based on the Meade and Pasquill formulas, are shown in Figs. 4-8 to 4-10.

The curves illustrate the principal general features of all equations that describe the diffusion of stack effluents. The concentration maxima occur at increasing distances with increasing values of h and with increasing degrees of stability. The maximum concentration decreases in inverse proportion to h^2.

The diffusion equation can be usefully employed to predict the maximum ground-level concentration that will occur downwind from a stack. If one solves Eq. (4-2) for χ_{max}, one obtains

$$\chi_{max} = \frac{2Q}{e\pi\bar{u}h^2}\frac{\sigma_z}{\sigma_y} \tag{4-3}$$

where e is the base of the natural logarithms (2.72). The downwind distance from the stack to χ_{max} is usually equivalent to 15–30 stack heights.

It is seen that χ_{max} is inversely proportional to the wind speed and the square of stack height and directly proportional to the rate of emission.

Note that the concentration downwind from the stack depends on emis-

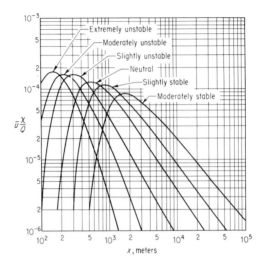

Fig. 4-9. Values of $\bar{u}\chi/Q$ as a function of downwind distance for a source located at a height of 30 m (Hilsmeier and Gifford, 1962).

sion rate from the source rather than on concentration of the contaminant in the effluent air. The concentrations downwind cannot be reduced by simply increasing the quantity of air being discharged. The concentration in the plume is determined by the mass rate of emission and the volume of the plume. The latter is a function of meteorological parameters and is ordinarily unaffected by the volume of exhaust air which is an insignificant addition to the mass (or volume) of the turbulence cells into which the contaminant will be dispersed.

EFFECTIVE STACK HEIGHT

Pollutants are frequently emitted to the atmosphere at an elevated temperature and with considerable vertical velocity. The buoyancy resulting from the combined effects of the temperature and velocity of the exhaust gas sometimes results in an *effective* stack height somewhat higher than the actual height. The problem of combining the various meteorological parameters together with the effluent temperature and velocity is a complicated one because of the way in which the various parameters interact. For example, while the effect of a temperature inversion is to suppress vertical motions and might, therefore, be expected to reduce the effective stack height, the inversion reduces turbulent mixing to an extent that results in a more persistent temperature differential between the plume and the surrounding atmosphere.

Many investigators have attempted to develop theoretical and empirical methods of computing the effective stack height, taking into consideration the amount of heat in the plume and the ambient vertical temperature gradient (Slade, 1968).

Briggs (1969) has reviewed the various methods of calculating the buoyancy effect of a heated plume and concluded that all the proposed methods have defects, but that most of the experimental data reported in the literature fit the following equation:

$$\Delta h = 1.6 F^{1/3} u^{-1} x^{2/3} \tag{4-4}$$

where

Δh = effective stack height increment (m)
$F = 3.7 \times 10^{-5} Q_H$
Q_H = heat emission due to efflux of stack gasses (cal/sec)
x = downwind distance with a maximum value of 10 stack heights (m)

EFFECT OF BUILDINGS AND TERRAIN ON PLUME DISPERSION

One must, of course, be cautious to avoid applying these formulas in topographical situations in which normal flow patterns are apt to be per-

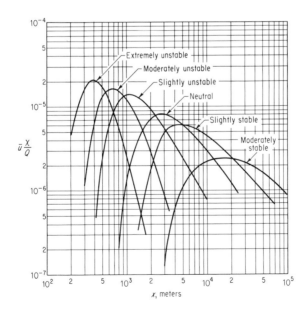

Fig. 4-10. Values of $\bar{u}\chi/Q$ as a function of downwind distance for a source located at a height of 100 m (Hilsmeier and Gifford, 1962).

turbed. Buildings close to the stack may result in the conditions shown in Fig. 4-11, which is a wind tunnel photograph of the effect of local structures on the dispersion of a plume. Micrometeorology in the vicinity of buildings is too complex for analytical solutions, and in many instances the question of how high a stack should be in order to avoid the down-wash in the lee of a building is best answered by wind tunnel tests (Strom, 1968).

It is helpful to understand the general characteristics of the wake downwind of a building. This is shown in Fig. 4-12 (Halitsky, 1968), which depicts the flow near a sharp-edged building. Three distinct zones exist: the displacement zone, the wake, and the cavity. The displacement zone is the volume in which the air is deflected around the solid building. Immediately downwind of the building is a region of torroidal circulation known as the cavity or eddy zone in which it is possible for the concentrations to accumulate in high concentrations. Beyond the cavity is the true wake, which is a region of high turbulence within which the contaminant spreads throughout. The area of the wake may be taken as twice the projected area of the building, and the average wind velocity is $\bar{u}/2$. From this we can estimate the average concentration on the assumption that the contaminant is mixed uniformly across the area of the wake, in which

Fig. 4-11. Perturbation of stack plumes by buildings. Note the effect of increased stack height in eliminating the downwash in the lee of the building (Professor Gordon Strom).

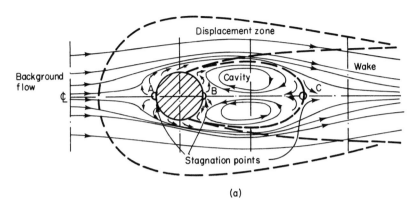

(a)

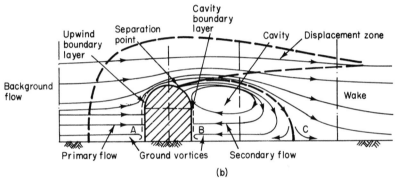

(b)

Fig. 4-12. Flow around a rounded building. (a) Flow in a horizontal plane near the ground; (b) flow in the longitudinal center plane (Halitsky, 1968).

case

$$\bar{\chi} = Q/Au \tag{4-5}$$

in which

A = area of the building projected to the airstream

Mountains and valleys also tend to distort flow under certain conditions, as shown in Fig. 4-13. There are also occasional situations in which inversions in combination with terrain features result in meteorological isolation of an area in the manner shown in Figs. 4-14 and 4-15. In Fig. 4-14, an inversion above a valley floor is restricting dilution of atmospheric contaminants. This was apparently the cause of the sudden onset of acute respiratory distress among the inhabitants of Donora, Pennsylvania dur-

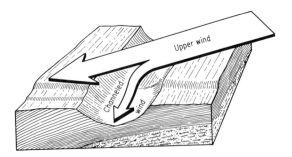

Fig. 4-13. Wind channeling by valley walls (U.S. Weather Bureau, 1955.

ing a 3-day inversion in 1948 (Schrenk *et al.*, 1949). The exact cause of this disaster, which contributed to the death of 20 people, has not been definitely established, but it is believed to have been caused by the sudden buildup of irritant gases emitted from the stacks of factories in the valley.

DIFFUSION OF ACCIDENTAL RELEASES IN THE LOWER ATMOSPHERE

Accidental releases may involve instantaneous discharges of radioactive contaminants to the lower atmosphere. When this happens, the manner in which dispersion occurs is dependent not only on the meteorological parameters but also on the temperature of the cloud and the total time during which the release occurs. The downwind concentrations resulting from a sudden release of a given magnitude will depend on the buoyancy of the cloud, which in stable air is altered by the relative density of the air which is entrained. According to Briggs (1969) rise of the cloud is approximated by

$$\Delta h = 2.66 \left(\frac{Q_i}{C_p \rho \, \partial\theta/\partial z} \right)^{1/4} \tag{4-6}$$

where

Q_i = instantaneous heat release (cal)
C_p = specific heat at constant pressure
ρ = density of air
$\partial\theta/\partial z$ = potential temperature gradient ($^\circ$K m^{-1})

During neutral conditions the rise of the cloud will be greater, but there has been insufficient experimentation to warrant the presentation of an equation in this text. Cloud rise in unstable air is limited only by the depth of the unstable layer.

The height at which cloud stabilization occurs, as determined from the

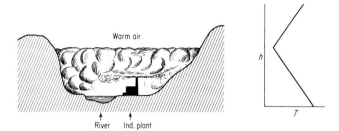

Fig. 4-14. Fumigation of valley floor caused by an inversion layer that restricts diffusion from a stack.

above equation, may be substituted for h in the diffusion equation for the continuous-point source.

As it is frequently not practical to specify the length of time during which the release will take place, one may modify formula (3-2) by substituting $\bar{Q}$, the total number of curies released in the incident, for Q, curies discharged per unit time. This results in the substitution of $\bar{\chi}t$, curie seconds per cubic meter (Ci sec/m^3) in place of χ. This is a convenient form in which to express the results of the accident, because when the result is expressed in curie seconds, one may proceed directly to a computation of the intergal dose from the radioactive contaminants.

DISPERSION OF AEROSOLS

The Sutton equations were derived for gaseous effluents but may be applied to plumes in which the contaminants are in the form of aerosols, provided their particle sizes are such that the settling rates are insignificant compared to the scale of vertical motion due to turbulent mixing. Baron, Gerhard, and Johnstone (1949) have shown that the Sutton formulas

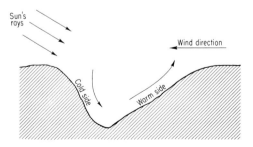

Fig. 4-15. Atmospheric overturn caused by uneven solar heating of valley walls.

can be used with particle diameters up to 20 μm when applied to the continuous release from stacks.

Deposition of Particulate Effluents

The mechanisms by which particulates will deposit on surfaces vary depending on whether the cloud passage is associated with precipitation. Dry deposition may result from gravitational settling or impaction on surfaces deflecting turbulent atmospheric flow. When precipitation occurs below the rain-forming level, the dust is washed to the surface by falling raindrops (washout). At higher altitudes, dust particles may serve as nuclei for condensing raindrops (rainout). This phenomenon is apparently responsible for removal of most submicron particles from the atmosphere.

The descent of particles through the atmosphere follows certain well-known physical laws (Drinker and Hatch, 1954; Dalla Valle, 1949) which govern the resistance to the motions of particles moving in viscous media. The force resisting the motion of a particle falling through air is

$$R = \tfrac{1}{2}\rho_a v^2 A C_d \qquad (4\text{-}7)$$

where

R = resistance (g sec/cm^2)
ρ_a = density of air (g/cm^3)
v = velocity of particle (cm/sec)
A = cross-sectional area of particle (cm^2)
C_d = drag coefficient

The drag coefficient is not constant for all conditions but varies systematically with the Reynolds number:

$$\text{Re} = v\, d\rho/\mu \qquad (4\text{-}8)$$

where

Re = Reynolds number
μ = viscosity of air (dyne/cm^2)
d = particle diameter (cm)
ρ = density of particle (g/cm^3)

For Reynolds numbers greater than 10^3, C_d is reasonably constant, and for spheres it has an average value of 0.44 (Drinker and Hatch, 1954). In this zone of turbulent motion, the viscosity of the air has no effect, and the resistance R varies with d^2 and v^2.

$$R = k\rho_a d^2 v^2 \qquad (4\text{-}9)$$

where $k = 0.055\ \pi$ for spheres.

For Reynolds numbers less than 3.0, in which region the flow is streamlined (laminar), $C_d = 24/\text{Re}$, and for spheres the resistance is defined by the classic equation developed by Stokes.

$$R = 3\pi\mu \, dv \tag{4-10}$$

For Reynolds numbers between 3 and 10^3, an intermediate zone, the relationships between C_d and Re become so complex that a mathematical solution has not been derived. An approximation adopted from Lapple and Sheppard (1940) is given by Drinker and Hatch.

A free falling particle will reach a terminal velocity when the resistance owing to its motion through the air is just balanced by the force, F_g dynes, due to gravitational attraction. Thus

$$F_g = R = \tfrac{1}{6}\pi d^3(\rho - \rho_a)g \tag{4-11}$$

where

ρ = density of particle
ρ_a = density of air
g = gravitational constant

From these relationships, one can obtain three equations for the terminal velocities of spherical particles describing flow in the streamline, intermediate, and turbulent regions.

Streamline motion $\left(d < \dfrac{115}{\rho^{1/3}}\right)$:

$$v = 0.003\rho d^2 \tag{4-12}$$

Intermediate motion $\left(\dfrac{115}{\rho^{1/3}} < d < \dfrac{2130}{\rho^{1/3}}\right)$:

$$v = 0.34\rho^{2/3}d \tag{4-13}$$

Turbulent motion $\left(d > \dfrac{2130}{\rho^{1/3}}\right)$:

$$v = 16\rho^{1/2}d^{1/2} \tag{4-14}$$

The settling velocities for quartz particles as a function of particle size are given in Fig. 4-16. Such particles will fall at slightly lower velocities than spheres because of their irregular shape. A useful rule of thumb is that a 10 μm mineral particle having a density of about 2.5 falls at a rate of about 1 fpm at sea level. The rate of fall of particles of other diameters (d) can be approximated from the relationship of $v = d^2/100$.

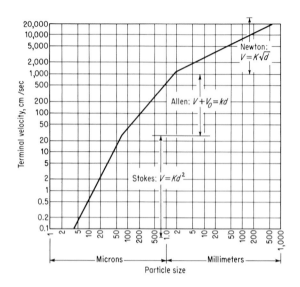

Fig. 4-16. Terminal velocity for quartz particles ($d = 2.6$) in air at sea level. From "Industrial Dust" by P. Drinker and T. Hatch. Copyright 1954, McGraw-Hill. Used with permission of McGraw-Hill Book Company.

When a particle's size becomes so small that its diameter is less than the mean free path of the gas molecules, the terminal velocity will increase over the values calculated from Stoke's equation. This may be corrected by using Cunningham's modification of Stoke's equation [Eq. (4-15)].

$$v_c = V_s\left(1 + \frac{1.7\lambda}{d}\right) \tag{4-15}$$

where

v_c = corrected velocity
V_s = velocity calculated from Stoke's law
λ = mean free path of gas molecules (cm)

Since λ is about 10^{-5} cm at sea level, uncorrected terminal velocities may be used at sea level for particles above 0.5 μm. However, a fivefold correction is required for particles of 0.05 μm at sea level.

IMPACTION OF DUST ON SURFACES

When a turbulent mass of air flows across a solid surface, suspended particles may be impacted against the surface. The efficiency of impaction is defined as the ratio of the number of particles deposited on the surface to those originally contained in the volume of air diverted. When a wind flows around a cylinder, the efficiency of impaction increases as the velocity increases and the diameter of the cylinder decreases (Chamberlain, 1955).

Theories have been developed which explain the manner in which impaction occurs as a function of particle size and velocity when regular shapes such as cylinders and planes are involved, but it is not practical to apply these to the irregular surfaces one finds in practice. Chamberlain modified the Sutton diffusion formulas by inserting into the equation a term to allow for the diminution in concentration due to surface deposition. The Chamberlain modification, which is applicable, provided the settling velocities of the particles are inappreciable (i.e., for particle diameters less than about 20 μm), employs the concept of velocity of deposition v_g which is defined as

$$v_g = \frac{\text{amount deposited per m}^2/\text{sec}}{\text{volumetric concentration per m}^3 \text{ above surface}}$$

In the absence of a theoretical basis for estimating v_g, a number of investigators have undertaken field observations designed to provide measurements of the velocity of deposition for a variety of surfaces. These data have been well summarized by Van der Hoven (1968), who found it difficult to draw any general conclusions from the data except that chemically active aerosols such as ^{131}I deposit more readily than inactive materials such as ^{137}Cs or nonradioactive fluorescent particles. Deposition on vegetation surfaces such as grasses and bushes was more rapid than on bare surfaces. However, the data show wide variability, which up to the present time have not been related to meteorological parameters. The data gathered for ^{131}I by Chamberlain in the United Kingdom (1960) and by Hawley and his associates in Idaho Falls (1964) show that the ^{131}I deposition velocity varies over an order of magnitude from about 0.2 to 3.7 cm/sec depending on the type of surface and other factors. Chamberlain and Chadwick (1966) concluded that the velocity of deposition of ^{131}I under experimental conditions was about 1 cm/sec on the average, but 0.5 cm/sec for radioiodine contained in fallout.

Chamberlain also considered the effect of washout of plumes by rain. He assumed that raindrops remove a constant fraction of the plume throughout its full height below the height at which the rain is formed. A *washout factor* (Λ) is thus substituted for the factor that accounts for deposition velocity, giving

$$\chi = \frac{2Q}{u\pi C_y C_z x^{2-n}} e^{-\Lambda x/u} \tag{4-16}$$

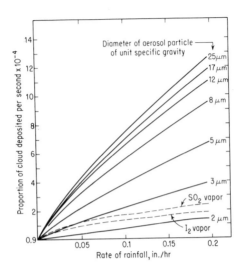

Fig. 4-17. Percentage removal of particles according to particle size and rate of rainfall (U.S. Weather Bureau, 1955, after Chamberlain, 1955).

in which Λ is the proportion of the cloud deposited per second, as determined from Fig. 4-17. This equation is applicable to continuous ground-level sources.

Similarly, the rate of deposition per unit area can be shown to be

$$D = \frac{\Lambda Q_0 e^{-\Lambda/u}}{u \pi^{1/2} C_y x^{(2-n)/2}} \qquad (4\text{-}17)$$

The inherent uncertainties in Eq. (4-17) can be eliminated if one considers only the maximum values of deposition (Chamberlain, 1953), thereby limiting consideration to the worst possible case. This washout rate for the Gaussian plume model, assuming that the rain-forming layer is above the plume, is

$$D = \frac{\Lambda Q}{u \sigma_y (2\pi)^{1/2}} \exp - \left(\frac{y^2}{2\sigma_y^2} \right) \qquad (4\text{-}18)$$

WIND PICKUP OF DEPOSITED PARTICLES

When radioactive particles have been deposited on the ground, there may be a risk that the particles can become reairborne as the result of wind action. This is another aspect of the general problem of airborne radioactivity of which little is known.

Healy and Fuquay (1958) have reviewed the literature on atmospheric

TABLE 4-3

ESTIMATED AIR CONCENTRATIONS OVER AN INFINITE PLANE FROM ONE PARTICLE OF A GIVEN SIZE PER SQUARE METER AT VARIOUS WING SPEEDS[a]

Particle diameter (μm)	Estimated terminal velocity (m/sec)	Particle concentration (particles/m³)				
		1(m/sec)	5(m/sec)	10(m/sec)	20(m/sec)	30(m/sec)
1.5	4.9×10^{-5}	2.7×10^{-3}	6.8×10^{-2}	0.27	1.08	2.4
8	1.1×10^{-3}	2.3×10^{-5}	5.6×10^{-4}	2.3×10^{-3}	9.1×10^{-3}	2.0×10^{-2}
14	4.6×10^{-3}	3.1×10^{-6}	7.8×10^{-5}	3.1×10^{-4}	1.2×10^{-3}	2.8×10^{-3}
78	0.11	2.3×10^{-8}	5.8×10^{-7}	2.3×10^{-6}	9.3×10^{-6}	2.1×10^{-5}
160	0.34	3.7×10^{-9}	9.2×10^{-8}	3.7×10^{-7}	1.5×10^{-6}	3.3×10^{-6}
600	1.5	2.2×10^{-10}	5.6×10^{-9}	2.2×10^{-8}	8.9×10^{-8}	2.0×10^{-7}
1000	2.4	8.3×10^{-11}	2.1×10^{-9}	8.3×10^{-9}	3.3×10^{-8}	7.5×10^{-8}
2000	4.9	2.4×10^{-11}	6.0×10^{-10}	2.4×10^{-9}	9.6×10^{-9}	2.2×10^{-8}

[a] Adopted from Healy and Fuquay (1958).

erosion of soils, have considered the problem theoretically, and have also conducted experiments in the field. This has enabled them to estimate the atmospheric concentrations of particles of various diameters that would be encountered over an infinite plane having a specified deposition of particles. Their data are given in Table 4-3.

Dose Calculations from Radioactivity in the Atmosphere

In Chapter 2, we reviewed the methods used by the ICRP and NCRP for computing the maximum permissible dose to the critical organ. The MPC's for air and water derived in this way are intended primarily for chronic exposure and although they can be used for acute episodes in the manner noted, they, in themselves, are insufficient for this purpose.

In the event of an emergency involving the release of radioactive materials to the atmosphere, it may be necessary to obtain knowledge of the following.

1. The external β- and/or γ-radiation dose from the passing cloud.

2. The external γ-radiation dose from a given ground deposit.

3. The potential thyroid dose from ^{131}I which may be expected from a given level of ground deposition.

4. The dose to the lung and other critical organs from inhalation of the passing cloud.

5. The potential dose from other radionuclides that may be incorporated into food chains or drinking water.

The direct radiation dose from the cloud and the dose from inhalation will be considered here. The doses due to conditions 4 and 5 were discussed in Chapter 3. For a presentation of the more detailed derivations of the dose calculations and the more exact forms required to fit a variety of conditions, the reader is referred to the excellent work of Healy and Baker (1968) from which the following is summarized.

The procedure for calculation of the integral γ dose at any point downwind of a given source for any combination of meteorological parameters is a formidable one if a solution is desired that takes into consideration all the interacting effects of time, cloud dimensions, and fission-product spectra on dose at the ground. Certain approximations seem justifiable if one admits to the great uncertainties involved in estimating the amount of radioactivity released in any given situation, as well as possible errors in assigning the various constants in the diffusion equations. Hendrickson and Strenge (1970) have discussed the sources of uncertainty in cloud dose calculations.

β Dose from Passing Cloud

Assuming the radius of the cloud to be greater than the range of the β particles, the dose rate to an individual located at the center of the cloud will be

$$R' = \frac{\bar{E}\ (1.6 \times 10^{-6})\,(3.7 \times 10^{10})\ \chi}{(1293)\,(100)} \tag{4-19}$$

$$= 0.46\bar{E}\chi$$

where

$R' =$ dose rate (rad/sec)
$\bar{E} =$ average β energy (MeV/disintegration)
$1.6 \times 10^{-6} =$ number of ergs/MeV
$3.7 \times 10^{10} =$ disintegration rate per curie (disintegration/sec/Ci)
$1293 =$ density of air (g/m³)
$100 =$ absorbed energy per gram of tissue (erg/g)
$\chi =$ concentration of β-emitting radionuclides (Ci/m³)

and the integrated dose delivered by the passing cloud will be

$$R_\infty = 0.46\bar{E}\ \overline{\chi t} \tag{4-20}$$

in which

$R_\infty =$ dose (rad)
$\overline{\chi t} =$ product of concentration $\times$ time (Ci sec/m³) calculated by substituting $Q_\Sigma =$ total curies released during incident for $Q = $ Ci/sec in Eq. (4-2)

If the cloud contains several nuclides, some of which have half-lives comparable or shorter than the time required for cloud passage, the concentration will be changing according to an exponential or power function, and Eq. (4-20) must be altered accordingly.

In most cases the above equation will overestimate the dose by as much as a factor of 2 owing to the short range of the β particles and the fact that the dose is not delivered isotropically.

External γ Dose from Passing Cloud

Although more rigorous solutions to the problem may be found (Slade, 1968), a more simplified approach to the problem is based on the assumption that the individual is standing on the ground immersed in an infinite medium through which radioactivity is uniformly dispersed during the period in which one receives the total dose measured in curies per cubic meter times time. The dose estimate may be simplified by neglecting back-scatter from the ground. This error, which tends to reduce the dose esti-mate, offsets the error in the opposite direction introduced by the premise of a cloud of infinite dimensions. (U. S. Atomic Energy Commission, 1957). The dose from a cloud containing 1 Ci sec/m³ is given by

$$R = \frac{\frac{1}{2}\overline{\chi t}\ (3.7 \times 10^4 d/\text{sec}/\mu\text{Ci})\ (\bar{E})\ (1.6 \times 10^{-6}\ \text{erg/MeV})}{(100\ \text{erg/g rad})\ (0.0012\ \text{g/cm}^3)}$$

$$= 0.25\overline{\chi t}\bar{E} \tag{4-21}$$

where

R = dose (rads)

$\bar{E}$ = average γ quanta energy (MeV)

$\overline{\chi t}$ = product of concentration $\times$ time (Ci sec/m³) calculated by sub-stituting Q_Σ = total curies released during incident for Q = Ci/sec in Eq. (4-2)

The dose from the passing cloud of fission products, as from a nuclear explosion or reactor release, can be expected to be of a low order in com-parison with the dose received by inhalation (particularly inhalation of the iodine) or with the γ dose caused by large-scale deposition of the cloud on surfaces.

DOSE TO THE LUNGS

When radioactivity is released from a reactor core that has been operating for more than a few days prior to the release, the atmosphere contamina-tion will be caused largely by radioiodine. Under these conditions the criti-cal organ is the thyroid, and the dose to the lung will not be governing. However, if the release is from some other source in which the effects of insoluble long-lived radionuclides must be considered, the dose to the lung may be of interest.

If one assumes an exposure, $\overline{\chi t}$ (μCi sec/ml), in which the period of ex-posure is short in relation to the half-life of the radionuclide, then the dose to the lungs may be estimated as

$$R = \frac{\overline{\chi t}v_t(f)\ (3.7 \times 10^4)\ (3600 \times 24)\ (1.6 \times 10^{-6}E)\ (\text{RBE})}{100m}$$

$$\times \int_0^t \exp - 0.693[(120 + T_r)/120T_r]t\ dt$$

where

$\overline{\chi t}$ = exposure (μCi sec/ml)
R = dose to lungs to time t (rem)
v_t = volume of air inhaled per given unit of time = 115 ml/sec
f = fraction of inhaled dust retained in the lower lungs at the end of 1 day = 0.12
m = mass of lungs (g)
E = effective energy of particulate radiation (MeV)
t = days
T_r = radiological half-life (days)
T_b = biological half-life (days)

$$T = \frac{T_r T_b}{T_r + T_b} = \frac{120 T_r}{120 + T_r}$$

where the values for f, v_t, and T_b are the ICRP values for the standard man. From the above,

$$R = 130\left(\frac{T_r}{120 + T_r}\right)\overline{\chi t}\, E(\text{RBE})\left[1 - \exp-\,0.693\left(\frac{(120 + T_r)}{120 T_r}\right)t\right] \quad (4\text{-}22)$$

Thyroid Dose from Inhaled Radioiodine

If the curie seconds of exposure to ^{131}I is known, the thyroid dose can be estimated from the following conversion factor:

$$1 \text{ Ci sec/m}^3 = 330 \text{ rem} \quad (4\text{-}23)$$

This equivalence assumes that the ^{131}I is inhaled in a period which is short in relation to the effective half-life of ^{131}I (7.6 days).

Tropospheric and Stratospheric Behavior

The emissions we have discussed thus far from near-surface sources are diffused measurably over scales of distances measured in tens of miles. In contrast, when the debris originates from nuclear or thermonuclear explosions, it is spread throughout the atmosphere on a global scale.

Our knowledge of stratospheric dispersion has been obtained from gas and dust samples obtained by aircraft or balloons penetrating to an altitude of about 115,000 ft. Studies have been made of the distribution of ozone and water vapor and of a wide variety of radioactive substances including debris from high-yield nuclear explosions. The transport of naturally occurring radionuclides such as ^{7}Be, ^{32}P, and ^{14}C, all which are induced by cosmic-ray bombardment of the upper atmosphere, has also been studied. On two occassions, tracers have been incorporated into nuclear weapons exploded at unusually high altitudes. [102]Rh was injected into the stratosphere by an explosion at an altitude of 43 km above Johnston Island in the Pacific in August, 1958. The cloud from this nuclear explosion was believed to have risen to about 100 km. In July, 1962, an explosion conducted about 400 km above Johnston Island contained a known amount of [109]Cd, which has been measured by investigators in many parts of the world.

Until comparatively recently the isothermal character of the stratospheric profile was assumed to imply a high degree of stability in this portion of the atmosphere. However, during the past 10 years, mainly as the result of studies of the debris from thermonuclear explosions, it has been learned that the lower stratosphere has a greater degree of turbulence than had formerly been believed to exist.

The first significant stratospheric measurements were of the movement of stratospheric water and ozone by Brewer (1949) and Dobson (1956). Their observations were studied by Stewart et al. (1957), who developed a model of stratospheric–tropospheric exchange that is consistent with the observed pattern of nuclear weapons fallout as will be discussed in Chapter

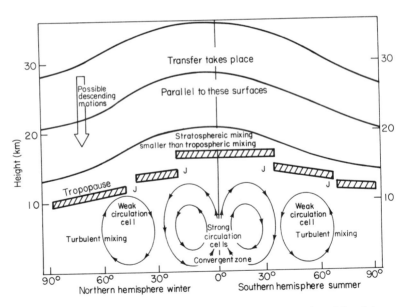

Fig. 4-18. Schematic cross section displaying characteristics of meridional transport ("J" locates typical jet stream positions) (United Nations Scientific Committee on the Effects of Atomic Radiation, 1964).

14. According to this model, air enters the stratosphere in the tropical regions where it is heated and rises to an altitude of about 30 kilometers at which level it begins to move toward the poles. As shown in Fig. 4-18, the tropopause is lower in the polar regions than at the equator, and tropopause discontinuities in the temperature regions facilitate transfer from the stratosphere to the troposphere. The westerly jet streams occur at these discontinuities with velocities of 100 to 300 km/hr. The rate of transfer from the lower stratosphere is most rapid in the winter and early spring.

The mean residence time of stratospheric debris depends on altitude, time of year, and latitude. Debris injected into the lower polar stratosphere by Russian thermonuclear explosions has been found to have a mean residence time of less than 6 months, whereas in tropical latitudes the residence time has been found to be from 2 to 3 years for the middle stratosphere and 5 to 10 years for debris injected at 100 km.

Sedimentation is evidently a significant factor in the thin stratospheric atmosphere since Feeley (1965) has shown that ^{14}C injected as carbon dioxide has a longer residence time than ^{90}Sr injected in particulate form (see Chapter 14).

Aerosols introduced into the troposphere are distributed by the planetary winds and deposit on the surface of the earth mainly by the scavenging of rain. The remarkable correlation of ^{90}Sr deposition with rainfall is demonstrated on the Olympic Peninsula by Hardy and Alexander (1962) and will be discussed in Chapter 14. The mean residence time of dust injected into the troposphere is about 30 days on the average, but this can vary from 5 days for dust in the lower rain-bearing region of the troposphere to 40 days in the higher initial altitudes (UNSCEAR, 1964). Rainfall removes aerosols from the troposphere primarily by droplet formation around the particle (rainout) and also by a scrubbing action (washout).

There have been suggestions that ocean spray is effective in scavenging dust near the ocean–atmosphere interface, and that this phenomenon might explain reports that fallout of ^{90}Sr into the oceans is higher than on land. However, field studies conducted by Freudenthal (1970a) conclude that ocean spray is not significant in this respect.

Chapter 5

The Food Chain from Soil to Man

In order to understand the manner in which radioactive contamination may pass from the soils and plants to man, it is useful to have at least a passing acquaintance with some of the physical and chemical properties of the soils and the mechanisms by which radioelements pass from the soils through the food chain. Most of the food consumed by human beings is grown on land and, except for elements like carbon and oxygen which may be obtained from the atmosphere, it is the soil that nourishes the complicated terrestrial ecological system of which man is a part.

As would be expected, radionuclides that occur naturally in soil are incorporated metabolically into plants and ultimately find their way into the bodies of animals, including man (Chapter 7). Artificially produced radionuclides introduced into soil behave in a similar manner, and worldwide contamination of the food chains has taken place during the past as a result of nuclear weapons testing. In addition to root uptake, direct deposition may occur on foliar surfaces, and when this happens the contaminants may be absorbed metabolically by the plants or may be transferred directly to animals that consume the contaminated foliage. Foliar deposition is potentially a major source of food-chain contamination by both radioactive and nonradioactive substances (Russell, 1965; Russell and Bruce, 1969).

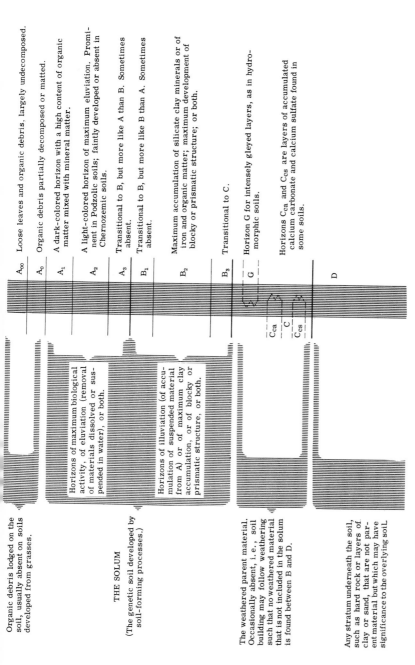

A_{00} Loose leaves and organic debris, largely undecomposed.

A_0 Organic debris partially decomposed or matted.

A_1 A dark-colored horizon with a high content of organic matter mixed with mineral matter.

A_2 A light-colored horizon of maximum eluviation. Prominent in Podzolic soils; faintly developed or absent in Chernozemic soils.

A_3 Transitional to B, but more like A than B. Sometimes absent.

B_1 Transitional to B, but more like B than A. Sometimes absent.

B_2 Maximum accumulation of silicate clay minerals or of iron and organic matter; maximum development of blocky or prismatic structure; or both.

B_3 Transitional to C.

G Horizon G for intensely gleyed layers, as in hydromorphic soils.

Horizons C_{ca} and C_{cs} are layers of accumulated calcium carbonate and calcium sulfate found in some soils.

D

Organic debris lodged on the soil, usually absent on soils developed from grasses.

Horizons of maximum biological activity, of eluviation (removal of materials dissolved or suspended in water), or both.

THE SOLUM

(The genetic soil developed by soil-forming processes.)

Horizons of illuviation (of accumulation of suspended material from A) or of maximum clay accumulation, or of blocky or prismatic structure, or both.

The weathered parent material. Occasionally absent, i. e., soil building may follow weathering such that no weathered material that is not included in the solum is found between B and D.

Any stratum underneath the soil, such as hard rock or layers of clay or sand, that are not parent material but which may have significance to the overlying soil.

Fig. 5-1. The principal horizons in a hypothetical soil profile. Not all horizons are present in any single profile (U.S. Department of Agriculture, 1957).

Some Properties of Soils

Soils (U.S. Department of Agriculture, 1957; Hillel, 1971) consist of
mineral and organic matter, water, and air arranged in a complicated
physiochemical system that provides the mechanical foothold for plants
in addition to supplying their nutritive requirements.

Vertical profiles through soils reveal horizontal layers (horizons) which
differ in their physical characteristics and which, in part, determine the
kinds and amounts of vegetation that a soil will support. Broadly speaking,
three types of horizons may be identified. The uppermost, which may be
from 1 to almost 2 ft in thickness, is the surface soil in which most of the
life processes take place. The second horizon is the subsoil, extending to
about 3 ft below the surface. Still further below the surface, to a depth of
about 5 ft, is a layer of loose and partly decayed rock which is the parent
material of the soils. This, in brief, is a description of the basic soil profiles
which are conventionally designated the A, B, and C horizons, which can
be differentiated further, as shown in Fig. 5-1.

The inorganic portion of surface soils may fall into any one of a number
of textural classes, depending on the percentage of sand, silt, and clay.
Sand consists largely of primary minerals such as quartz and has a particle
size ranging from 50 μm to about 2 mm. The silt consists of particles in the

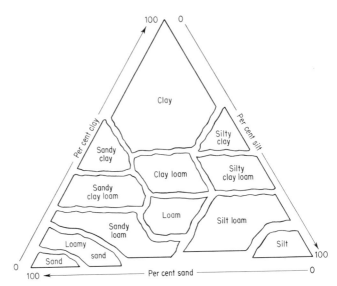

Fig. 5-2. The textural classes of soils according to the percentages of clay, sand, and
silt (U.S. Department of Agriculture, 1957).

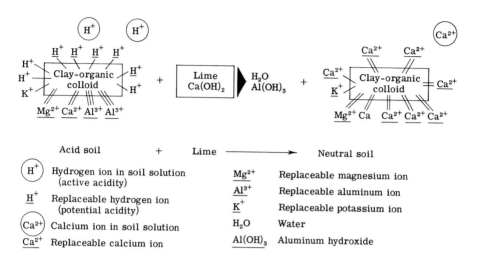

Fig. 5-3. The cation-exchange reactions when an acid soil is limed (U.S. Department of Agriculture, 1957).

range of 2 to 50 μm, and the clay particles are smaller than 2 μm in diameter. Figure 5-2 illustrates the textural classes into which soils may be differentiated, depending on the percentages of each of the three principal constituents.

The important physiochemical processes by which soils provide the nourishment for plants are controlled largely within the clay fraction of the soil. An essential characteristic of the platelike particles of secondary aluminum silicates comprising the clay is the abundance of negative surface charges. The resultant ability of the clay particles to attract ions, especially positive ions, to their surfaces is one of the most important properties of soils.

Most of the nutrient ions in the soils are not dissolved in the soil water, but are adsorbed on the surfaces of soil particles. A much greater reservoir of nutrient elements can be held in this way than can be dissolved in the soil water. As a matter of fact, dissolved nutrients would not remain long in soil but would be leached from it were it not for the extraordinary ability of the soil clays to bind the soil elements in ionic form. Since it is estimated that a total surface of 1 ft^2 of loam will have a total surface area of more than 500,000 ft^2, the opportunities for ion exchange are very great.

Cations in water solution are exchanged with cations adsorbed on the surface of the clays. Thus, most soils tend to become acidic after a period of time because of the replacement of adsorbed cations by hydrogen ions, and for this reason such soils must be limed from time to time. This procedure replaces the hydrogen ions with ions of calcium and magnesium, as is illustrated schematically in Fig. 5-3.

The ability of a given soil to adsorb cations is given quantitatively by its exchange capacity, customarily expressed as the milliequivalents of cations required to neutralize the negative charge of 100 g of soil at pH 7. Montmorillonite clay has such excellent cation-exchange properties that soils rich in it may have cation-exchange capacities of about 100 mEq/100 g compared to less than 10 mEq/100 g for predominantly kaolin-type soils. Organic matter derived from the decay of plant material can furnish the major part of the exchange capacity of many soils.

Sandy soils, as would be expected, tend to have a low exchange capacity which can be increased by adding organic matter.

Behavior of Radionuclides in Soils

Uptake of a long-lived radionuclide by plants depends to a considerable degree on whether it remains within reach of the roots of plants and the extent to which it is chemically available.

When a radionuclide is added in soluble form, it can adsorb on clays, precipitate as an oxide or hydroxide, chelate with soil organics, or, somewhat improbably, remain in solution (Schultz, 1965). The manner in which the radionuclide is distributed among these various fractions will determine how long it will remain at the site of deposition and the extent to which it will be available for uptake by plants.

Very little is known about the rates at which the various elements will migrate through different soils under varying conditions of pH and moisture. Information of this type is needed in order to permit one to forecast the behavior of accumulations of fission products underground, where the rate of vertical and horizontal movement will determine the rate at which nearby water supplies become contaminated. Knowledge of such movements nearer the surface would make it possible to predict the rate at which radionuclides are leached beyond the root zones of plants.

In general, the extent of penetration of an ion depends on its valence. A divalent cation will ordinarily be bound more tightly by soil than the monovalent cations, but other factors may be overriding. Thus, ^{90}Sr, which in ionic form is divalent, will penetrate more readily than monovalent ^{137}Cs because strontium ions are relatively abundant in soil compared to cesium ions. The latter ions are relatively rare in soil, and monovalent ^{137}Cs is actually bound in soils more readily than divalent strontium. In situations where there is abundant moisture and a deficiency of soil potassium, Broseus (1970) observed that ^{137}Cs introduced by fallout is more readily available than otherwise for uptake by forage.

We shall see in Chapter 12 that large volumes of radioactive solutions at Hanford have been stored on and under the ground in trenches and cribs. Studies of the migration of the individual fission products toward the water table below the stored wastes have been made (Pearce *et al.*, 1960) and have shown that, as would be expected, the fission products are adsorbed on the soil particles at different rates. The individual radionuclides form an adsorption cone in which the elements are sequentially removed in an order dependent on their relative affinities for the adsorption media. The soils at Hanford are sandy with relatively low cation-exchange capacity (5–10 mEq/100 g).

Field studies in the Soviet Union (Spitsyn *et al.*, 1960) showed that ^{90}Sr migrated at a rate of 1.1 to 1.3 cm/day through soils that had moderately high exchange capacity and that were permeated with ground water. This is an annual movement of less than 5 m, despite the fact that the soil was permeated with water. Since the average life of a ^{90}Sr atom is about 40 years, the mean distance that would be traversed by the isotope before its decay would be less than 200 m under the given conditions. The total amount of ^{90}Sr would diminish to 0.1% of the original in 10 half-lives (280 years), by which time the distance traveled would be 1400 m. The capacity of the soils to store fission products in ionic form is, thus, seen to be substantial.

Field and laboratory studies by Spitsyn *et al.*, Alexander *et al.* (1960), and the United Kingdom Agricultural Research Council (1961) all indicate that the ^{90}Sr fallout from nuclear weapons debris is normally held tightly in the upper few inches of soil. Alexander (1967) has studied the downward movement of ^{90}Sr and ^{137}Cs in fields having a variety of soil types and vegetation characteristics. At all sites studied, the two nuclides were contained mainly in the upper 4 in. of soil. At one site that was examined in more detail than the others, both nuclides were present in appreciable amounts in the leaf litter. From estimates of the cumulative fallout at the time of sampling, it was concluded that only 50% of the deposit could be accounted for in leaf litter and soil and that the remaining half was presumably contained in the vegetative stand. The greater part of the fallout, after only a few years, is thus contained in the biological cycle: soil—plant—litter—microbial utilization—soil (Ritchie *et al.*, 1970).

Alexander's measurements confirm that ^{137}Cs is more tightly bound by soil than [90]Sr. For purposes of forecasting future doses from ^{90}Sr deposited on soil by radioactive fallout from nuclear weapons tests, the United Nations Scientific Committee on the Effects of Atomic Radiation has assumed that ^{90}Sr is leached away from the root zone of cattle fodder at a rate of 2% per year (UNSCEAR, 1969).

Although some generalizations concerning the relative degree of fixation of the principal radionuclides in soil are possible, their behavior is so dependent on factors such as rates and amounts of rainfall, drainage, and extent of tillage that more general quantitative forecasts are not practical at this time (Schuffelen, 1961).

Uptake from Soils

Although almost every element can be identified in all soils in more or less amounts, 16 elements are considered to be necessary for the growth and reproduction of plants. These are carbon, hydrogen, oxygen, nitrogen, phosphorus, sulfur, potassium, calcium, magnesium, iron, manganese, zinc, copper, molybdenum, boron, and chlorine. All are derived from soil except carbon, hydrogen, and oxygen, which may be supplied by the atmosphere.

As a general rule, radioisotopes present in soil will pass into the root system in the same manner as nonradioactive isotopes of the same cations. The element may or may not be required for normal metabolism, and some elements like iodine, cobalt, uranium, and radium are known to be present in plants although they serve no known metabolic function. It may be assumed that if a cation is present in the soil it will probably be present in the tissues of plants grown on the soil.

The extent to which a radionuclide is absorbed from the soil by plants depends on the chemical form of the nuclide, the metabolic requirements of the plant, and physiochemical factors in the soil. According to Nishita et al. (1961), the relative uptake of various radioelements from soils is Sr $\gg$ I > Ba > Cs, Ru > Ce > Y, Pm, Zr, Nb > Pu. In Table 5-1 are assembled numerous data showing the relative tendency of the various nuclides to be concentrated from soil by crop plants (Menzel, 1965). The relative concentration factor, expressed as parts per million in dry plant material/part per million in dry soil is shown to vary from factors of 10 to 1000 for elements like potassium, rubidium, phosphorus, sodium, and lithium to less than 0.01 for lead, plutonium, zirconium, and yttrium.

Radioisotopes of elements ordinarily present in soil and normally utilized in plant metabolism are absorbed in a manner independent of their radioactive properties. Thus, ^{45}Ca in soil becomes part of the pool of available calcium in the soil, and the plant will not differentiate significantly between ^{45}Ca and the stable isotopes of calcium. By available calcium is meant that portion of the total soil calcium which exists in ionic form and is, therefore, available for transport to the root system. The roots may be

TABLE 5-1

RELATIVE CONCENTRATION OF ELEMENTS IN THE FIRST CROP PLANTS GROWN
AFTER THE ELEMENTS HAVE BEEN APPLIED IN WATER-SOLUBLE FORM AND
MIXED INTO SURFACE SOIL[a,b]

10–1000 Strongly concentrated	1–100 Slightly concentrated	0.1–10 Not concentrated	0.01–1 Slightly excluded	<0.01 Strongly excluded
K	Mg	Ba	Cs	Sc
Rb	Ca	Ra	Be	Y
N	Sr	Si	Fe	Zr
P	B	F	Ru	Ta
S	Se	I		W
Cl	Te	Co		Ce
Br	Mn	Ni		Pm
Na	Zn	Cu		Pb
Li	Mo			Pu
				Sb

[a] Menzel (1965).

[b] Relative concentration factor $\left(\dfrac{\text{ppm in dry plant material}}{\text{ppm in dry soil}}\right)$.

unable to distinguish between chemical congeners. Thus, plants grown in simple inorganic solutions containing calcium and strontium are unable to discriminate between the two (Food and Agriculture Organization, 1960).

The cations adsorbed on the surfaces of the soil particles constitute a reservoir from which many of the nutritive requirements of the plant may be supplied. They pass to the vicinity of plant root tips by means of the soil water which contains cation concentrations in equilibrium with the adsorbed solid phase.

The roots of plants are located at soil depths characteristic of the species: less than 1 ft in the case of spinach, in contrast to alfalfa and asparagus roots which penetrate to 10 ft or more. The effectiveness of root structures as an exchange surface is illustrated by the observation that the roots of a single winter rye plant had a surface ara of 6870 ft² (Wadleigh, 1957).

Foliar Deposition of Radionuclides

Radioactive substances can bypass the soil and pass directly to the food chain by foliar deposition (Russell, 1965). The radionuclides may then pass directly to grazing animals or man as superficial contamination or they may be absorbed metabolically from the plant surface. Based on data obtained in the United Kingdom during the period when fallout was occurring from weapons tests, Russell concluded that of the ^{90}Sr then present in cows' milk only 20% was due to direct uptake from soil, whereas the balance was due to deposition directly on the surface of herbage. The same investigator (1966) concluded that virtually all the ^{137}Cs detected in foods results from foliar deposition. He noted that a cow in summer months may consume herbage each day which is equivalent to the ground cover over 50 m²,* and that 25% of recently deposited fallout may be contained in this herbage. Others, using United States data, have shown that foliar absorption accounts for a somewhat smaller fraction of the ^{90}Sr found in milk (see Chapter 14).

The question of how to apportion the ^{90}Sr content of cow's milk between that which originated from foliar deposition and that which is taken by the plant from the soil occupied a good deal of interest during the period when fallout was accumulating from nuclear weapons tests in progress up through about 1962. The importance of this arose from the fact that in order to make future forecasts of human exposure from expected deposition of ^{90}Sr, it is necessary to know to what extent a given dose from drinking milk is due to fresh fallout which will disappear after weapons tests are stopped and how much was due to ^{90}Sr that will remain in the soil. A number of investigators (Tajima, 1956; Russell, 1965) attempted to fit proportionality factors to equations of the following type:

$$C = p_r F_r + p_d F_d \qquad (5\text{-}1)$$

where

C = 12 month mean ratio of ^{90}Sr to calcium in milk (pCi/g Ca)
F_r = the annual deposit of ^{90}Sr (mCi/km²/year)
F_d = the cumulative deposits of ^{90}Sr (mCi/km²)
p_r = the proportionality factor for the rate-dependent component of C
p_d = the proportionality factor for the deposit-dependent component of C

In order to account for the effect of fallout during the previous growing year on the ^{90}Sr content of milk from cows fed stored feed for part of the year, Bartlett and Russell (1966) proposed inclusion of a third term, the

---

* This figure must of necessity be quite variable, depending on the type and yield of vegetation, as well as other factors. It will be noted elsewhere that Garner (1960) used a figure of 160 m²/day.

"lag factor," thus

$$C = p_r F_r + p_r F_d + p_l F_l \tag{5-2}$$

in which

$F_l =$ fallout during the last 6 months of the previous year

$p_l =$ "lag" proportionality factor

This refinement could be significant during periods when the rate of fallout is varying significantly, but Aarkrog (1971) did not find it necessary to utilize this additional term in order to relate the observed rates of fallout in Denmark during the period 1962–1968, when the rate of global fallout reached a maximum just prior to the ban on atmospheric testing of nuclear and thermonuclear weapons.

During the period of relatively heavy fallout in the early 1960's (see Chapter 14), direct uptake from soil apparently accounted for as little as 10% of the ^{90}Sr seen in milk. The worldwide average soil uptake component for ^{90}Sr contamination of milk was 0.3 (UNSCEAR, 1964, 1966), and the proportionality factor for the rate-dependent component was 0.8.

The potential hazard from fallout in farm areas may depend to a major extent on whether direct contamination of the plant surfaces is occurring. In the case of relatively short-lived isotopes like ^{131}I, food-chain contamination via the roots is unlikely because radioactive decay would reduce the opportunity for the isotope to pass from the soil into the plants. Root uptake is a leisurely process compared to the immediate contamination of foods that occurs when the plant surfaces become contaminated by direct fallout.

The significance of surface contamination varies with the growing season, since the possible risk owing to direct contamination of crops is obviously greater just before a harvest or when active grazing by stock animals (particularly milk-producing animals) is in progress. Conversely, the danger may be lowest in winter months when there are no standing crops, although it is possible that even during these months direct fallout on the basal structure of grasses in permanent pastures may be stored until the following spring when contaminants may be absorbed by the growing plants. Retention of this type will be greatest for plants that develop a "mat" of basal parts, old stems, and surface roots (Russell, 1965).

The significance of foliar contamination of plants will also depend on the structure of the plants and the role of the various parts of the plant in relation to the dietary habits of man. The inflorescences of wheat plants have a shape that tends to maximize entrapment of fallout particles. It is probably for this reason that wheat was found to be a major source of ^{90}Sr from weapons testing fallout in Western diets (U.S. Atomic Energy

Commission, 1960b). It has been reported that cereals generally are subject to relatively higher foliar retention. The influence of dietary practices is illustrated by the fact that white bread has been shown to contain less ^{90}Sr than whole-wheat bread which is made from unsifted flour that contains the brown outercoat as well as the inner white portion of the wheat grain.

Foliar contamination can be removed by radioactive decay, volatilization, leaching by rain, other weathering effects, and by dying and dropping of plant parts. Chamberlain (1970) has examined the data from a number of investigators and found the field loss during the growing season to be about 0.05 per day, not considering radioactive decay. Chadwick and Chamberlain (1970) found the rate of field loss to be highly variable. The half-time of removal of strontium sprayed on crops varied from 19 days in the summer to 49 days in winter. New growth contained less than 1% of the initial deposit.

Hansen et al. (1964) in a 3-year study of the ^{131}I and ^{90}Sr content of milk from dairy herds grazing on well-fertilized and badly fertilized fields found a 50% difference in the level of contamination. The radionuclide content of the milk from cattle pastured in well-fertilized fields was lower, presumably because the faster growing grass diluted the contamination present as foliar deposition.

Metabolic Transport through Food Chains

From the foregoing we can identify root uptake and foliar deposition as the two ways by which fallout can contaminate crops that are eaten by man or that serve as food for stock animals. Much remains to be learned about the ways in which each radionuclide behaves in passing from the root to the eatable portion of the plant, through the body of the stock animal, and into the milk, meat, or eggs consumed by man. A quantitative understanding of the food-chain transport mechanisms is required if one wishes to predict the dose to man from a given deposition of a radionuclide in the environment or a given concentration in foodstuffs.

It is sometimes possible to forecast the behavior of a radioelement from knowledge of its chemical congeners. Comar et al. (1956) have called attention to the possibility that in studying the metabolism of radionuclides in relation to their congeners the extent to which discrimination takes place at any metabolic step may be described by the observed ratio OR, which may be expressed as follows:

$$OR_{\text{sample-precursor}} = \frac{C_e/C_c \text{ sample}}{C_e/C_c \text{ precursor}}$$

where C_e and C_c are the concentrations of the element and its congener, respectively.

To understand the use of the observed ratio, consider the ingestion by a cow of herbage containing given concentrations of calcium and its radioactive congener ^{90}Sr and suppose one wishes to predict the concentration of ^{90}Sr that will appear in the milk. This can be accomplished by using the OR[milk-herbage], which describes the ability of the cow to discriminate metabolically in favor of calcium and against ^{90}Sr in the production of milk. It proves convenient to report analytical data in food-chain studies as the concentration of contaminant contained by its congener. Thus, since this particular OR is known to be about 0.1, we would expect the cow's milk to contain 10 pCi ^{90}Sr/g Ca, if the herbage on which the cow feeds contains 100 pCi ^{90}Sr/g Ca.

Of the many artificial radionuclides that have contaminated soils and plants, ^{90}Sr, ^{137}Cs, and ^{131}I have been studied most thoroughly. The principal factors that influence the passage of these nuclides from soil to man will be discussed here. The reader is referred to the Appendix for more detail about these and certain other nuclides and to Chapter 14 for a discussion of the observations made of the behavior of the nuclides following fallout from nuclear weapons testing.

STRONTIUM-90 (U. S. ATOMIC ENERGY COMMISSION, 1972)

Most discussions of ^{90}Sr in food chains use the ratio pCi ^{90}Sr/g Ca in which 1 pCi ^{90}Sr/g Ca is defined as the strontium unit (SU). This practice continues despite the fact that the calcium content of plants has been found to vary considerably, and contamination reported as pCi/g food is sometimes less variable than when related to the unit mass of calcium (UNSCEAR, 1969). However, all things considered, this ratio is useful in following the ^{90}Sr from one biological step to the next.

However, the ratio may be meaningless when applied to soil under practical conditions. This is because the ^{90}Sr will not normally be homogeneously mixed throughout the soil, and there is no way of expressing the strontium–calcium ratio of the nutrients to which the roots are exposed. Under laboratory conditions where the ^{90}Sr was well mixed with the soil, Fredrikson *et al.* (1958) showed that the observed ratio from plant to soil varied only within the narrow range of 0.7 to 0.8 over a wide degree of variation in the amount of soil calcium. The discrimination against strontium was apparently unaffected as long as the soil was not oversaturated with calcium ions. This would indicate that only a small degree of differentiation takes place at the soil–root interface, but it is difficult to determine in practice exactly how the ^{90}Sr is distributed through the soil, and the use of the OR at the level of soil is, therefore, not practical.

It has been shown by Roberts and Menzel (1961) that a portion of the ^{90}Sr may become unavailable to plants as a result of reactions in the soil. Studies over a 3-year period showed this fraction to be variable from about 5 to 50%.

The OR$_{human\ bone-diet}$ are the most important observed ratios insofar as ^{90}Sr is concerned. Measurements of the OR$_{human\ bone-diet}$ have been reported from seven countries (UNSCEAR, 1969) and have been shown to vary from 0.13 to 0.33. To what extent this range is due to differences in analytical techniques or to real differences in dietary or metabolic factors cannot be said. In the United States, the OR$_{human\ bone-diet}$ is 0.18 and ranges from 0.15 (Chicago) to 0.22 (San Francisco). There appears to be no systematic difference between dairy and nondairy foods.

The observed ratio in passing from plants to milk by way of the cow has been shown (Comar and Wasserman, 1960) to be about 0.1. Thus, the overall observed ratio (OR$_{human\ bone-plant}$) would be about 0.25 if the plants are consumed directly and about 0.025 if the calcium is consumed via milk. In short, the lactating animal has been shown to be a strontium decontaminator with an efficiency of about 90%. Lough *et al.* (1960) have shown the observed ratio in passing from diet to human milk is also about 0.1, the lactating human thus having the same decontaminating effectiveness as the lower animals.

The overall effect of metabolic differentiation between strontium and calcium in passing from soil to human bone can be summarized for milk and vegetable diets as follows, starting with soil containing 1 pCi ^{90}Sr/g Ca:

$$
\begin{array}{llll}
1\ \text{SU} & 1\ \text{SU*} & 0.13\ \text{SU} & 0.032\ \text{SU} \\
\text{in} & \rightarrow \text{in} & \rightarrow \text{in} & \rightarrow \text{in} \\
\text{Soil} & \text{Plant} & \text{Milk} & \text{Human bone}
\end{array}
$$

$$
\begin{array}{lll}
1\ \text{SU} & 1\ \text{SU*} & 0.25\ \text{SU} \\
\text{in} & \rightarrow \text{in} & \rightarrow \text{in} \\
\text{Soil} & \text{Plant} & \text{Human bone}
\end{array}
$$

The net ^{90}Sr/calcium ratio in human bone will thus vary, depending on whether the dietary calcium is derived from dairy foods or other sources. The people of the United States derive almost all their calcium from milk, but this is not true of other countries. The dietary sources of calcium in various countries are given in Table 5-2. In Chile, the population receives

---

* This holds only when the strontium and calcium are uniformly mixed throughout the root zone of the plant.

TABLE 5-2

GENERAL SOURCES OF NATURAL[a] CALCIUM IN THE FOOD SUPPLIES OF SELECTED COUNTRIES (Expressed as Percentages of Estimated Total Calcium Content). CALCULATIONS BASED ON FOOD BALANCE SHEETS, 1954–1956 AVERAGE[b]

| Country | Percent of total calcium supply | | | | | | | | | Total supply of Ca (g per capita/day) |
	Cereals	Starchy roots	Pulses, nuts	Vegetables	Fruits	Meat, poultry	Eggs	Fish	Milk, cheese	
Austria	6	2	1	10	3	1	1	—	76	0.90 0.10
Denmark	7	3	1	9	2	1	1	1	76	0.95 0.10
Finland	6	2	—	2	1	1	1	1	87	1.3 0.10
Germany	6	4	1	7	3	1	1	1	76	0.85 0.10
Italy	10	1	4	22	4	1	2	1	57	0.60 0.10
Norway	6	2	1	4	1	1	1	2	82	1.1 0.10

[a] Excluding additions of calcium salts of mineral origin to flour or other foods.
[b] Food and Agriculture Organization (1960).

only 8% from dairy products, compared to 77% in the United States and 87% in Finland. The cereals, pulses, and vegetables are the dominant sources of calcium in the Far Eastern Countries.

In many parts of the world the calcium of metabolic origin in foodstuffs is supplemented with mineral calcium, a practice which would tend to lower the ratio ^{90}Sr/Ca. Calcium carbonate is added to the maize used in the preparation of tortillas in Mexico and the flour used in bread making in the United Kingdom (Food and Agriculture Organization, 1960).

Several methods have been proposed by which the uptake of ^{90}Sr from soils can be reduced (Menzel, 1960). These include the application of lime, gypsum, fertilizer, and organic matter. According to Menzel, these techniques are only moderately effective and could not be expected to reduce the uptake of ^{90}Sr by more than 50% in productive soils. This is a modest diminution in uptake considering the rather large amounts of amendments required. For example, at the levels of exchangeable calcium ordinarily found in productive soils, it would require several tons of lime per acre to effect a measurable change in strontium uptake .However, unproductive soils having low cation-exchange capacity and low exchangeable calcium have relatively large uptakes of ^{90}Sr which can be reduced appreciably by the addition of calcium in available form. In the United Kingdom, the ^{90}Sr uptake in herbage was shown to increase manyfold as the amount of exchangeable calcium diminished from 3 to 4 g/kg of soil to less than 2 g/kg (United Kingdom Agricultural Research Council, 1961).

CESIUM-137

It has been well established by several investigators that cesium is so tightly bound by the clay minerals of the soil that root uptake is slight, and foliar absorption is, therefore, the main portal of entry of ^{137}Cs to the food chains.

Cesium is a congener of potassium, but the Cs/K ratio is not as constant in biological systems as is the ratio Sr/Ca. The uptake of cesium from soil has been shown to be inversely proportional to the potassium content of soils in which there is a potassium deficiency (Nishita et al., 1961; Menzel, 1964), and Broseus (1970) has shown that this inverse dependence on the potassium content of soil explains the high cesium content of milk grown in certain parts of the island of Jamaica.

Although cow's milk is the largest contributor of ^{137}Cs to the United States adult diet, other foods including grain products, meat, fruit, and vegetables contribute about two-thirds of the dietary cesium intake (Gustafson, 1969). Measurements made by Gustafson on representative United States diets from 1961 through 1968 are given in Table 5-3.

TABLE 5-3

PARTITION OF ^{137}Cs IN UNITED STATES ADULT DIET[a]

Year	Milk (%)	Grain products (%)	Meat (%)	Fruit (%)	Vegetables (%)
1961	31	17	12	20	15
1962	38	17	13	15	14
1963	39	21	22	8	6
1964	34	26	21	8	5
1965	28	23	26	9	4
1966	25	30	23	11	5
1967	24	28	17	8	7
1968	31	19	19	10	9

[a] Gustafson (1969).

Wilson *et al.* (1969) studied the transport of ^{137}Cs from the atmosphere to milk and concluded that root uptake was so slight that it could be neglected in any model designed to forecast the dose to humans from ^{137}Cs in fallout. For reasons not known, the crude fiber content of the forage was found to influence uptake by the cow with transfer coefficients (the concentration of ^{137}Cs in milk/^{137}Cs intake per day) varying from 0.0025 for alfalfa and corn silage to 0.01 for mixed grain.

Wilson and his associates developed a model for predicting the ^{137}Cs content of milk from cattle fed on stored feed. It was assumed that for the first 6 months in any given year the cows are fed stored feed that had been exposed to the previous year's fallout and are fed on feed contaminated by the current year's fallout during the second 6 months. The mean concentration in picocuries per liter was then calculated for each of the half-years using the following equation:

$$C_m = C_a B \tag{5-4}$$

where

C_m = the average concentration in cow's milk (pCi/liter)
C_a = the average concentration in air (pCi/m^3)
B = a coefficient obtained by multiplying the fallout contamination factor and transfer coefficient shown in Table 5-4

This technique was shown to be remarkably useful for forecasting the ^{137}Cs content of milk from seven milk sheds across the nation from 1962 to 1967, although for reasons that were not understood by the investigators,

TABLE 5-4

SUMMARY OF INFORMATION FOR MODEL OF ^{137}Cs IN MILK FOR DRY LOT HERD[a]

	Dry weight intake (kg/day)	Fallout contamination factor (m³/kg)	Feed to milk transfer coefficient (day/liter)
Hay	14	9100	0.0025
Grain	7	1470	0.010
Silage	5	6740	0.0025

[a] Wilson et al. (1969).

the model did not work well for two milk sheds in Tampa, Florida and Seattle, Washington. The excellent correlation between the observed ^{137}Cs concentration in air during the growing season May–July and the mean quarterly ^{137}Cs content of milk is shown in Fig. 5-4, from which the Tampa and Seattle data are excluded. The ^{137}Cs levels in Tampa have been anomalously high, a fact that has been attributed to the Pangola grass prevalent in the area (Porter et al., 1967).

When ^{137}Cs is ingested by man, about 80% is deposited in muscle and about 8% in bone (UNSCEAR, 1969; Spiers, 1968). The half-life in adults

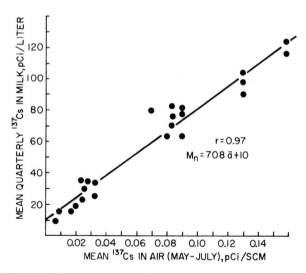

Fig. 5-4. Correlation between surface air activities of ^{137}Cs and quarterly averaged ^{137}Cs in milk from seven milksheds across the nation from 1962 to 1967 (Wilson et al., 1969).

depends on body weight, sex, and dietary habits and has been shown to vary between 50 and 200 days. Turnover is much more rapid in children, among whom the half-life has been found to be on the order of 10 days for newborn infants. A burden of 1 pCi/g K produces an absorbed dose of 0.02 mrad/year.

One must be careful that any generalization about pathways to man is not negated by some special dietary consideration. In the case of ^{137}Cs, it has been shown that Laplanders and other residents of the far north are subject to relatively high ^{137}Cs intake owing to their dependence on reindeer which feed on lichens that have a tendency to concentrate a variety of trace substances present in the atmosphere (Miettinen, 1969). About 25% of the ^{137}Cs contained in lichens is absorbed by reindeer (Holleman et al., 1971). As a result of this phenomenon, the Finnish Lapps contain about 50 times the ^{137}Cs body burden of Finns in the southern part of the country.

IODINE-131

Because of the short half-life of ^{131}I (8 days), it is not a significant environmental contaminant insofar as uptake from the soils is concerned. The decay rate is relatively rapid in relation to the growing time of a crop, and any significant contamination by means of root uptake would, for this reason, be improbable. On the other hand, radioiodine deposited on the surfaces of plants can be ingested directly by cattle and pass in this way to milk or other dairy. Since the time is usually short from the collection of milk to its consumption, the possibility of iodine contamination of fresh milk must be considered. If the milk is processed into powered form, radioiodine contamination will be less of a problem because longer storage time will permit decay of the isotope. However, cottage cheese reaches the consumer almost as quickly as fresh milk and may, therefore, be contaminated with short-lived nuclides such as radioiodine.

Mathematical models for the transfer of radioiodine to food are handicapped by the fact that since radioiodine has a short half-life, contamination must of necessity take place over a relatively short period of time. Thus, transfer from air to forage is apt to be variable because of meteorological factors that tend to average out in the case of longer-lived nuclides such as ^{137}Cs or ^{90}Sr. Field observations of the various constants and coefficients needed for mathematical modeling have been found to be very variable. Chamberlain found the values of deposition velocity (v_g) to range from 0.1 to 0.4 cm/sec following the Windscale accident (Chamberlain and Dunster, 1958). Other measurements of the radioiodine deposition velocity in the United Kingdom have ranged from 0.4 to 2.5 cm/sec (Chamberlain, 1960; Chamberlain and Chadwick, 1966). Hawley et al. (1964) found the deposition velocity to be 0.6 cm/sec on the average for radioiodine released experimentally at the National Reactor Test Station in Idaho, but the individual observations varied from 0.17 to 1.1 cm/sec.

Soldat (1963), reporting on studies at Hanford Laboratories, found that over a 2-year growing season the ratio pCi/kg grass:pCi/m³ air was about 4200. He found the ratio pCi/liter milk:pCi/kg grass to be about 0.15. Various estimates have been made of the extent to which the human thyroid dose is increased by the grass—cow—milk pathway compared to inhalation. A widely used factor of 700 has been justified by Burnett (1970) and is intermediate among various proposed factors.

Chamberlain (1970) concluded from an analysis of data from various investigators that the ¹³¹I foliar deposition data conformed to the concept that vegetation behaves as a filter, with deposition dependent on the density of the foliage.

Once the radioiodine deposits on foliage, it is fixed to about the same extent as other substances and is removed by weathering and dying of plant parts at a rate of about 5% per day, which with the 8-day radioactive half-life of ¹³¹I gives an effective half-life of removal from grass ranging from 3.5 days in the Idaho experiments to about 6 days at Windscale. An effective half-life of about 5 days is usually used for purposes of risk calculation. Soldat (1965) found that radioiodine in milk reached a peak 3 days following an accidental release.

According to Garner (1960), a milk cow consumes about 20 kg grass/day on the average, and an average vegetative stand for an acre of pasture is about 500 kg. It follows that a cow grazes an area of about 160 m²/day. It can be assumed that about 5% of the ingested radioiodine will be secreted in cow's milk (Lengemann, 1966).

The fraction of ingested iodine that reaches the human thyroid following ingestion is usually taken as 0.3 and the effective half-life for elimination from the thyroid as 7.6 days (ICRP, 1960). However, thyoridal radioiodine uptake can be blocked almost completely by 50–200 mg of stable iodine, as the iodide or iodate (Blum and Eisenbud, 1967). Substantial dose reduction will result from administration of KI as long as 2 hours after exposure.

Chapter 6

The Aquatic Environment

The streams, ponds, lakes, rivers, bays, seas, and oceans provide many opportunities for human exposure to liquid radioactive wastes (IAEA, 1971a). The principal routes are apt to be via contamination of food or water, although under special circumstances other pathways may become important, such as contamination of fishing gear or beaches used for recreation.

The contemporary general awareness of the importance of minimizing all types of pollution, together with the special potential risks of radioactive pollution, especially by long-lived nuclides, has resulted in very conservative policies in regard to disposal of radioactive liquid wastes. However, the nuclear energy industry, like many others, needs water for cooling or other purposes, and some degree of contamination can result when the water is returned to the source from which it was originally taken. The quantities of radioactive liquid wastes can be reduced to whatever extent is desired by adding separation steps to remove progressively larger fractions of the contaminants, but it is not possible to remove them completely (see Chapter 12). The physical and chemical techniques by which contamination is removed from waste water are capable of removing any desired fraction, but some residual contamination must be released. Intelligent waste management requires that the ecological behavior of the radionuclides released to the aquatic environment be understood quantitatively

137

so that the dose commitments to man, wild life, and stock animals can be estimated and controlled to whatever required extent.

The doses that cause injury to plant and animal life are very much higher than would be permitted under any circumstances in which human safety must be considered. Whereas the maximum permissible dose to human inhabitants near nuclear energy installations is limited to a few millirem per year, the doses associated with perceptible injury to the lower life forms are higher by many orders of magnitude (Auerbach et al., 1971; Templeton et al., 1971), a conclusion which is supported by many reports in the literature. As an example, Auerbach cites the work of Donaldson who found no effects following irradiation of chinook salmon at dose rates up to 5.0 rem/day beginning immediately after fertilization of the egg and continuing for 80 to 100 days until the fish were completely formed. Many experiments were conducted in which the number of exposed fish ranged from 96,000 to 256,000, and their viability was tested by allowing the salmon to go to sea and then return to the hatching tanks after 4 years. It was concluded that there were no demonstrable effects on the fish and that, if anything, some of the irradiated fish showed higher-than-average viability and weight on their return to the hatching tanks.

Because warm water is discharged by the condenser cooling systems of power stations, the question is often raised of possible synergistic ecological effects of temperature and ionizing radiation. No such synergism has been demonstrated over a wide range of temperatures or dose rates, except that in some cases organisms grown in warm water have been shown to absorb radionuclides as much as 50% faster owing to increased growth rates (Harvey, 1970). Ophel and Judd (1966) administered ^{131}I and ^{90}Sr to goldfish (Carassius auratus) and found no impairment of ability to withstand near-lethal temperatures at doses of 10^4 rad to bone and 10^5 rad to the thyroid. Angelovic et al. (1969) have reported that the estuarine fish (Fundulus heteroclitus) has lowered tolerance to heat and salinity when irradiated, but the administered doses were very high, in the range of 2000–6000 rem. The doses at which effects have been observed in irradiated aquatic organisms are very much higher than would be permitted in practice.

Diffusion within Aquatic Systems

We have seen in Chapter 4 that the concentration of an atmospheric pollutant downwind of a source of known strength can be calculated conveniently if a few readily obtainable meteorological data are available. Except under conditions of unusual topography, the same diffusion equations can be used anywhere in the world, and if the source strength is known, one need only measure wind direction and velocity and estimate the degree of atmospheric stability to approximate the downwind concentration within reasonable limits of uncertainty.

Unfortunately, a generalized approach to the dispersion of pollutants introduced into a body of water is not possible in the state of our knowledge. Physical mixing is highly dependent on depth of water, type of bottom, shoreline configuration, tidal factors, wind, the temperature and depth at which the pollutant is introduced, and other factors. Each stream, river, bay, lake, sea, and ocean has its own mixing characteristics that vary from place to place and from time to time. Hydrologists have developed useful diffusion equations that are valuable for the specific situations for which they are intended, but the equations are usually based on a combination of theoretical calculations, model measurements, and field observations that are applicable only to the given locality.

In the aquatic environment the mixing problem is further complicated by the fact that in addition to the badly understood diffusion processes the fate of a pollutant is dependent on other physical and biological processes (Fig. 6-1 and 6-2). If a pollutant is a suspended solid, it can settle to the bottom, be filtered by certain organisms, or become attached to plant surfaces. Pollutants in solution can adsorb on suspended organic and inorganic solids or can be assimilated by the plants and animals. The suspended solids, dead biota, or excreta settle to the bottom and become part of the organic substrate that supports the benthic community of organisms. The

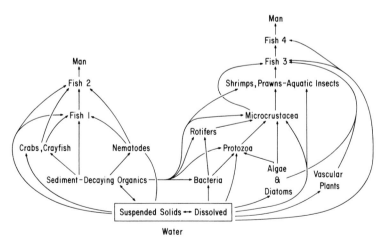

Fig. 6-1. Possible routes of trace metal transfer through an estuarine ecosystem (Steven Jinks).

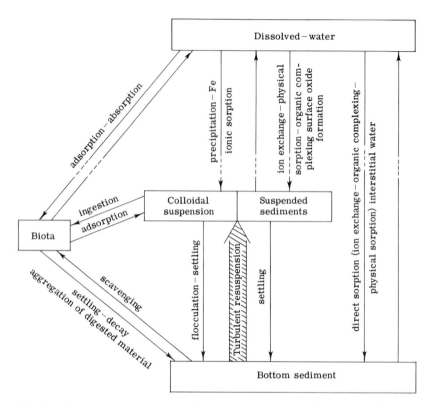

Fig. 6-2. The basic components of the aquatic ecosystem. The complexities of the pathways among the biota are illustrated in Fig. 6-1 (Leland Hairr).

sediments more often act as a sink for pollution, but they may also become a source, as when they are resuspended during periods of increased turbulance. The sediments may also serve as a source of pollution when the concentration of contaminant in the overlying water becomes sufficiently reduced. Lentsch *et al.* (1972) has shown that the role of estuarine sediments as a source or sink for manganese seems to be related to salinity and may oscillate with the tidal cycle under estuarine conditions.

Mechanisms also exist for the transport of radioelements from the aquatic to the terrestrial environment, as when man eats marine organisms or where birds take fish from the sea and then deposit droppings along the shore.

In time, it is likely that sufficient quantitative information will become available about these mechanisms to permit one to estimate with some degree of certainty the capacity of a given river, estuary, or ocean to receive radioactive nuclides without injury to the ecosystem of which man is a part. For the time being, however, much remains to be learned, and the uncertainties that exist are likely for some time to result in a highly conservative approach to the discharge of radionuclides into the aquatic environment. Proper liquid-waste management practices often require that extensive ecological studies be conducted in each locality where a radioactive waste outfall is to be located.

Mixing Characteristics in Receiving Water

THE OCEANS

The oceans (Sverdrup *et al.*, 1942) cover an area of 3.6 × 10⁸ km² and, with an average depth of 3800 m, they contain a total volume of 1.37 × 10⁹ km³ (Revelle and Schaefer, 1957). Bordering the oceans are the continental shelves which skirt most of the coastlines to a depth of about 150m and in some places extend seaward for more than 150 km. The ocean bottoms were largely unmapped until development of the echo sounder in World War II made it possible to obtain precise topographical records of the ocean floor.

The near-surface waters of the oceans, to a depth which varies geographically from 10 to 200 m, is a region in which rapid mixing occurs as a result of wind action. Because of thorough mechanical mixing, the vertical gradients of temperature, salinity, and density are nearly uniform.

About 75% of the oceans is cold, deep water at a temperature of 1°–4°C and a salinity of 3.47% (IAEA, 1960), but between the surface water and the deep water is an intermediate zone characterized by decreasing temperature and increasing salinity and density with depth. The fact that the salinity is increasing with depth reduces vertical motions, and the intermediate zone, therefore, tends to restrict exchange between the surface waters and the deep waters (Revelle *et al.*, 1956; Pritchard *et al.*, 1971). The intermediate zone may be as much as 1000 m in depth and is known as the *thermocline* or *pycnocline*.

The characteristic currents of the ocean surface layer are due primarily to wind action and tend to coincide with the surface wind patterns. The movement of surface water has been shown to be as much as 90 miles/day in the Florida current and 41 miles/day in the Kuroshio current of the western Pacific (National Academy of Sciences—National Research Council, 1957b). Radioactivity from Bikini Atoll was found to drift westward at a rate of about 9 miles/day after tests of nuclear weapons in 1954 (Miyake and Saruhashi, 1960). The major surface currents of the oceans are shown in Fig. 6-3.

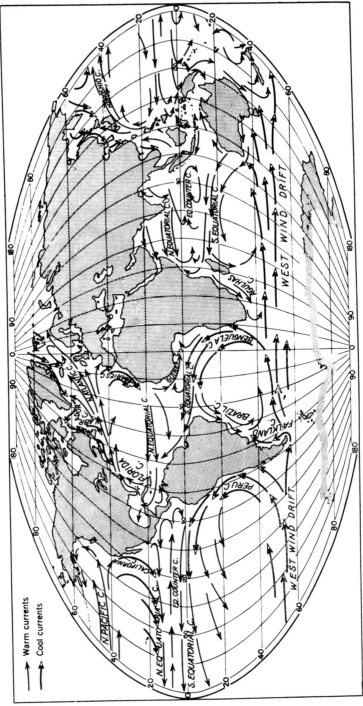

Fig. 6-3. The principal ocean currents of the world. From "Introduction to Meteorology" by S. Pettersen. Copyright 1958, McGraw-Hill. Used with permission of McGraw-Hill Book Company.

A number of observations of the manner in which radioactive substances diffuse vertically and horizontally in the mixed layer have been made in connection with the United States weapons testing program in the Marshall Islands and in the Irish Sea where the British have conducted extensive studies of the fate of radioactive contamination from Windscale.

Studies in the Pacific in 1958 of mixing of radioactive fallout indicated a holdup of radioactivity in the vicinity of the thermocline at the end of 48 hr (Lowman, 1960). Figure 6-4 gives the distribution of activity as a function of depth for 6, 28, and 48 hr after fallout occurred and illustrates the conspicuous peaking which was found at the end of 48 hr in the vicinity of thermocline. No explanation was given for the apparent rise of activity between 150 to 300 m in depth.

After the Marshall Island tests of 1954, three extensive surveys were made of the spread of radioactivity in the northern Pacific. The general course of the contamination was initially in a westerly direction to the region of the Asiatic mainland where the contamination turned north into the Kuroshio current. The data from these surveys have been summarized by Miyake and Saruhashi (1960), as shown in Fig. 6-5 and Table 6-1.

Fission products introduced into surface waters near Bikini in the Marshall Islands (Folsom and Vine, 1957) were found to have moved 120 miles in 40 days and to have diffused during this time only to a depth of 30 to 60 m. The horizontal area was found to be about 40,000 km². The

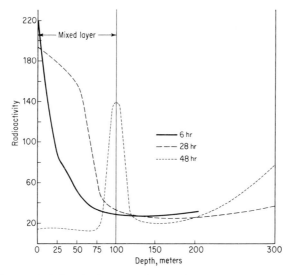

Fig. 6-4. The vertical distribution of fission products in the ocean at various times after fallout of debris from a nuclear explosion (Lowman, 1960).

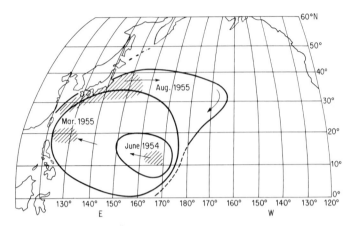

Fig. 6-5. The horizontal dispersion of nuclear weapons debris in the Western Pacific Ocean after tests by the United States in the Marshall Islands in 1954 (Miyake and Saruhashi, 1960).

dilution was such that had 1000 Ci been introduced the average concentration at the end of 40 days would have been 1.5×10^{-10} μCi/ml.

It is known that there are many deep-sea trenches having depths greater than 7 km, and there is much interest in the mechanisms of diffusion at these depths since it has been proposed that they be used for the storage of radioactive wastes (Bogorov and Kreps, 1958).

The rate of movement of the bottom water has not been mapped, but Koczy (1958, 1960) undertook to estimate rates of vertical diffusion from measurements of radium and other substances dissolved in ocean waters. Dissolved ^{238}U in ocean waters gives a rise by radioactive decay to ^{230}Th which precipitates rapidly to the bottom sediments. The decay of ^{230}Th results in the formation of ^{226}Ra which tends to return into solution at the ocean bottom. Koczy took the vertical gradient of dissolved radium as a measure of the rate of movement of bottom water toward the surface.

From various data, Koczy (1960) has drawn the model of vertical diffusion shown in Fig. 6-6. Dissolved substances released from the ocean floor diffuse slowly through a friction layer 20 to 50 m in depth, where diffusion rates are of the order of molecular diffusion. Mixing is most rapid (3 to 30 cm²/sec) just above the friction layer and decreases rapidly with height above the ocean floor to a level about 1000 m below the surface where a secondary minimum ($\sim 10^{-2}$ cm²/sec) is thought to exist. Diffusion rates then increase as one approaches the mixed layer where diffusion coefficients ranging from 50 to 500 cm²/sec are found.

Koczy estimated the vertical velocity in the Atlantic Ocean to be between

TABLE 6-1

⁹⁹Sr Contamination of North Pacific Ocean, 1954–1955 (in pCi/liter)[a]

Source	Period	Contaminated area (>0.1 pCi/liter) (km²)	Maximum activity observed (location and depth)	Maximum averaged activity of water column	Averaged activity in the contaminated area	Total amount of ⁹⁹Sr (MCi)
1st Shunkotsu expedition	May 31–June 30, 1954	4.2×10^6	194[b] 11°05'N 160°12'E 80 m	45[b] (down to 100 m)	7[b] (down to 100 m)	$2.6 + x$
Operation troll	March 9–April 13, 1955	1.9×10^7	3.9 21°30'N 124°45'E 0 m	1.8 (down to 400 m)	0.4 (down to 400 m)	3
NORPAC expedition	August–September, 1955	2.7×10^7	0.5 28°25'N 129°E 0 m	0.5 (down to 500 m)	Contaminated area (>0.1 pCi/liter) 0.2 (500 m)	3

[a] Miyake and Saruhashi (1960).
[b] Except near the test site.

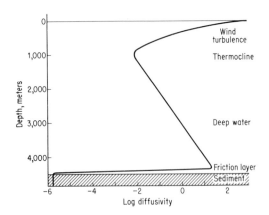

Fig. 6-6. Vertical diffusion from the ocean deeps according to Koczy (1960).

0.5 and 2 m/year at depths ranging from 750 to 1750 m. If these values apply at greater depths, a radioactive solution placed at a depth of 3000 m would not appear in the surface water for more than 1000 years. Using these data, he estimated that 10^9 Ci of ^{90}Sr could be deposited at a depth of 3600 m below the thermocline without exceeding the maximum permissible concentration (MPC) for ^{90}Sr in the mixed layer.

The vertical motions of the oceans have been more recently summarized by Pritchard *et al.* (1971) using vertical profiles of ^{14}C concentrations. A vertical velocity of about 6.6 m/year was found to be typical in the northeast Pacific Ocean to a depth of about 1000 m.

Pritchard (1960) also undertook to calculate the rate at which a radioactive substance would diffuse to the surface if released from the ocean depths. He used a simplified model in which the total amount of radioactivity released in 1 year was assumed to diffuse instantaneously to a height of 5 m. His source was assumed to be at a depth of 4000 m and to ascend at a rate of 5 m/year. Each year the 5 m layer of contamination is assumed to rise by 5 m and its position occupied by a new contaminated layer. No exchange between the 5 m layers is assumed, but the horizontal diffusion does occur, so that as the radioactivity rises, each layer has a greater horizontal spread and, therefore, a lower concentration than the layer behind it.

The horizontal diffusion velocity for the deep ocean is not known but is assumed by Pritchard to lie between 2×10^{-3} and 2×10^{-4} m/sec, and horizontal spread of the material over most of the area of the North Atlantic would occur inbetween 40 and 400 years, which is somewhat less than the 600 years required to rise from a depth of 4000 m to a depth of 1000 m.

Based on Pritchard's model, at the end of 600 years the radioactivity from 1 year's release would be contained in a layer 5 m thick having a horizontal area of 3×10^{13} m². In proceeding to calculate the concentration at the base of the intermediate layer, Pritchard allowed for the fact that water at a depth of 1000 m is returned to the deep layers when it reaches high latitudes. On remixing with the fresh contamination at the bottom, it is, therefore, adding a measure of residual activity.

Using this model, Pritchard estimated that 5×10^{11} Ci of ⁹⁰Sr could be placed on the bottom each year at a depth of 4000 m without exceeding a concentration of 10^{-9} μCi/ml at the base of the 1000 m layer. An alternative estimate, based on the assumption that rapid mixing occurs in the bottom 1000 m, reduces Pritchard's estimate to 9×10^{8} Ci, which is in close agreement with the figure suggested by Koczy.

The above estimates are based on relatively few data. Since there is no information concerning the variability in vertical movement, it is possible that at some time or in some place the pollution could reach the surface much more quickly. For example, it is known that there are ocean areas in which upwelling carries water from the deep water to the surface in vertical currents having velocities on the order of a few meters per day (Wooster and Ketchum, 1957), and although it is thought that upwelling is limited to depths of less than 500 ft, such a phenomenon would greatly accelerate vertical transports in those places where upwelling occurs.

COASTAL WATERS

The relatively shallow coastal waters are of special interest insofar as the likelihood of radioactive contaminants being introduced into the ocean is of concern. In the absence of sufficient knowledge with which to compute dilution, a number of simplifying assumptions can be made which enable one to calculate the maximum concentrations which would be found at various distances downstream from a source of radioactivity. These computations, which are due to Reid (National Academy of Sciences—National Research Council, 1959), neglect radioactive decay, assume the diffusion process to be Fickian, and neglect the depletion of the radioactivity in sea water owing to adsorption by bottom solids or by the uptake of marine organisms. The maximum concentration of a contaminant (C_m) along the axis of a current at a distance from a source is given by

$$C_m = \frac{Q}{2D\sqrt{2\pi KUx}}$$

TABLE 6-2

MAXIMUM CONCENTRATION C_m AT DISTANCES x FROM SUSTAINED GROSS SOURCE
WITH RATE OF SUPPLY 100 Ci/YEAR[a]

x (km)	C_m (μCi/ml)
1	2×10^{-7}
2	1.6×10^{-7}
4	1×10^{-7}
10	7×10^{-8}
20	5×10^{-8}
40	3×10^{-8}
100	2×10^{-8}

[a] National Academy of Sciences—National Research Council (1959).

where

$$Q = \text{Ci/year}$$
$$D = \text{depth of water (m)}$$
$$K = \text{diffusivity coefficient}$$
$$U = \text{current velocity (m/day)}$$
$$x = \text{distance from source (m)}$$

For a continuous source of 100 Ci/year (274 mCi/day), a uniform depth
of water of 30 m with a current velocity of 10 cm/sec and a diffusivity coefficient of 1 cm²/sec, the values of C_m at various distances from the source
are given in Table 6-2. The values in this table are thought to be high by
a factor of at least 10 and probably 100 because of the simplifying assumptions that were made.

ESTUARIES

The estuarine waters, in which tidal action brings about a mixing of salt
and fresh water, are of special importance because of their high biological
productivity. Not only are shellfish frequently harvested in estuarine
waters in large quantities, but they serve also as the nursery grounds for
many species which later move to off-shore waters where they are harvested
by man (Odum, 1963).

Pritchard (1958) has divided estuaries into four types based on the mixing patterns. In type A, the sea water flows upstream as an underlying
wedge having clearly defined surfaces. Vertical exchange across the interface between the fresh and salt water is greatly restricted.

In type B estuaries, there is sufficient turbulent mixing to blur the sharp interface between the freshwater and salt water. The salt water from the ocean flows in toward the head of the estuary below fresh water flowing from the river to the ocean.

In type C estuaries, there is so much vertical mixing that the salinity gradient disappears in the vertical axis. A horizontal gradient develops instead, in which for the Northern Hemisphere the seaward net flow of relatively freshwater occurs toward the right bank of the estuary and the saltier ocean water flows in from the ocean near the estuary's left bank, looking toward the sea.

In type D estuaries, turbulent mixing is so complete that the gradients disappear in both the horizontal and vertical directions.

Estuary flow is so complicated that there is little prospect of the dispersion mechanism ever yielding to a generalized theoretical approach in the foreseeable future. Each estuary has its own physical characteristics which must be studied in detail on an individual basis. An example of the kinds of studies that can be taken is the study of diffusion and convection in the Delaware River Basin (Parker *et al.*, 1961). In order to understand

Fig. 6-7. Scale model of the Delaware River, one of several models of rivers, estuaries, and bays by which flow characteristics are studied by the United States Army Corps of Engineers at their laboratory in Vicksburg, Mississippi (United States Army Corps of Engineers).

the effects of the release of contaminants from the first nuclear-powered merchant ship, N.S. Savannah, a scale model of the Delaware Basin was constructed at the United States Army Waterways Experiment Station at Vicksburg, Mississippi (Fig. 6-7). The model was 750 ft long and 130 ft wide. The expected dilution was studied using dyes. Figure 6-8 illustrates the type of information obtained from a study of the dilution of dye over a period of 58 tidal cycles (about 1 month). In this particular study, which involved instantaneous injection of a given dose, the concentration at the end of 58 tidal cycles remained at approximately 1% of the maximum concentration during the initial tidal cycle. The conditions of the experiment were conservative: There was no radioactive decay, sedimentation, or biological uptake. Thus, only the dimunition owing to mixing was measured.

The Vicksburg station also operates a model of the Hudson estuary which has been used in conjunction with theoretical studies, field observations, and other scale models located at Worcester, Massachusetts to yield

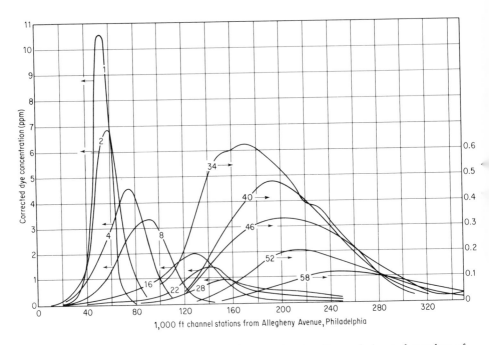

Fig. 6-8. Longitudinal distribution of contaminants after a designated number of tidal cycles in the Delaware River near Philadelphia (Parker *et al.*, 1961).

quantitative relationships applicable to pollution studies in the Hudson (Howells and Lauer, 1969).

RIVERS

Several excellent studies have been undertaken, but there is as yet far too few data about the fate of radioactive substances in rivers and estuaries to permit one to generalize quantitatively. Thus, it is important that the mixing characteristics of each river be ascertained to permit intelligent decisions to be made in regard to waste management practices.

The most extensive set of observations in the United States is probably on the Columbia River, where five large plutonium-producing reactors at Hanford used river water for cooling. The water was purified before entering the reactor, but traces of induced radioactivity, notably ^{32}P and ^{65}Zn, were present in the effluents, and studies of these two isotopes have been made in the water, muds, and biota of the river.

An example of the manner in which dilution of the Hanford effluents takes place is shown in Fig. 6-9 (Foster, 1959). This pattern is characteristic for the Columbia River only at a particular location and at a particular time. Techniques are not available by which one can forecast the concentrations downstream of any given point of discharge with any degree of certainty. Norton (1957) attempted to develop a general mathematical relationship to describe the turbulent mixing of contaminants in the Columbia River, but was unable to obtain a satisfactory solution to the differential equations encountered in the theoretical development. Using an empirical approach, an expression was developed that fit his field data, but this is of limited application.

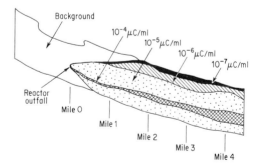

Fig. 6-9. Horizontal mixing of radioactive reactor effluents in the Columbia River (Foster, 1959).

Biological Uptake of Radionuclides

In an aquatic ecosystem, the supply of basic compounds such as carbon dioxide and of elements such as oxygen, calcium, hydrogen, and nitrogen is contained either in solution or is held in reserve in the bottom sediments (Odum, 1963). These nutrients are absorbed and metabolized through the utilization of solar energy by two main types of food-producing organisms: rooted or large floating plants and minute floating plants called phytoplankton. Since food production by the higher plant forms is limited to relatively shallow water, it is the phytoplankton that have the main responsibility for converting the mineral resources of the aquatic environment into food for the higher organisms.

The phytoplankton serve as food for the smaller zooplankton, which, in turn, serve as the basic nourishment of several higher trophic levels. The phytoplankton also serve as food for certain filter-feeding fish and bottom-dwelling animals (IAEA, 1960). Sedimentation of the excrement of aquatic animals and the action of organisms which decompose dead plants and animals eventually complete the cycle by returning the nutrient elements to abiotic forms in which they again become available to the phytoplankton (Lowman et al., 1971).

Although certain elements like carbon, oxygen, and hydrogen are abundant in aquatic organisms, other elements such as copper, manganese, zinc, and iron are present in varying amounts that may or may not be essential to life processes. The aquatic organisms sequester these trace elements to a varying degree, and this phenomenon must be taken into consideration when any decision is made as to the rate at which any given radionuclide can be released to any body of water.

The ratio of the concentration of an element in the organism to the concentration in the water is known as the concentration factor (Polikarpov, 1966). In fresh water, where the mineral constituents are usually present in very much lower concentrations than in sea water, the concentration factors tend to be higher than in the oceans. Moreover, concentration factors in the marine environment are apt to be more uniform than in freshwater because of the very much greater variability of the mineral constituents of the latter, which vary not only from river to river and lake to lake but also change from day to day and season to season, depending on the amounts of rainfall and other factors. This is illustrated in Tables 6-3 and 6-4 in which the concentration factors reported by various investigators are given for the main groups of edible organisms in the freshwater and marine environments.

A comparison of the two tables shows that remarkable differences exist in the concentration factors in freshwater and marine organisms. In each

TABLE 6-3

REPORTED CONCENTRATION FACTORS (CF) FOR VARIOUS CLASSES OF FRESHWATER ORGANISMS[a]

Element	Group	CF Range	Mean CF	Ref.[b]
Cs	Plants	80–4,000	907	1,2,3,4,5
	Fish	120–22,000	3,680	2,3,6,7,8,9,10
Sr	Plants	80–410	200	11,12
	Fish	0.85–90	14	11,12,13,14,15,16
Mn	Plants	1,300–600,000	150,000	1,6,17,18,19
	Crustacea	1,700–250,000	125,000	1,7,18,20
	Molluscs	1,100–1,600,000	∼300,000	1,19,20,21
	Fish	0.1–400	81	1,6,7,19,20
Co	Plants	300–30,000	6,760	1,19,22
	Molluscs	300–85,000	32,408	19,22
	Crustacea	—	—	
	Fish	60–3,450	1,615	1,19,22
Zn	Plants	140–15,000	3,155	6,18,19,22
	Molluscs	30–140,000	33,544	19,22
	Crustacea	300–4,000	1,800	7,18,22
	Fish	10–7,600	1,744	1,6,7,19,22
Fe	Plants	40–45,000	6,675	6,18,19,22
	Molluscs	20–80,000	25,170	19,22
	Crustacea	60–1,800	930	18,22
	Fish	0.1–1,225	191	6,19
I	Plants	10–200	69	22,23,24
	Molluscs	60–1,000	320	22,24
	Crustacea	—	—	
	Fish	0.5–25	9	22,23,24
Ce	Plants	200–35,000	3,180	7,22,24
	Molluscs	400–1,500	1,100	22,24
	Crustacea	300–1,000	600	22
	Fish	2–160	81	22,24
K	Fish	340–18,000	4,400	7,25,26
Ca	Plants	64–720	350	20
	Fish	0.5–470	70	7,20

[a] Jinks and Eisenbud (1972).

[b] References: (1) Lentsch et al. (1970); (2) Harvey (1964); (3) Pendleton and Hanson (1958); (4) Pendleton (1962); (5) Wlodek (1967); (6) Feldt (1971); (7) Bryan et al. (1966); (8) Lentsch et al. (1970); (9) Gallegos et al. (1970); (10) Kolehmainen et al. (1968); (11) Ophel and Judd (1969); (12) Ophel (1963); (13) Nelson (1967); (14) Agnedal (1967); (15) Templeton and Brown (1964); (16) Reed and Nelson (1969); (17) Howells and Bath (1969); (18) Jaakkola et al. (1967); (19) Merlini et al. (1967); (20) Vinogradov (1953); (21) Harrison (1969); (22) Polikarpov (1966); (23) Kohlehmainen et al. (1967); (24) Beninson et al. (1966); (25) Nelson (1969); (26) Gustafson (1969).

TABLE 6-4

REPORTED VALUES OF CONCENTRATION FACTORS (CF) FOR VARIOUS CLASSES OF MARINE ORGANISMS[a]

Element	Group	CF Range	Mean CF	Ref.[b]
Cs	Plants	17–240	51	1,2
	Molluscs	3–28	15	1,3,4
	Crustacea	0.5–26	18	1,4
	Fish	5–244	48	1,4
Sr	Plants	0.2–82	21	1,4,5
	Molluscs	0.1–10	1.7	4,5
	Crustacea	0.1–1.1	0.6	4,5
	Fish	0.1–1.5	0.43	1,4,5
Mn	Plants	2,000–20,000	5,230	4,5,6
	Molluscs	170–150,000	22,080	3,4,5,6
	Crustacea	600–7,500	2,270	1,4
	Fish	35–1,800	363	4,5,6
Co	Plants	60–1,400	553	1,4,5,6
	Molluscs	1–210	166	4,5
	Crustacea	300–4,000	1,700	4,5,7
	Fish	20–5,000	650	4,5,7
Zn	Plants	80–2,500	900	1,4,5
	Molluscs	2,100–330,000	47,000	1,4,5,8
	Crustacea	1,700–15,000	5,300	1,4,5
	Fish	280–15,500	3,400	1,4,5
Fe	Plants	300–6,000	2,260	3,4,5,6
	Molluscs	1,000–13,000	7,600	4,5
	Crustacea	1,000–4,000	2,000	4,5
	Fish	600–3,000	1,800	5
I	Plants	30–6,800	1,065	3,4,5,6
	Molluscs	20–20,000	5,010	4,5
	Crustacea	20–48	31	3,4,5
	Fish	3–15	10	4,5
Ce	Plants	120–4,500	1,610	1,3,4
	Molluscs	100–350	240	4
	Crustacea	5–220	88	1,4
	Fish	0.3–538	99	1,3,4
K	Plants	4–31	13	1,3,6
	Molluscs	3.5–10	8	1,3,4,6
	Crustacea	8–19	12	1,3,4,6
	Fish	6.7–34	16	1,4

TABLE 6-4 (continued)

Element	Group	CF Range	Mean CF	Ref.[b]
Ca	Plants	1.8–31	10	4,6
	Molluscs	0.2–112	16.5	4,6
	Crustacea	0.5–250	40	4,6
	Fish	0.5–7.6	1.9	4,6
Pu	Plants	—	1,000	4
	Molluscs	—	286	9
	Fish	0.1–5	2.55	4,9
Rb	Plants	12–42	23	6
	Molluscs	11–27	17	4
	Crustacea	8.9–17.3	13	4
	Fish	7.6–23.8	17	4
Ru	Plants	15–2,000	448	1,3,4
	Molluscs	1–3.6	2.2	4
	Crustacea	1–100	38	1,4
	Fish	0.4–26	6.6	1,4
Zr–Nb	Plants	170–2,900	1,119	1,4
	Molluscs	8–165	81	4
	Crustacea	1–100	51	1,4
	Fish	0.05–247	86	1,4

[a] Jinks and Eisenbud (1972).
[b] References: (1) Bryan *et al.* (1966); (2) Smales and Salmon (1955); (3) Polikarpov (1967); (4) Polikarpov (1966); (5) Ichikawa (1961); (6) Vinogradov (1953); (7) Welander (1969); (8) Seymour (1966); (9) Wong and Noshkin (1970).

case the wide ranges of concentration factors may be due in part to species differences or to differences in the trace element content of the water, particularly in the freshwater environments. Another factor which may introduce variability is that in some instances the organisms sampled may not have been in equilibrium with the water. Not the least possible influence is that many of the data are based on field measurements, and some of the results may have been affected by errors in the analytical procedures. This possibility is particularly relevant to analysis of water samples owing to the difficulties in obtaining representative samples and reliable results in situations where the concentration of the nuclide may be near the limit of detectability. The importance of this factor is illustrated by a world-wide intercomparison of analytical results of sea water samples submitted to 44 laboratories for radiochemical analysis by the International Atomic Energy Agency (IAEA, 1971b).

The significance to man of the presence of radionuclides in marine and freshwater foods depends on the part of the organism in which the radionuclide is located. When it concentrates in an organ that is eaten by man, it is of more significance than if it occurs in the portion that is not eaten. Thus, although it is known that clams, oysters, and scallops concentrate ^{90}Sr, as do certain crabs (Chipman, 1960), this element is stored in the shell which is not ordinarily consumed. On the other hand, ^{65}Zn and ^{60}Co are known to concentrate in the eatable tissues of seafood.

No conclusion can be drawn about the relative hazard of a given rate of discharge of radioactive substances without a quantitative understanding of the physical characteristics of the body of water, its ecology, and the pathways of human exposure.

In British studies of the fate of radioactive effluents from Windscale (Dunster, 1958; Preston and Jefferies, 1969), it was found that the factor which limited the quantity of waste discharged into nearby coastal waters was the accumulation of radioactivity in seaweed harvested in Cumberland about 20 km from the point of discharge. The few local inhabitants who regularly consume substantial quantities of seaweed limited the maximum amount of radioactivity that could be discharged into this particular environment.

Beasley et al. (1971) have shown that ingestion of many radionuclides, both natural and artificial, can be greatly increased by consumption of marine protein concentrates. This may become an increasingly important factor in the future.

Another example in which local factors influence the significance of contamination of a river is to be found at Hanford, where Columbia River water irrigates thousands of square miles of land downstream from the Hanford reservation, resulting in ^{65}Zn contamination of farm produce (Davis et al., 1958). The concentration in beef and cows' milk was particularly noteworthy.

Tvären, a Baltic bay with a capacity of 409×10^3 m^3 and a surface area of 17.5 km^2, has been studied by Agnedal and Bergström (1966) to identify the human exposure pathways and to provide estimates of the maximum permissible release of radioactive wastes from nearby nuclear facilities. Fish consumption habits, sun-bathing, and commercial fishing techniques were evaluated. It is of interest that the average fish consumption for the local population was found to be 30 kg/year, that the local fishermen spend a maximum of 3000 hr/year in contact with their gear, but that the maximum time at sun-bathing was taken as 100 hr/year. The release limits were calculated to be 500×10^6 Ci/year of tritium, 5600 Ci/year ^{106}Ru, 3800 Ci/year ^{60}Co, 63,000 Ci/year ^{90}Sr, and 28,000 Ci/year ^{239}Pu.

Role of the Suspended Solids and Sediments

The sediments play a predominant role in aquatic radioecology by serving as a repository for radioactive substances which they pass by way of the bottom-feeding biota to the higher trophic levels (Duursma and Gross, 1971).

Figure 6-10 (Wrenn *et al.*, 1972) illustrates the relatively high degree of concentration of ^{137}Cs that takes place in Hudson River sediments and the effect of the sediments in stabilizing the concentration of ^{137}Cs in fish. These data were collected in the northern (fresh water) reaches of the Hudson River estuary beyond the range of influence of a nuclear power reactor located at Indian Point. Shown in Fig. 6-10 are: (a) the concentration of this nuclide in river water, sediments, aquatic plants, and fish and (b) the rate of fallout accumulation of ^{137}Cs. It is remarkable that although the rate of fallout and the concentration in water diminished from 1964 through 1971, the concentration in the bottom sediments changed very little (and actually seems to have increased), whereas the fish dropped only slightly. The aquatic plants seem to follow changes in the ^{137}Cs content of the sediments.

Pollution introduced into a body of water reaches the bottom sediment primarily by adsorption on suspended solids that later deposit on the bottom. The deposited remains of biota that have absorbed or adsorbed pollutants may also be an important source.

In studies of the Clinch River below Oak Ridge, Tennessee, the amounts of radioactivity contained by the suspended solids were found to be variable (Parker *et al.*, 1966), which is not surprising considering that the load of solids varies from place to place in a river and also varies with time. In addition to being dependent on the burden of suspended solids, the tendency of the sediments to remove pollutants will depend on their ion-exchange capacity as well as the temperature and salinity of the overlying water. A materials balance study of the Clinch River revealed that of all the radioactive materials released over a 20-year period the sediments contained 21% of the ^{137}Cs, 9% of the ^{60}Co, 0.4% of the ^{106}Ru, about 25% of the rare-earth radionuclides, and about 0.2% of the ^{90}Sr (Pickering *et al.*, 1966). The distribution of ^{137}Cs in cores several feet deep were found to be well correlated with the annual releases to the river. This was found not to be so for ^{60}Co. In other river systems the sediments might be less quiescent owing to stirring during periods of high run-off.

Lentsch *et al.* have studied the behavior of ^{54}Mn in that portion of the Hudson River near the Indian Point reactor, a source of ^{54}Mn. The manganese content of several species of rooted aquatic plants has been found to

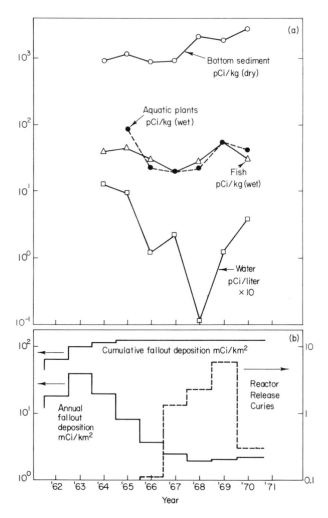

Fig. 6-10. [137]Cs in the upper reaches of the Hudson River, 1962–1971 (Wrenn *et al.*, 1972).

be proportional to the dissolved manganese concentration in water, but this in turn has been shown to be highly variable owing to the periodic intrusion of salt water which was found to release manganese bound in the river sediments. In this investigation, the [54]Mn was found to behave similarly to stable manganese present in the river system (Lentsch *et al.*, 1972).

Chapter 7

Natural Radioactivity

The phenomenon of natural radioactivity was first discovered by Becquerel just before the end of the nineteenth century, and the classic experiments, which were subsequently undertaken by the Curies and others, have had profound effects generally on science, technology, and society.

It soon became apparent that natural radioactivity was a useful tool with which to study the structure and properties of matter, both on a microscopic and macroscopic scale. The phenomenon of natural radioactivity has been used, for example, to reveal the structure of the atomic nucleus, to estimate the age of the earth, and to measure the rates of sediment formation on ocean bottoms.

With the advent of man's utilization of nuclear energy, studies of the natural levels of radioactivity have become necessary in order to understand fully the environmental influences of the radioactivity produced by the atomic energy industry. Only by having knowledge of the amounts of natural radioactivity and the manner in which it varies can an intelligent interpretation of monitoring data be ensured, whether in the vicinity of a reactor site, the oceans, the atmosphere, or in the tissues of man.

Traces of naturally occurring radioactivity can be demonstrated in all substances, living and nonliving. In addition, ionizing radiations originating beyond the earth continually bombard its surface and subject all inhabitants to another source of exposure. The natural radioactivity to which man is exposed is thus of both terrestrial and extraterrestrial origin.

About 340 nuclides have been found in nature, of which about 70 are radioactive and are found mainly among the heavy elements. All elements having an atomic number greater than 80 possess radioactive isotopes, and all isotopes of elements heavier than number 82 are radioactive. The relative abundances of the isotopes now present on earth are derived from the isotopic ratios produced when the universe was formed. Alpher and Herman (1953) call attention to the surprising uniformity of the ratios of elements not only on earth but in the atmospheres and on the surfaces of the planets, in meteorites, in interstellar matter, and even in extragallactic nebulas. This observation, together with the observed constancy in the relative abundance of the isotopes everywhere on earth, has led to the inference that constant isotopic distribution is a fundamental property of the universe. Alpher and Herman and Rankama (1954, 1963) have reviewed the relationships between this inference and the several hypotheses by which the origin of the universe is explained. However, minor shifts in isotopic ratios do occur gradually in nature as the result of physiochemical geological processes that are sensitive to slight differences in mass.

When the matter of which the universe is now formed first came into being, a relatively large number of isotopes must have been radioactive, but in the several billion years during which the world is believed to have existed, all the shorter half-lived isotopes would have disappeared. The radionuclides which now remain are those that have half-lives comparable to the age of the universe. Radioisotopes with half-lives of less than about 10^8 years would have become undetectable in the 30 or so half-lives since their creation, whereas radionuclides having half-lives of greater than 10^{10} years have decayed very little up to the present time. Kohman (1959) divides the residual radioactivity of the earth into the primary radionuclides, whose half-lives are sufficiently long that they have survived the interval since their creation and the secondary radionuclides which are derived by radioactive decay from the primaries.

We will see that in most places in the world the natural radioactivity varies within relatively narrow limits, but that in certain localities there are wide deviations from normal levels owing to the presence of abnormally high concentrations of radioactive minerals in the local soils.

Naturally Occurring Radioactive Substances

The naturally occurring radionuclides can be divided into those that occur singly (Tables 7–1a and 7–1b) and those that are components of

TABLE 7-1a

SINGLY OCCURRING NATURAL RADIONUCLIDES PRODUCED BY COSMIC RAYS[a]

Radio-nuclide	Half-life	Average atmospheric production rate (atoms/cm² sec)	Tropospheric concentration (pCi/kg air)	Principal radiations and energies (MeV)	Observed average concentrations in rainwater (pCi/liter)
^{3}H	12.3 years	0.25	3.2×10^{-2}	β^- 0.0186	—
^{7}Be	53.6 days	8.1×10^{-3}	0.28	γ 0.477	18.0
^{10}Be	2.5×10^6 years	3.6×10^{-2} [b]	3.2×10^{-8}	β^- 0.555	—
^{14}C	5730 years	2.2 (²)	3.4	β^- 0.156	—
^{22}Na	2.6 years	5.6×10^{-5}	3.0×10^{-5}	β^+ 0.545, γ 1.28	7.6×10^{-3}
^{24}Na	15.0 hr	—	—	β^- 1.4, γ 1.37, 2.75	0.08–0.16
^{32}Si	~650 years	1.6×10^{-4}	5.4×10^{-7}	β^- 0.210	—
^{32}P	14.3 days	8.1×10^{-4}	6.3×10^{-3}	β^- 1.71	"a few"
^{33}P	24.4 days	6.8×10^{-4}	3.4×10^{-3}	β^- 0.246	"a few"
^{35}S	88 days	1.4×10^{-3}	3.5×10^{-3}	β^- 0.167	0.2–2.9
^{36}Cl	3.1×10^5 years	1.1×10^{-3}	6.8×10^{-9}	β^- 0.714	—
^{38}S	2.87 hr	—	—	β^- 1.1, γ 1.88	1.8–5.9
^{38}Cl	37.3 min	—	—	β^- 4.91, γ 1.60, 2.17	4.1–67.6
^{39}Cl	55.5 min	1.6×10^{-3}	—	β^- 1.91, γ 0.25, 1.27, 1.52	4.5–22.5

[a] Perkins and Nielsen (1965).
[b] Korff (1971).

TABLE 7–1b

SINGLY OCCURRING NATURAL RADIONUCLIDES OF TERRESTRIAL ORIGIN[a]

Radio- nuclide	Abun- dance (%)	Half-Life (years)	Principal radiations: energy (MeV) and yield (%)	Specific activity (elemental) (pCi/g)
^{40}K	0.012	1.26×10^{9}	β^- 1.33, 89% γ with EC 1.46, 11%	855
^{50}V	0.25	6×10^{15}	γ with β^- 0.78, 30% γ with EC 1.55, 70%	3.0×10^{-3}
^{87}Rb	27.9	4.8×10^{10}	β^- 0.28, 100%	2.4×10^{4}
^{115}In	95.8	6.0×10^{14}	β^- 0.48, 100%	4.98
^{123}Te	0.87	1.2×10^{13}	EC	2.11
^{138}La	0.089	1.12×10^{11}	β^- 0.21, 80% γ with EC (0.81, 1.43), 70%	20.7
^{142}Ce	11.07	$>5 \times 10^{16}$	(α)	5.6×10^{-3}
^{144}Nd	23.9	2.4×10^{15}	α 1.83	0.25
^{147}Sm	15.1	1.05×10^{11}	α 2.23	3.5×10^{3}
^{148}Sm	11.27	$>2 \times 10^{14}$	—	1.37
^{146}Sm	13.82	$>1 \times 10^{15}$	—	0.33
^{152}Gd	0.20	1.1×10^{14}	α 2.1	4.3×10^{-2}
^{156}Dy	0.052	$>1 \times 10^{18}$	—	1.2×10^{-6}
^{174}Hf	0.163	2×10^{15}	α 2.5	1.68×10^{-3}
^{176}Lu	2.6	2.2×10^{10}	β^- 0.43 γ 0.089, 0.203, 0.306	2.4×10^{3}
^{180}Ta	0.012	$>1 \times 10^{12}$	—	0.239
^{187}Re	62.9	4.3×10^{10}	β^- 0.003	2.8×10^{4}
^{190}Pt	0.013	6.9×10^{11}	α 3.18	0.36

[a] Data assembled from Lederer et al. (1967).

three distinct chains of radioactive elements: (1) the uranium series which originates with ^{238}U (Table 7–2); (2) the thorium series which originates with ^{232}Th (Table 7–3); and (3) the actinium series which originates with ^{235}U (Table 7–4). The singly occurring radionuclides are of both cosmic and terrestrial origin.

The three families of radioactive heavy elements are all found in the earth's crust and account for much of the radioactivity to which man is exposed. A fourth family, the neptunium series which originated in the parent element ^{241}Pu, is known to have existed at one time, but ^{241}Pu has a half-life of only 14 years and, therefore, survived very briefly after its original formation. Other members of that series also have relatively short half-lives. The only surviving member of the neptunium family is the

TABLE 7–2

NUCLIDES OF THE URANIUM SERIES AND THEIR PRINCIPAL EMISSIONS

Isotope	Relative isotopic abundance (%)	Half-life	Radiation	Energy (MeV)	Percent yield
^{238}U	99.28	4.5×10^9 years	α	4.20	75
				4.15	23
			γ	0.048	23
^{234}Th	—	24 days	β	0.192	65
				0.100	35
			γ	0.092	4.0 (doublet)
^{234m}Pa	—	1.2 min	β	2.29	98
				1.53	<1
				1.25	<1
			γ (IT)	0.39	0.13
			γ	0.817	4
^{234}U	0.0057	2.5×10^5 years	α	4.77	72
				4.72	28
			γ	0.093	5
^{230}Th	—	8.0×10^4 years	α	4.68	76
				4.61	24
				4.51	0.35
			γ	0.068	0.6
				0.253	0.02
^{226}Ra	—	1622 years	α	4.78	94.3
				4.59	5.7
			γ	0.186	4
				0.26	0.007
^{222}Rn	—	3.8 days	α	5.48	100
			γ	0.510	0.007
^{218}Po	—	3.05 min	α	6.0	99+
^{214}Pb	—	26.8 min	β	0.72	100
			γ	0.053	1.6
				0.242	4
				0.295	19
				0.352	36
^{218}At	—	1.5–2.0 sec	α	6.70	94
				6.65	6
^{214}Bi	—	19.7 min	β	3.26	19
				1.51	40
				1.00	23

TABLE 7-2 (continued)

Isotope	Relative isotopic abundance (%)	Half-life	Radiation	Energy (MeV)	Percent yield
				1.88	9
			α	5.52	0.008
				5.45	0.011
				5.27	0.001
^{214}Po	—	1.64×10^{-4} sec	α	7.68	100
			γ	0.799	0.014
^{210}Tl	—	1.3 min	β	1.9	56
				1.3	25
				2.3	19
			γ	0.296	80
				0.795	100
				1.31	21
^{210}Pb	—	22 years	β	0.015	81
				0.061	19
			γ	0.0465	4
^{210}Bi	—	5.0 days	β	1.17	99+
			α	5.0	5×10^{-5}
^{210}Po	—	138 days	α	5.3	100
			γ	0.80	0.0011
^{206}Tl	—	4.2 min	β	1.51	100
^{206}Pb	25.2	Stable	—	—	—

nearly stable nuclide ^{209}Bi which has a half-life estimated to be about 2×10^{17} years (Rankama, 1963).

The transuranic elements that have been produced artificially (Seaborg, 1958) must have existed in nature at one time, but their half-lives are so short that these elements would have disappeared long ago. However, some transuranic elements should be produced continuously as a consequence of the availability of naturally occurring neutrons for capture by the uranium isotopes in the earth's crust. Thus, ^{239}Pu has been detected in pitchblende in a ratio to ^{238}U of 10^{-11} to 10^{-13}, and ^{237}Np has been identified in uranium minerals in a ratio to ^{238}U of 1.8×10^{-12} (Rankama, 1954).

In nature, ^{235}U and several other nuclides including ^{238}U, ^{234}U, ^{232}Th, and ^{226}Ra undergo fission spontaneously and are also fissioned by neutrons originating from cosmic rays, from (a,n) reactions with light nuclei, or from fission of other nuclei. None of these processes results in significant radioactivity. The half-life of ^{235}U owing to spontaneous fission is between

TABLE 7-3

NUCLIDES OF THE THORIUM SERIES

Isotope	Relative isotopic abundance (%)	Half-life	Radiation	Energy (MeV)	Percent yield
^{232}Th	100	1.4×10^{10} years	α	4.01	76
				3.95	24
			γ	0.055	24
^{228}Ra	—	6.7 years	β	0.055	100
^{228}Ac	—	6.13 hr	β	2.18	10.0
				1.85	9.6
				1.72	6.7
				1.11	53.0
				0.64	7.6
				0.46	13
			γ	0.058	53.0
				0.129	5.2
				0.184	<1
^{228}Th	—	1.9 years	α	5.42	71
				5.34	28
			γ	0.083	1.6
^{224}Ra	—	3.64 days	α	5.68	94
				5.45	6
				5.19	0.4
			γ	0.241	3.7
^{220}Rn	—	55 sec	α	6.28	100
			γ	0.50	0.07
^{216}Po	—	0.16 sec	α	6.77	100
^{212}Pb	—	10.6 hr	β	0.33	~88
				0.57	~12
			γ	0.176	<1
				0.238	47
				0.300	3.2
^{212}Bi	—	60.5 min	β	2.25	64
			α	6.086	9.8
				6.047	25.1
			γ's with β	1.81	1.0
				1.61	1.8
				1.03	2
				0.83	13
				0.72	7
			γ's with α	0.040	2
				0.288	0.5
				0.46	0.8
					(complex)

TABLE 7–3 (continued)

Isotope	Relative isotopic abundance (%)	Half-life	Radiation	Energy (MeV)	Percent yield
^{212}Po	—	3.04×10^{-7} sec	α	10.55	<1
				8.785	99
^{208}Tl	—	3.1 min	β	1.80	100
			γ	2.61	100
				0.86	12
				0.58	86
				0.51	23
^{208}Pb	52.3	Stable	—	—	—

10^{15} and 10^{16} years, which means that decay by this process proceeds at a rate less than 10^{-7} of that due to α emission. Others of the heavy nuclides undergo spontaneous fission with half-lives that range from 10^{14} to 10^{20} years (Rankama, 1963).

URANIUM

The uranium normally found in nature consists of three isotopes having mass numbers of 234, 235, and 238. ^{238}U, the parent of the uranium series, is present in the amount of 99.28% and is in equilibrium with its great granddaughter ^{234}U which is present in the amount of 0.0058%. ^{235}U present in the amount of 0.71% is the parent isotope of the actinium series and is the principal nuclide utilized in the fission process.

Uranium is found in most rocks and soils. Typical concentrations are listed in Table 7–5 in which it is shown that the acid igneous rocks contain concentrations of the order of 3 ppm, about 100 times greater than the ultrabasic igneous rocks but considerably less than the phosphate rocks of Florida and elsewhere which contain as much as 120 ppm and which have been considered as a commercial source of uranium (Clegg and Foley, 1958). The high uranium content of phosphate rocks is reflected in correspondingly high uranium concentrations in commercial phosphate fertilizers. Spalding and Sackett (1972) have found that the uranium content of North American rivers is higher than in the past, which they attribute to increased run-off of phosphate fertilizers.

The uranium content of air in New York State has been found to range from 0.10 to 1.47 ng/m³ (0.035 to 0.47 fCi/m³), with a significant corre-

TABLE 7-4

Isotope	Relative isotopic abundance (%)	Half-life	Radiation	Energy (MeV)	Percent yield
^{235}U	0.72	7.1×10^8 years	α	4.32	3
				4.21	5.7
				4.58	8 (doublet)
				4.5	1.2
				4.4	57
				4.37	18
			γ	0.110	2.5
				0.143	11.0
				0.163	5.0
				0.185	54.0
				0.205	5.0
^{231}Th	—	25.64 hr	β	0.302	52
				0.218	20
				0.138	22
			γ	0.026	2
				0.085	10 (complex)
^{231}Pa	—	3.25×10^4 years	α	5.00	24
				4.94	22
				5.02	23
				5.05	10
			γ	0.027	6
				0.29	6 (complex)
^{227}Ac	—	21.6 years	β	0.046	99
			α	4.95	1.2 (doublet)
			γ	0.070	0.08
^{227}Th	—	18.2 days	α	6.04	23
				5.98	24
				5.76	21
				5.72	14 (doublet)
			γ	0.050	8
				0.237	15 (complex)
				0.31	8 (complex)

TABLE 7-4 (continued)

Isotope	Relative isotopic abundance (%)	Half-life	Radiation	Energy (MeV)	Percent yield
^{223}Fr	—	22 min	β	1.15	99+
			α	5.35	4×10^{-3}
			γ	0.050	40
				0.080	13
				0.234	4
^{223}Ra	—	11.4 days	α	5.75	9
				5.71	54
				5.61	26
				5.54	9
			γ	0.149	10 (complex)
				0.270	10
				0.33	6 (complex)
^{219}Rn	—	4.0 sec	α	6.82	81
				6.55	11
				6.42	8
			γ	0.272	9
				0.401	5
^{215}Po	—	1.77×10^{-3} sec	α	7.38	99+
			β	—	2.3×10^{-4}
^{211}Pb	—	36.1 min	β	1.36	92
				0.95	1.4
				0.53	6
			γ	0.405	3.4
				0.427	1.8
				0.832	3.4
^{215}At	—	$\sim 10^{-4}$ sec	α	8.00	100
^{211}Bi	—	2.16 min	α	6.62	84
				6.28	16
			β	—	0.27
			γ	0.35	14
^{211}Po	—	0.52 sec	α	7.45	99
				6.89	0.5
			γ	0.57	0.5
				0.90	0.5
^{207}Tl	—	4.79 min	β	1.44	100
			γ	0.870	0.16
^{207}Pb	21.7	Stable	—	—	—

TABLE 7-5

AVERAGE URANIUM CONCENTRATION IN VARIOUS ROCKS[a]

Rock type	Uranium concentration (ppm)
Acid igneous	3.0
Intermediate igneous	1.5
Basic igneous	0.6
Ultrabasic igneous	0.03
Meteorites	0.003
Phosphate rock (Fla.)	120
Phosphate rock (N. Africa)	20–30
Bituminous shale (Tenn.)	50–80
Normal granite	4
Limestones	1.3
Other sedimentary rocks	1.2

[a] Lowder and Solon (1956).

lation with suspended particulates (McEachern *et al.*, 1971). The soils and coal flyash are the most likely sources.

Uranium occurs in traces in many commercial products. Pre-World War II samples of steel analyzed by Welford and Sutton (1957) contained uranium in the range of 0.01 to 0.2 ppm. Surprisingly, photographic emulsions and other photographic materials have been shown to contain from 0.2 to 1 ppm of uranium (F. A. Smith and Dzuiba, 1949).

Because the uranium isotopes are α emitters, they of themselves do not contribute to the γ-ray background, and they are present in too low a concentration to contribute significantly to the internal α dose delivered to human beings; but, as would be expected from the fact that uranium occurs in soils and fertilizers, it is possible to demonstrate the presence of uranium in food and human tissues. Welford (1960), in a study of 26 subjects with no occupational or other known exposure to uranium, found that their urinary excretion varied from 0.03 to 0.3 μg/liter of urine. Welford and Baird (1967) estimate the annual uranium dietary intake in New York, Chicago, and San Francisco to be about 500 μg/year (175 pCi/year). Based on measurements in the United Kingdom, Hamilton (1972) estimated the uranium content of the standard man to be 100–125 μg, in equilibrium with a daily intake of about 1 μg U.

Uranium is a substance of enormous economic and military importance and has been the subject of intensive geological study during the past 30 years. For discussions of the occurrence of uranium in economically significant quantities, the reader is referred to texts that deal specifically with this subject (Faul, 1954).

RADIUM-226

From the point of view of the ionizing radiation dose delivered to man, the isotope ^{226}Ra and its daughter products are of special interest. Referring to Table 7-2, it is seen that ^{226}Ra is an α emitter that decays, with a half-life of 1622 years, to ^{222}Rn, a 3.8-day nuclide of the noble gas.

The decay of radon is followed by the successive disintegration of a number of short-lived α- and β-emitting progeny. After six decay steps, producing isotopes ranging in half-lives from 1.6×10^{-4} sec to 26.8 min, the isotope ^{210}Pb is produced, having a half-life of 22 years. This isotope is followed by three additional decay steps in which the longest-lived isotope produced is ^{210}Po (half-life 138 days). The radioactive series ends in stable ^{206}Pb.

Radium, being an α emitter, does not add directly to the γ activity of the environment but does so indirectly through its γ-emitting descendants.

Radium-226 Content of Rocks and Soils

^{226}Ra is present in all rocks and soils in amounts that vary with the type of rock. Igneous rocks tend to contain somewhat higher concentrations than sandstones and limestones. Rankama and Sahama (1950) give a mean concentration of 0.42 pCi/g in limestone and 1.3 pCi/g in igneous rock, as listed in Table 7-6. Evans and Raitt (1935), who measured the ^{226}Ra content of rock and soil specimens from the scene of cosmic-ray

TABLE 7-6

AVERAGE RADIUM, URANIUM, THORIUM, AND POTASSIUM CONTENTS
IN VARIOUS ROCKS[a]

Type of rock	^{226}Ra (pCi/g)	^{238}U (pCi/g)	^{232}Th (pCi/g)	^{40}K (pCi/g)
Igneous	1.3	1.3	1.3	22
Sedimentary				
Sandstones	0.71	0.4	0.65	8.8
Shales	1.08	0.4	1.1	22
Limestones	0.42	0.4	0.14	2.2

[a] Adopted from UNSCEAR (1958).

TABLE 7–7

Average Radium Content of Various Rocks and Soils[a]

No. of specimens	Classification	Specimen, average value (pCi/g)
1	Quartz-mica schist	0.20
1	Quartzite	0.20
4	Limestones	0.29
1	Sandstone	0.32
2	Glacial sand and rubble	0.38
4	Gravels	0.41
2	Soils	0.73
1	Kaibab limestone	0.97
5	Granites	1.02
2	Peruvian larvas	2.06
23	All specimens	0.70

[a] Evans and Raitt (1935).

observations made by Millikan and his co-workers, give the values shown in Table 7–7, which are somewhat lower than the average values given by Rankama and Sahama.

Radium-226 in Water

The ^{226}Ra concentration of ocean bottom water is said to be remarkably uniform and is thought to originate from sediments in which it is formed at a uniform rate from ^{230}Th (Koczy, 1958). The concentration in bottom water was found to vary from 0.08 pCi/liter in the Indian and Atlantic Oceans to 0.15 pCi/liter in the Pacific. The ^{226}Ra content of ocean surface water is reported by Koczy to be of the order of 10^{-4} pCi/liter.

The literature contains many references to the radium content of water from rivers, streams, and wells of various depth. Many of the older measurements are of uncertain quality and are, therefore, inconclusive. However, more recent data published by Hursh (1953) indicate that the ^{226}Ra content of public water supplies is highly variable and in some places may actually approach the maximum permissible concentration (MPC) for continuous consumption.

The data published by Hursh are listed in Table 7–8, where it is seen

TABLE 7-8

²²⁶Ra CONTENT OF PUBLIC WATER SUPPLIES IN THE UNITED STATES[a]

City supply	Source	Ra concentration, 10^{-1} pCi/liter H_2O	
		Raw water	Tap water
Atlanta, Ga.	Chattahoochee R.	0.17	0.09
Baltimore, Md.	Gunpowder R.	0.20	0.08
Birmingham, Ala.	Cahaba R. and L. Purdy	0.24	0.17
Bismarck, ND.	Missouri R.	2.43	0.26
Boise, Idaho	Shallow wells 75%⎫ Deep wells 25%⎭	1.03	0.96
Boston, Mass.	Nashua R.	0.14	0.17
Buffalo, N.Y.	L. Erie	0.35	0.28
Charleston, S.C.	Edisto R.	1.81	1.41
Charleston, W.Va.	Elk R.	0.41	0.45
Cheyenne, Wyo.	Surface and wells	0.50	0.34
Chicago, Ill.	L. Michigan	0.24	0.29
Cincinnati, Ohio	Ohio R.	0.61	0.33
Cleveland, Ohio	L. Erie	0.33	0.23
Dallas, Tex.	Garza and Bachman L.	0.85	0.32
Denver, Colo.	South Platte R.	0.77	0.47
Detroit, Mich.	Detroit R.	0.26	0.18
Indianapolis, Ind.	Fall Cr. and White R.	1.37	0.86
Joliet, Ill.	Deep wells	65.40	57.9
LaVerne, Calif.	Colorado R.	1.00	0.35
Los Angeles, Calif.	Owens Valley Aqu.	0.09	0.08
Louisville, Ky.	Ohio R.	0.84	0.41
Miami, Fla.	Shallow wells	4.78	1.68
Memphis, Tenn.	Shallow and deep wells	2.11	1.70
New Orleans, La.	Mississippi R.	4.25	0.16
New York, N.Y.	Catskill supply	0.15	0.18
New York, N.Y.	Croton supply	0.20	0.18
Oklahoma City, Okla.	N. Canadian R.	1.06	0.42
Omaha, Neb.	Missouri R.	17.70	0.54
Philadelphia, Pa.	Delaware R.	0.48	0.44
Phoenix, Ariz.	Along Verde R.	0.27	0.18
Pierre, S.D.	Shallow wells along Missouri R.	0.62	0.15
Pittsburgh, Pa.	Allegheny R.	37.00	1.41
Portland, Ore.	Bull Run R.	0.14	0.01
Raleigh, N.C.	Walnut Cr.	0.22	0.25
Richmond, Va.	James R.	0.33	0.23
Sacramento, Calif.	Sacramento R.	0.18	0.15
Salt Lake City, Utah	Cottonwood Cr.	0.34	0.50

TABLE 7-8 (continued)

City supply	Source	Ra concentration, 10^{-1} pCi/liter H_2O Raw water	Tap water
San Francisco, Calif.	Calaveras Res.	0.18	0.07
St. Louis, Mo.	Mississippi R.	10.8	0.28
Tacoma, Wash.	Green R.	0.02	0.00
Washington, D.C.	Potomac R.	0.33	0.27
Wichita, Kans.	Deep wells	2.27	0.75

[a] Hursh (1953).

that if we exclude Joliet, Illinois the tap-water values range from 0.0 to 0.17 pCi/liter with an average value of 0.42×10^{-10} Ci/ml. Drinking water from deep wells in southern Illinois is exceptionally high and has been the subject of detailed study by Stehney (1956) and Samuels (1964), who have attempted to relate the levels of radium observed in the Joliet water supply and the skeletal burden of people living in the area. According to Samuels, about 1 million people in northern Illinois and southern Iowa drink water having a ^{226}Ra concentration greater than 3 pCi/liter, and about 50,000 people drink water having a concentration greater than 10 pCi/liter. We will see later that the normal dietary intake of ^{226}Ra is about 2 pCi/day, and that this results in an equilibrium body burden of about 30 pCi. Individuals who consume daily 1 liter of water containing 10 pCi/liter might be expected to receive a bone dose about five times normal.

According to Morris and Klinsky (1962), the Zeolite water-softening process is capable of removing as much as 99% of the ^{226}Ra in raw water.

Radium-226 in Food

Radium is chemically similar to calcium and is, therefore, absorbed from the soil by plants and is passed up the food chain to man. Because the radium in food originates from soil and the radium content of soil is known to be variable, there is considerable variability in the radium content of foods. In addition, it is reasonable to expect that chemical factors such as the amount of exchangeable calcium in the soil will determine the rate at which radium will be absorbed by plants. We have seen in Chapter 6 that this is true for strontium, which also is related chemically to calcium.

One of the earliest attempts to estimate the radium content of food was undertaken by Mayneord and his associates (1958, 1960). This group made

α-radiation measurements of ashed samples of foods and differentiated
the thorium from uranium series by counting the double α pulses from
the decay of ^{216}Po and ^{212}Po, whose disintegrations are separated by only
the 0.158-sec half-life of ^{212}Po. These early measurements served to approximate the total ^{226}Ra and ^{228}Ra concent of foods (Table 7-9) and
were highlighted by the fact that Brazil nuts were found to be extra-
ordinarily radioactive. This was later investigated by Penna Franca *et al.*
(1968), who showed that the phenomenon is due to the tendency of the
Brazil nut tree (*Bertholletia excelsa*) to concentrate barium, which is also a
chemical congener of radium. Penna Franca found the radium content of
Brazil nuts to range between 273 and 7100 pCi/kg, with only 3 out of 15
samples assaying less than 1000 pCi/kg. The radioactivity was about
equally divided between ^{226}Ra and ^{228}Ra and was not related to the radium
or barium content of the soil in which the tree is grown.

Shleien (1969) reported on the ^{226}Ra content of total diets collected in
11 cities in the United States and found the mean values to range from
0.52 to 0.73 pCi/kg. The radium concentration of Brazil nuts is thus of
the order of 1000 times greater than the radium concentration in the
average diet in the United States.

TABLE 7-9

ACTIVITY OF FOODS[a]

Foodstuff	Maximum α activity observed per 100 g (pCi)
Brazil nuts	1400
Cereals	60
Teas	40
Liver and kidney	15
Flours	14
Peanuts and peanut butter	12
Chocolates	8
Biscuits	2
Milks (evaporated)	1–2
Fish	1–2
Cheeses and eggs	0.9
Vegetables	0.7
Meats	0.5
Fruits	0.1

[a] Mayneord *et al.* (1958).

TABLE 7–10

^{226}RA IN NEW YORK CITY DIET[a]

Diet category	$\dfrac{\text{kg}}{\text{year}}$	$\dfrac{\text{g Ca}}{\text{year}}$	1966 $\dfrac{\text{pCi}}{\text{kg}}$	1966 $\dfrac{\text{pCi}}{\text{year}}$	1968 $\dfrac{\text{pCi}}{\text{kg}}$	1968 $\dfrac{\text{pCi}}{\text{year}}$	Average $\dfrac{\text{pCi}}{\text{kg}}$	Average $\dfrac{\text{pCi}}{\text{year}}$
Dairy products	200	216.0	0.25	50	0.30[b]	60	0.25	50[c]
					0.19[b]	38		
Fresh vegetables	48	18.7	0.50	24	1.6	77	1.1	53
					1.6	77		
Canned vegetables	22	4.4	0.65	14	0.68	15	0.67	15
Root vegetables	10	3.8	1.4	14	1.2	12	1.3	13
Potatoes	38	3.8	2.8	106	1.7	65	2.3	87
Dry beans	3	2.1	1.1	3.3	0.98	2.9	1.0	3
Fresh fruit	59	9.4	0.43	25	0.20	12	0.32	19
Canned fruit	11	0.6	0.17	1.9	0.15	1.7	0.16	1.8
					0.16	1.8		
Fruit juice	28	2.5	0.42	12	0.90	25	0.66	18
Bakery products	44	53.7	2.8	123	1.7	75	2.3	101
Flour	34	6.5	1.9	65	2.3	78	2.1	71
Whole grain products	11	10.3	2.2	24	2.7	30	2.5	28
Macaroni	3	0.6	2.1	6.3	1.4	4.2	1.8	5.4
Rice	3	1.1	0.76	2.3	3.3	9.9	2.0	60
Meat	79	12.6	0.01	0.8	0.02	1.6	0.02	1.6
Poultry	20	6.0	0.76	15	0.10	2.0	0.44	8.8
					0.11	2.2		
Eggs	15	8.7	6.1	92	14	210	10	150
					14	210		
Fresh fish	8	7.6	0.67	5.4	1.1	8.8	0.89	7.1
Shellfish	1	1.6	0.80	0.8	0.90	0.9	0.85	0.9
Yearly intake		370		584.8		680.2		639.6
Daily intake pCi/g Ca				1.6		1.8		1.7

[a] Fisenne and Keller (1970).
[b] Two different samples.
[c] Average of three samples.

Fisenne and Keller (1970) estimated the ^{226}Ra intake of inhabitants of New York City and San Francisco to be 1.7 and 0.8 pCi/day respectively, but this twofold difference was not reflected by differences in the ^{226}Ra content of human bone from the two cities. The New York City value was 0.036 pCi/g Ca, and the San Francisco value was 0.031 pCi/g Ca. Studies of this kind involve highly sophisticated and sometimes uncertain food and bone sampling techniques that may involve errors that can obscure differences of a factor of 2.

The ^{226}Ra content of the New York City diet is shown in Table 7–10. The relatively large contribution from eggs seems curious and has not been explained. The radium content of eggs reported by Fisenne and Keller in New York is about three to five times the values reported for San Francisco. The daily intake of ^{226}Ra by infants in the first year of life has been estimated to average 0.6 pCi/day in New York City (Engelmann, 1961).

It is evident from all the above-mentioned data that the uptake from drinking water is not likely to be significant unless the concentration of ^{226}Ra is greater than 1 pCi/liter, a level which seems to be reached in only a few localities.

Radium-226 Content of Human Tissues

A number of investigators in various parts of the world have undertaken to estimate the total body content of ^{226}Ra and the dose delivered by this nuclide and its radioactive progeny. The data presented in Table 7–11, as assembled from various investigators throughout the world by the United Nations (UNSCEAR, 1966), indicate that in most normal situations the ^{226}Ra content of bone varies from about 0.010 to 0.015 pCi/g of ash, corresponding to a total body burden of 30 to 40 pCi of ^{226}Ra in an adult skeleton of 2800 g of ash. According to Hursh *et al.* (1960), the skeletally deposited radium is about 78% of the radium in the total body.

The dose delivered by radium and its daughter products to the various portions of the skeletal tissue is not easily estimated because of uncertainty as to the fraction of radon retained and because of difficulties associated with calculating a dose that results from α emissions in loci that are not exactly known on the microscopic scale in which the absorption of α particles occurs.

The methods of estimating dose to the skeletal tissues from the radium series have been carefully described and evaluated by Spiers (1968). Based on his methods, the United Nations concluded that a skeleton burden of 30–40 pCi of ^{226}Ra produces a dose to the osteocytes of 6 mrem/year and a dose to the bone marrow of 0.3 mrem/year (UNSCEAR, 1966). This comparatively large difference is due to the fact that the osteocyte dose is produced by α particles for which a quality factor of 10 applied, whereas the bone marrow dose is due largely to the β activity of the radium decay products (United Nations, 1958). These estimates do not include the dose from the radium daughters, which are summarized in Table 7–19.

TABLE 7–11

²²⁶Ra in Human Bone as Reported by Various Investigators, 1962[a]

Location of area	pCi/g ash	pCi/g Ca	Total[b] in the skeleton (pCi)
Normal areas			
Central America			
United States			
Puerto Rico	0.006	0.017	17
Europe			
Federal Republic of	0.013	0.040	36
Germany			
United Kingdom	0.008–0.02	—	—
North America			
United States			
Illinois	0.012[c]	—	32
New England	0.014	—	39
New York, N.Y.	0.012	0.032	32
Rochester, N.Y.	0.010; 0.017	—	28.48
San Francisco, Calif.	0.0096	0.026	27
High level areas			
Asia			
India			
State of Kerala	0.096	—	~270
(monazite area)	(0.03–0.14)	—	—
North America			
United States			
Illinois	0.037[d]	—	~100
Illinois	0.028[d]	—	78

[a] Adopted from United Nations UNSCEAR, (1966).
[b] Skeleton of 7000 g fresh weight yielding 2800 g ash was assumed.
[c] In people consuming water with "normal" levels of ²²⁶Ra.
[d] In people consuming water with elevated ²²⁸Ra concentration.

Thorium-232

The thorium content of various rocks, as reported by Faul, indicates a range of 8.1 to 33 ppm for igneous rocks, with a mean value of 12 ppm. Limestone, as would be expected, contains only about 1 ppm. Rankama and Sahama report that the concentration in sandstone is 6 ppm intermediate between limestone and the igneous rocks. The thorium content of igneous rocks is, thus, about four times the uranium content, but since the specific activity of ²³²Th is 0.11 pCi/g compared to 0.33 pCi/g for ²³⁸U, the radioactivity owing to the two nuclides is more nearly 1:1.

The characteristics of the thorium series are basically different from that of the uranium series in a number of respects.

1. ^{228}Ra has a shorter half-life than ^{226}Ra (5.8 years compared to 1620 years).

2. ^{228}Ra is a β emitter that decays to α-emitting ^{228}Th, with a half-life of 1.9 years. ^{228}Th, in turn, decays through a series of α emitters including the noble gas ^{220}Rn (thoron), which has a half-life of only 54 seconds compared to 3.8 days for ^{222}Rn, the decay product of ^{226}Ra. The thoron, thus, has less opportunity to diffuse from its place of formation.

3. The insolubility of ^{228}Th prevents its uptake by vegetation. The solubility of ^{228}Ra in soil is comparable with ^{226}Ra, but the dose rate to an organism from assimilated ^{228}Ra, a β emitter, is time dependent because of the ingrowth of α-emitting ^{228}Th and its short-lived descendants.

4. In the ^{228}Ra chain, there is no long-lived "stopping" nuclide comparable to ^{210}Pb ($T_{1/2} = 22$ years). The longest-lived nuclide beyond ^{228}Th is ^{212}Pb with a half-life of 10.6 hr.

These differences affect the relative dose from the thorium and uranium series. The dosimetry and radiochemistry of the throium series tends to be complicated by these characteristics (Fresco et al., 1952).

Because of its relative insolubility and low specific activity, ^{232}Th is normally present in biological materials only in insignificant amounts. Petrow and Strehlow (1967) found the ^{232}Th content of human bone to range from 0.006 to 0.01 μg/g of ash, in reasonable agreement with measurements previously made by Pavlovskaya (1960) and subsequently by Lucas et al. (1970). The principal biological significance of ^{232}Th is due to its daughters, which will be discussed separately.

RADIUM-228 (MESOTHORIUM)

Although ^{228}Ra frequently occurs in soil and water in approximately a 1:1 ratio to ^{226}Ra, there is surprisingly little information about its occurrence in foods or in human tissues. Dudley (1959) calls attention to the likelihood that in infants the ratio of ^{226}Ra to ^{228}Ra is about 1, but that because of the 6.7 half-life of ^{228}Ra, this ratio would be expected to increase in older persons. Systematic ^{228}Ra measurements in food and water have not been made on a scale comparable to ^{226}Ra, but such data as do exist suggest that under normal circumstances the ^{228}Ra content of food, water, and human tissues is from one-half to one-fourth of the ^{226}Ra content (UNSCEAR, 1966).

^{228}Ra is of special interest in those areas of the world noted for abnormally high concentrations of soil thorium. This is discussed later in this chapter.

RADON-222 AND THORON RADON-220

When ^{226}Ra decays by α emission, it transmutes to its daughter ^{222}Rn, an inert gas having a half-life of 3.8 days. Similarly, ^{224}Ra which is a descendant of ^{232}Th decays by α emission to 54-sec ^{220}Rn, commonly known as thoron.

Each of the two gaseous isotopes diffuses into the atmosphere to some extent. The 3.8-day radon isotope has a greater opportunity to escape to the atmosphere before undergoing decay than does thoron, which has a half-life of only 54 sec and, therefore, has a correspondingly smaller probability of diffusing from its place of birth to the atmosphere. It has been estimated that ^{222}Rn diffuses from the soil at an average rate of 1.4 $\pm$ 0.73 p Ci/m²/sec (Pearson and Jones, 1965).

The atmospheric concentrations of these noble gases and their daughter products depend on many geological and meteorological factors, some of which have not been thoroughly studied. The average radon concentration in Washington, D. C. was shown by Lockhart (1964) to be more than 100 times greater than the average concentration in Little America and more than 12 times the values observed at Kodiak, Alaska. The same investigator who made measurements at these and other localities for many years reports wide variability from day to day. For example, the daily concentrations varied more than one-hundredfold in Washington, D. C. during 1957 (Lockhart, 1958). Lockhart estimated the ^{222}Rn content of surface air by measuring the ^{214}Pb content of atmospheric dust. He has shown that the ^{214}Pb is in secular equilibrium with ^{222}Rn. The ^{212}Pb radioactivity of dust is also a measure of the ^{220}Rn (thoron) content of the atmosphere, but these nuclides are not thought to be in secular equilibrium. The radon concentrations at various places as estimated by Lockhart by means of the ^{214}Pb measurements are given in Table 7–12.

The radon concentration inside buildings is somewhat higher and in round numbers may be taken as 0.5 pCi/liter on the average (UNSCEAR, 1962). The corresponding figure for ^{220}Rn may be taken as 0.02 pCi/liter.

It is likely that these variations are dependent on meteorological factors that influence the rate of emanation of the gases from the earth. Thus, the rate of emanation from soil may increase during periods of diminishing atmospheric pressure and decrease during periods of high soil moisture, owing to the high solubility of radon. It is also likely that the history of an air mass for several days prior to observation influences its radon and thoron content (Barreira, 1961). Passage of the air over oceans and precipitation would tend to reduce the concentration of these gases, whereas periods of temperature inversion might cause the concentrations to increase by limiting the volume of the atmosphere within which dilution can take

TABLE 7–12

SUMMARY OF MEASUREMENTS OF NATURAL RADIOACTIVITY IN THE GROUND-
LEVEL AIR[a]

Site	Period of observa-tion	Radioactivity (pCi/m³) ^{214}Pb	^{212}Pb	Activity ratio $^{214}Pb/^{212}Pb$
Wales, Alaska	1953–1959	20	0.16	125
Kodiak, Alaska	1950–1960	9.9	0.04	250
Washington, D.C.	1950–1961	122	1.34	91
Yokosuka, Japan	1954–1958	56	0.48	117
Lima, Peru	1959–1962	42	1.33	28
Chacaltaya, Bolivia	1958–1962	40	0.53	72
Rio de Janeiro, Brazil	1958–1962	51	2.54	20
Little America V, Antarctica	1956–1958	2.5	<0.01	(>250)
South Pole	1959–1962	0.47	<0.01	(>50)

[a] Lockhart (1964).

place. The gases can be expected to be present in greater amounts over igneous masses than over large bodies of water or over sedimentary formations.

The concentration of radon is reported by Lockhart to be 50 to 100 times greater than that of thoron in each of the five locations studied. Washington, D. C. is significantly higher than the four overseas locations, and Little America has the lowest average concentrations of both radon and thoron, possibly because the snow cover obstructs the emanations from the earth's crust. Low temperatures and proximity to the ocean may also be factors. Similar arguments might be made for the two Alaskan locations.

Other reports from Japan, Great Britain, the Soviet Union, Austria, and Sweden indicate that the average concentrations of radon in outdoor air may be taken to be in the range of 0.1 to 0.5 pCi/liter. A number of investigators, including Blifford et al. (1952), Hultqvist (1956), Gold et al. (1964), and Cox et al. (1970), have observed periodicity in hour-to-hour observations of the radon and thoron content of outdoor air. Maximum concentrations are observed in the early morning hours and minima in the late afternoon, when the values are about one-third the morning maxima (Gold et al., 1964). These diurnal variations could be the result of many interacting factors. Diffusion from soil could be increased in the afternoon owing to rising ground temperatures and the normal increase in atmospheric turbulence in late afternoon. These factors, which would tend to increase removal of radon from soil, could be more than offset by the increased atmospheric stability in the early morning hours.

Seasonal variations in the atmospheric radon concentration have also been noted. Systematic differences in soil moisture are an important factor, as is freezing. Both of these factors would tend to inhibit diffusion of radon.

Because the daughter products of radon and thoron are electrically charged when formed, they tend to attach themselves to the inert dusts that are normally present in the atmosphere. If the radioactive gases coexist with the dust in the same air mass for a sufficiently long time, the parents and their various daughters will achieve radioactive equilibria. From examination of Tables 7-2 and 7-3, which give the physical properties of the uranium and thorium series, we see that the growth of the radon daughters would approach an equilibrium in about 2 hr and that beyond that time further growth would be slow because of the presence of 22-year ^{210}Pb. We can ignore the activity due to ^{210}Pb and its daughters because the average time in which a particle is suspended in the atmosphere is very much shorter than the average life of ^{210}Pb. For all practical purposes, equilibrium with the radon daughters adsorbed on atmospheric dust is achieved after about 2 hr.

The thorium series below thoron (^{220}Rn) has no long-lived member. The equilibrium between thoron and its daughters will, therefore, be achieved at a rate governed by the time required for the buildup of ^{212}Pb (half-life 10.6 hr).

The adsorbed radioactive daughters, thus, have the effect of endowing the ordinary dusts of the atmosphere with apparent radioactivity. Wilkening (1952) has observed that the radon daughters tend to distribute themselves on atmospheric dust in a manner which depends on the particle size of the dust and that the bulk of the activity is contained on particles having diameters less than 0.035 μm.

Anderson and his associates (1954; Anderson and Turner, 1956) observed a close correlation between the concentration of radon daughters in the atmosphere and the concentration of suspended solids. The average concentration in London in clear weather, as determined from total air measurements of radon, was 2 pCi/liter. This was about 20 times the concentrations reported by earlier workers at the turn of the century, and it was suggested that burning coal is the source of radon. Analysis of British coal shows the concentration of ^{226}Ra to be 0.05 to 0.3 pCi/g of coal.*

* Fossil fuels as a source of atmospheric radioactivity are discussed further on page 197.

Prospero and Carlson (1970) have shown the ^{222}Rn content of North Atlantic air to be influenced by the amount of dust carried from North Africa by the trade winds. Blifford and his associates (1952) investigated the relationships between the concentrations of radon and its various decay products in the normal atmosphere and found, as would be expected, that the atmosphere is markedly depleted in the amounts of ^{210}Pb relative to the precursors of this isotope. This is because the inert dust of the atmosphere, the radon, and the radon daughters coexist long enough under normal circumstances for equilibria to be reached between radon and the more short-lived daughters. Since the longest half-life prior to ^{210}Pb is 26.8-min ^{214}Pb, equilibrium is reached in about 2 hr. The ^{210}Pb, which has a 22-year half-life, would take about 100 years to reach equilibrium. However, various mechanisms exist for removing dust from the atmosphere, and the ratio of ^{210}Pb to its shorter-lived ancestors was shown by Blifford to be indicative of the length of time the dust resides in the atmosphere. He concluded by this method of analysis that the mean life of the atmospheric dust to which the radon daughters are attached is 15 days.

Wilkening (1964) found that the atmospheric content of ^{222}Rn daughters is depleted during passage of a thunderstorm, which he attributes to the action of electric fields that, in his measurements, changed from a normal value of about 1.8 V/cm to -340 V/cm during the storms.

The natural radioactivity of atmospheric dust, owing primarily to the attached daughters of radon, can be demonstrated readily. When air is drawn through filter media, the radon daughters attached to the filtered atmospheric dust cause both the α and β activity of the filter media to rise. Curve A of Fig. 7-1 illustrates the manner in which this increase in α radioactivity occurs in the case of normal air containing 5×10^{-14} Ci/liter of radon in equilibrium with its daughter products. The rise in α activity increases for about 2 hr, at the end of which time the accumulated daughters decay at a rate which is compensated by the decay of newly deposited daughters. The radioactivity of the filter media will not increase beyond this equilibrium unless either the rate of air flow or the concentration of radon is increased. When air flow ceases, the α radioactivity of the filter will diminish as shown in curve B of Fig. 7-1 with an effective half-life of about 40 min.

We have seen in Chapter 2 that, depending on particle size, inhaled dust may be deposited either on the bronchial epithelium or in the alveolar tissue. The larger particles tend to deposit on the bronchial epithelium of the upper lung from which they are removed by ciliary action in a matter of hours. The dust particles deposited in the alveolar tissue are sufficiently small since they have survived passage through the bronchial tree, and

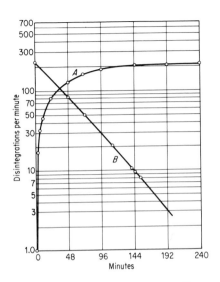

Fig. 7-1. Curve A: The buildup of the α activity on a filter paper through which is drawn 1 cfm of air containing 5×10^{-14} Ci/liter of radon in equilibrium with its daughter products up to but not including ^{210}Pb. Curve B: Decay of the accumulated α activity after cessation of flow.

these particles remain in the lung more persistently. It has been shown by Harley (1952), Shapiro (1956), Altshuler *et al.* (1964), and Jacobi (1964) that the dose delivered to the lung of an individual exposed to an atmosphere containing radon, thoron, and their daughter products is delivered primarily by the daughter products adsorbed on inert dust that accumulates in the lung.

When air is inhaled that contains radon or thoron in partial or total equilibrium with their daughter products, the inert gases will largely be exhaled immediately. However, a fraction of the dust particles will be deposited in the lung. The place of deposition and the manner of clearance from the lung will depend on the factors discussed in Chapter 2. With each breath, additional inert dust will be deposited, and radioactive equilibrium will be reached when the amount of activity deposited in unit time equals the amount of activity that is eliminated from the lung by the combination of physiological clearance and radioactive decay. In the case of radon in equilibrium with its daughters, the total energy dissipation in the lungs from the daughter products is about 500 times greater than that deriving from radon itself. Based on the models developed by Altshuler *et al.* and Jacobi, the dose to the basal cells of the epithelium of segmental and lobar bronchi could be as high as 100 millirads per year from the radon daughters normally present in the atmosphere. The dose from radon is delivered mainly by α particles. With a quality factor of 10, the dose to the basal cells of the bronchial epithelium could thus be a rem or more under normal circumstances.

As in the case of radon, Blanchard and Holaday (1960) have shown that the dose delivered to the lung by thoron is insignificant in comparison to the dose delivered by the thoron daughters, the dose from the thoron itself being about 0.001 of the dose from the daughter products with which it is in equilibrium. The dose to the lung from thoron and its daughters does not add significantly to the dose received from the radon series.

When air containing radon is inhaled, it is partially absorbed, and the decay products deliver a dose to the whole body that has been estimated at 3.0 mrem/year from a mean atmospheric concentration of 0.5 pCi/liter (UNSCEAR, 1966).

LEAD-210 (RADIUM-D) AND POLONIUM-210

^{210}Pb is a 22-year β emitter separated from its antecedent ^{222}Rn by six short-lived α or β emitters. We have seen that the longest radionuclide between ^{222}Rn and ^{210}Pb is ^{214}Pb, which has a half-life of only 26.8 min. The ^{210}Pb decays to 138-day ^{210}Po via the intermediate ^{210}Bi, which has a 5-day half-life. Thus, following the decay of 3.8-day radon in the atmosphere, ^{210}Pb is produced rapidly, but its long half-life provides assurance that very little will decay in the atmosphere before it precipitates to the earth's surface, mainly in rain or snow.

Because of these mechanisms, one would expect to observe radioactive disequilibria in the upper profiles of rocks and soils from which ^{222}Rn can diffuse. In addition, atmospheric transport of radon permits ^{210}Pb to achieve more widespread distribution than the ^{226}Ra from which it is derived, and because of the tendency of ^{210}Pb to return to the earth's surface, one would expect that broadleafed plants, presenting convenient surfaces to the precipitating ^{210}Pb, would be enriched in this radionuclide. This in fact is observed.

The ^{210}Pb/^{210}Po ratio will depend on the length of time the ^{210}Pb and its matrix coexist and whether the polonium is selectively removed from its site of production by chemical or biological mechanisms. Since ^{210}Po has a half-life of only 138 days, appreciable ingrowth in vegetation can take place during a single growing season, and additional buildup can occur during food storage, with equilibrium being reached in about 1 year. If ^{210}Pb is absorbed as such by man, there is ample time for ingrowth of ^{210}Po to occur in that fraction of the ^{210}Pb that is retained by the body.

The ^{210}Pb content of the atmosphere has been found to vary from 4.8 $\times$ 10^{-3} to 26 $\times$ 10^{-3} pCi/m^3, with the lowest values at island stations such as San Juan and Honolulu and the higher values in the interior of the United States (Magno *et al.*, 1970b). Since the mean residence time of dust suspended in the troposphere is 15 to 20 days, there is little time for ^{210}Po to be formed.

Jaworowski (1967) reported rainwater to contain from 1 to 10 pCi ^{210}Pb/liter with a mean of about 2 pCi/liter. In an area having 1 m of rainfall per annum, this would indicate a fallout of about 2 mCi/km^2/year. The United Nations (UNSCEAR, 1966) estimates that the total daily intake of ^{210}Pb is in the range of 1 to 10 pCi under normal circumstances.

Although Magno *et al.* found the ^{210}Pb concentration of air in the United States to be quite variable, the concentration in total diet was not significantly different between locations and averaged 0.80 pCi/kg. His values are in agreement with those reported by Morse and Welford (1971), which ranged from 0.70 to 1.0 pCi ^{210}Pb/kg in eight United States cities. Using the ICRP model for lead transfer from the gastrointestinal tract to blood, it can be estimated that about 0.2 pCi/day of ^{210}Pb reaches the bloodstream of average inhabitants of the United States. According to Morse and Welford, air and food contributed about equally to the blood level. Hill (1966) reported similar daily intakes for ^{210}Po, indicating that in most foods there has been sufficient time for this nuclide to reach equilibrium with ^{210}Pb.

There are two notable groups in which the dose from ^{210}Po is apt to be exceptionally high: cigarette smokers and residents of the northlands who subsist on lichen-eating reindeer.

Radford and Hunt (1964) were the first to report significant concentrations of ^{210}Po in cigarette smoke. A number of investigators (Kelley, 1965; Holtzman and Ikewicz, 1966; Ferri and Baratta, 1966; Little and Radford, 1967; Rajewsky and Stahlhofen, 1966) have studied this phenomenon, from which it can be concluded that the lungs and other tissues of cigarette smokers do have abnormally high concentrations of ^{210}Po. Based on α track measurements of autoradiographs of the lung tissuse of 13 smokers, Rajewsky estimated that the basal cells of the subsegmental bronchi of cigarette smokers may receive as much as 86 mrem/year and that the basal cells of their terminal bronchi may receive as much as 150 mrem/year. Holtzman and Ikewicz found about twice as much ^{210}Pb and ^{210}Po in the ribs of smokers compared to nonsmokers. The ^{210}Pb and ^{210}Po content of the lungs of smokers was about four times that of nonsmokers. Ribs from smokers contained 0.28 pCi ^{210}Pb/g of ash and 0.25 pCi ^{210}Po/g of ash. The lung contained 5.9 pCi ^{210}Po/kg wet. Based on these data,

Holtzman estimated that the dose to the total skeleton is elevated by about 30% in cigarette smokers. If the dose is calculated for the surface cells of bone, the increase is about 8%.

It is of interest that Hill (1966) found a close correlation between the ^{210}Po and ^{137}Cs content of human tissues from Canadian subjects, thus strengthening the suggestion that dietary habits that tend to favor broad-leafed vegetables or other foods subject to surface deposition may influence the ^{210}Po content of the tissues.

The tendency of lichens to absorb trace elements deposited from the atmosphere results in relatively high concentrations of ^{137}Cs and ^{210}Po in reindeer that feed on the lichens and men who feed on the reindeer (Beasley and Palmer, 1966; Blanchard and Moore, 1970). Kauranen and Miettinen (1969) found the ^{210}Po content of Lapps living in Northern Finland to be about 12 times higher than in residents of Southern Finland, where more normal dietary regimes exist. These investigators found the dose to the livers of Laplanders to be 170 mrem/year compared to 15 mrem/year for southern Finns. The comparative dose (haversian canals) was 27 and 11 mrem/year for bone, and 85 and 7 mrem/year for gonads, respectively.

Holtzman (1964) measured the ^{210}Pb and ^{210}Po content of Illinois potable waters known to be high in radium and found the ^{210}Pb and ^{210}Po to be significantly low relative to ^{226}Ra. This would seem to indicate a loss of ^{210}Pb owing to chemical precipitation, biological activity, or other factors.

POTASSIUM-40

Of the three naturally occurring potassium isotopes, only ^{40}K is unstable, having a half-life of 1.3×10^9 years. It decays by β emission to ^{40}Ca, followed by K capture to an excited state of ^{40}A, and γ-ray emission to the ^{40}A ground state. ^{40}K occurs to an extent of 0.01% in natural potassium, thereby imparting a specific activity of approximately 800 pCi/potassium. Representative values of the occurrence of potassium in rocks, as summarized by Kohman and Saito (1954), indicate a wide range of values, from 0.1% for limestones through 1% for sandstones and 3.5% for granite.

The potassium content of soils of arable lands is controlled by the use of fertilizers. Pertsov (1964) estimated that the quantity of ^{40}K added to the soils in the form of fertilizer was 1160 Ci in the United States in 1958.

A man weighing 70 kg contains about 140 g of potassium, most of which is located in muscle. From the specific activity of potassium, it follows that the ^{40}K content of the human body is of the order of 0.1 μCi. This isotope delivers a dose of about 20 mrem/year to the gonads and other soft tissues and about 15 mrem/year to bone. Because of its relative abundance and its energetic β emission (1.3 MeV), the radioactivity of ^{40}K is easily the predominant radioactive component in normal foods and human tissues. Seawater contains ^{40}K in a concentration of about 300 pCi/liter.

RUBIDIUM-87

Of the two rubidium isotopes found in nature, ^{85}Rb and ^{87}Rb, only the latter is radioactive, with a half-life of 4.8 $\times$ 10^{10} years. ^{87}Rb is a pure β emitter, and it is present in elemental rubidium in the amount of 27.8%, which endows this element with a specific activity of 0.02 pCi/g Rb. Pertsov (1964) quotes Vinogradov in listing the rubidium content of all but highly humic soils as about 0.01%. The ^{87}Rb content of ocean water has been reported to be 2.8 pCi/liter, with marine fish and invertebrates ranging from 0.008 to 0.08 pCi/g wet weight (Mauchline and Templeton, 1964).

The United Nations (UNSCEAR, 1966) estimates that the gonadal dose from ^{87}Rb is 0.3 mrem/year on the average.

RARE EARTHS

Naturally occurring radioactive rare earths have been detected in soils but in very low concentration. Pertsov (1964) again cited Vinogradov in reporting that the contributions of ^{142}Ce and ^{144}Nd to the α activity of six soil types is of the order of 10^{-6} pCi/g. The α activity owing to ^{147}Sm in the same soils was found to be about 10^{-3} pCi/g.

INDUCED RADIONUCLIDES

A number of radionuclides that exist on the surface of the earth and in its atmosphere have been induced by the interaction of cosmic rays with atmospheric nuclei. Among these the most important are tritium (^{3}H), ^{14}C, and ^{7}Be. Of lesser importance are ^{10}Be, ^{22}Na, ^{32}P, ^{33}P, ^{35}S, and ^{39}Cl. The properties of these isotopes and the extent to which they have been reported in various media were listed in Table 7–1.

^{14}C is formed by ^{14}N capture of neutrons produced in the upper atmosphere by cosmic-ray interactions. The incident cosmic-ray neutron flux is approximately 1 neutron/sec/cm^2 of earth's surface, and essentially all these neutrons disappear by ^{14}N capture (Anderson, 1953). The incident neutron flux integrated over the surface of the earth yields the natural rate of production of ^{14}C atoms, which has been estimated to be 1.6 atoms/cm^2/sec (UNSCEAR, 1964), and is believed to have been unchanged for at least 15,000 years prior to 1954, when nuclear weapons testing began to perturb the natural levels to a noticeable extent.

^{14}C exists in an equilibrium concentration in the carbon of living biological substances in a constant amount of 7.5 ± 2.7 pCi/g C. After death, the ^{14}C equilibrium is no longer maintained, and the concentration diminishes at a rate of 50% every 5600 years, thus making it possible to use the ^{14}C content of organic materials for the purpose of measuring age (Libby, 1952).

Because the ^{14}C originally present in coal and oil has decayed almost completely, the introduction into the atmosphere of carbon from these sources tends to reduce the specific ^{14}C activity of carbon. For this reason, the concentration of ^{14}C in atmospheric carbon tends to be lower in urban and industrial areas. Clayton, Arnold, and Patty (1955) and Lodge, Bien, and Suess (1960) have used the ^{14}C content of particulate atmospheric carbon to estimate the fractions of the dust originating from garbage incineration and combustion of fossil fuels. Incineration of food residues, textiles, paper, and other organic constituents of garbage will release smoke in which the ^{14}C is in contemporary equilibrium because such garbage is not likely to contain refuse so old that a significant fraction of the ^{14}C has decayed. On the other hand, because there has been essentially complete decay of the ^{14}C present in the organic matter from which fossil fuels have evolved, ^{14}C analysis makes it possible for one to estimate the relative contribution to the two sources of particulate carbon.

The total carbon content of the body is approximately 18% or 12.6 kg for a 70 kg man. The total ^{14}C body content is thus of the order of 0.1 Ci, but the dose is relatively small owing to the soft quality of the ^{14}C β particles (0.01 MeV). It is estimated that the dose from ^{14}C is 1.6 mrem/year to the skeletal tissues of the body and 0.7 mrem/year to the gonads (UNSCEAR, 1966).

Tritium, a radioactive isotope of hydrogen (^3H), is formed from several interactions of cosmic rays with gases of the upper atmosphere (Suess, 1958). Existing in the atmosphere principally in the form of water vapor, tritium precipitates in rain and snow. Like ^{14}C, it is produced in thermonuclear detonations, and the manner in which the atmospheric concentration of the isotope has increased since 1954 will be discussed in a subsequent chapter. The natural production rate of ^3H is thought to be about 0.5 atom/cm^3 sec, corresponding to an annual rate of about 2500 pCi/m^2 and an equilibrium global inventory of about 80 MCi (Harley and Lowder, 1971).

The natural concentration of tritium in lakes, rivers, and potable waters was reported to have been 5 to 20 pCi/liter prior to the advent of weapons testing. From these data, a total natural body tritium content of about 10 pCi can be derived. The total body dose from tritium of natural origin is estimated to be 0.003 to 0.006 mrem/year (UNSCEAR, 1962).

Other isotopes formed from cosmic-ray interactions with the atmosphere may be potentially useful as tracers for studying atmospheric transport mechanisms, but relatively few observations have been reported.

Natural Sources of External Ionizing Radiation

The dose received by human beings from external sources of ionizing radiation originates from cosmic rays and from γ-emitting radionuclides in the earth's crust. The United Nations (UNSCEAR, 1966) estimates the γ-radiation dose from external sources in "normal" areas to be 50 mrem/ year from terrestrial sources and 28 mrem/year from cosmic radiation.

Solon et al. (1958) and later Beck et al. (1966) made extensive measurements of the natural γ-radiation background in a number of cities throughout the United States. The Solon data tend to be about 30% higher than those reported by Beck, owing in all probability to the greater effect of fallout during the period when Solon's measurements were made. The techniques used by Beck enabled him to differentiate the effects of fallout (Beck et al., 1966; Beck and de Planque, 1968), which was not possible at the time Solon undertook his series of measurements.

The mean exposure in the 124 locations measured by Solon was 81 ± 20 mR/year, compared to Beck's mean of 61 ± 23 mR/year at 210 locations. Solon showed that his data were well correlated with barometric pressure, indicating the effect of cosmic sources of radiation. The data gathered by Beck at the principal cities where measurements were made are listed in Table 7–13.

TERRESTRIAL SOURCES OF EXTERNAL RADIATION

Hultqvist (1956) has derived simple formulas for calculating the ionization produced in air over rocks and soils having various concentrations of radioactive minerals. If his formula is modified so as to estimate the contribution to the dose rate (rads per year) from the various concentrations of radioactive materials in the ground, the following expressions are obtained (UNSCEAR, 1958):

$$\text{DRa} = 18.4 \times 10^{12} \times \text{SRa}$$

$$\text{DU} = 6.4 \times 10^{6} \times \text{SU}$$

$$\text{DTh} = 3.1 \times 10^{6} \times \text{STh}$$

$$\text{DK} = 13.3 \times 10^{2} \times \text{SK}$$

in which DRa, DU, etc., equal air dose rates in rads per year, measured

TABLE 7-13

TERRESTRIAL AND COSMIC γ-RADIATION LEVELS MEASURED IN THE UNITED STATES, 1965[a] (μR/hr)[b]

Location	^{40}K	^{238}U + Dtrs.	^{232}Th + Dtrs.	Total Terrestrial	Cosmic	Total
Aiken, S.C.	0.1	1.2	1.4	2.7	3.7	6.4
Dallas, Tex.	0.6	0.9	1.4	2.9	3.6	6.5
Reno, Nev.	1.3	1.0	1.8	4.1	5.5	9.6
Rapid City, S.D.	1.1	1.5	1.7	4.3	4.7	9.0
Spring Valley, Minn.	1.4	1.2	2.0	4.6	4.0	8.6
Salt Lake City, Utah	1.8	1.1	2.1	5.0	5.7	10.7
New Orleans, La.	1.9	1.2	2.6	5.7	3.5	9.2
Goleta Beach, Calif.	2.6	1.5	1.7	5.8	3.6	9.4
Largo, Md.	1.8	1.5	2.9	6.2	3.6	9.8
Pelham, N.Y.	2.2	1.3	3.0	6.5	3.6	10.1
Sundance, Wyo.	2.2	1.9	2.4	6.5	5.6	12.1
Argonne, Ill.	2.0	2.2	2.4	6.6	3.8	10.4
Sioux Falls, S.D.	2.1	1.8	3.0	6.9	4.1	11.0
Carlisle, Pa.	2.4	1.5	3.0	6.9	3.7	10.6
Chadron, Neb.	2.3	1.7	3.3	7.3	4.8	12.1
New York, N.Y.	2.2	1.4	4.2	7.8	3.6	11.4
Alamagordo, N.M.	2.8	2.1	3.3	8.2	5.0	13.2
Fort Collins, Colo.	2.5	1.8	4.0	8.3	5.8	14.1
Elko, Nev.	2.7	3.0	3.5	9.2	5.8	15.0
Bonny Doon, Calif.	7.8	1.4	1.8	11.0	4.1	15.1
Denver, Colo. (various locations)	(2.3–3.9)	(1.3–2.4)	(3.4–7.4)	(7.1–13.2)	(5.9–6.0)	(13.0–19.2)
Rolesville, N.C.	5.2	2.3	10.2	17.7	3.7	21.4

[a] Beck et al. (1966).
[b] 1 μR/hr = 8.2 mR/year.

over ground containing SRa, SU, etc., grams of radium, uranium, etc., per gram of rock or soil. When these equations are used to compute the dose over rocks having an average content of radioactive elements, the values obtained are those given in Table 7-14. The external γ radiation from radionuclides in the earth's crust is thus seen to be influenced by the kind of rock over which the measurements are made. The actual doses to

TABLE 7–14

γ-RADIATION DOSE RATES FROM RADIUM, URANIUM, THORIUM, AND POTASSIUM IN ROCKS[a]

Type of rock	Dose rate (mrem/year)			
	^{226}Ra	^{238}U	^{232}Th	^{40}K
Igneous	24	26	37	35
Sedimentary				
Sandstones	13	7.7	18	15
Shales	20	7.7	31	36
Limestones	7.7	8.4	4	4

[a] UNSCEAR (1958).

people cover a somewhat more narrow range owing to the fact that most people live on soil rather than rock. The soils tend to be less variable in their radioactive content owing possibly to the fact that the igneous rocks which are high in radioactive content weather more slowly and, therefore, contribute less to soils than the softer sedimentary rocks.

Beck and de Planque (1968) give the estimates shown in Table 7–15

TABLE 7–15

EXPOSURE RATES IN AIR DUE TO ^{40}K, ^{238}U, AND ^{232}TH IN THE SOIL[a] (μR/hr)

Detector height (m)	Exposure rate per 1% ^{40}K	Exposure rate per 1 ppm ^{238}U[b]	Exposure rate per 1 ppm ^{232}Th
0	1.68	0.667	0.314
1	1.65	0.654	0.307
3	1.60	0.635	0.298
5	1.56	0.618	0.290
10	1.47	0.583	0.274
30	1.22	0.476	0.226
100	0.72	0.274	0.133
300	0.19	0.068	0.036

[a] Beck and de Planque (1968).
[b] Exposure rate due to radon daughters in the atmosphere has been neglected but should not amount to more than a few percent of the total exposure rate even close to the interface.

of the exposure in air owing to γ radiation from the three main sources, potassium, uranium, and thorium. According to Kohman (1959), the similarity of dose rates from the various isotopes listed in Table 7–14 is a coincidence arising from the fact that the various isotopes happen to be present in rocks in amounts that are approximately inversely proportional to their specific activities.

Beck *et al.* (1966) also reported on a series of measurements of the natural γ dose rates at two communities in Westchester County, New York during 1963–1965. The individual observations at one location that averaged 6.9 $\pm$ 2.1 μR/hr ranged from 6.0 to 8.2 μR/hr. At the second location, which averaged 7.1 $\pm$ 2.1 μR/hr, the range was similar, from 6.0 to 8.3 μR/hr. These investigators thought the principal source of variation to be due to the effect of soil moisture, which can account for 30% by weight during wet periods. Whereas the soil moisture might be expected to attenuate the radiation from thorium and potassium, the uranium series might be expected to increase owing to the fact that the soil water would inhibit the diffusion of radon. However, an examination of the Beck data fails to show an inverse correlation of this type.

GALACTIC RADIATION

The primary radiations that originate in outer space and impinge isotropically on the top of the earth's atmosphere consist of 85% protons, 14% α particles, and about 1% of nuclei of atomic number between $Z = 4$ and $Z = 26$ (Langham, 1967). An outstanding characteristic of these radiations is that they are highly penetrating, with a mean energy of about 10^{10} eV and maximum energies of as much as 10^{19} eV. The primary radiations predominate in the stratosphere above an altitude of about 25 km.

It is know known that these radiations are mainly from galactic sources and that only a small fraction is normally of solar origin. However, the solar component becomes very significant following solar flares associated with sun-spot activity. Sun spots are known to follow an 11-year cycle.

The interactions of the primary particles with atmospheric nuclei produce electrons, γ rays, neutrons, and mesons. At sea level the mesons account for about 80% of the cosmic radiation and electrons account for about 20%. It has been estimated that 0.05% of primary protons penetrate to sea level (Myrloi and Wilson, 1951). With the development of high-altitude aircraft and manned space probes, the dose from primary cosmic radiations have attracted more interest in recent years. Sophisticated radiation measurements are now an important component of the scientific programs undertaken by satellites probing into space.

The dose from galactic radiation is affected by altitude and geomagnetic latitude. For the first few kilometers above the earth's surface, the galactic radiation doubles for each 2000 m increase in the altitude. For the first 3000 ft, however, the total dose rate actually decreases with altitude, owing to the fact that attenuation of the γ rays from terrestrial sources diminishes more rapidly than the increase in galactic radiation (Schaefer, 1971). The increase in dose rate at higher altitudes is shown in Fig. 7-2, in which it is seen that at polar latitudes rates in excess of 1 mrem/hr are received at altitudes of 60,000–80,000 ft, the upper limit of high-performance aircraft such as the SST. On rare occasions, once or twice during the 11-year cycle, a giant solar event may deliver dose equivalents in the range 1–10 rem/hr, with a peak of as high as 5 rem during the first hour (Upton, 1966). During a well-documented solar flare in February, 1956, dose rates well in excess of 100 mrem/hr existed briefly at altitudes as low as 35,000 ft (Schaefer, 1971). Under more normal circumstances, the occupants of an SST would receive a dose of 1 to 2 mrem/hr from galactic and solar radiation.

As one rises above the earth's atmosphere, the dose consists of two main components. One is the dose from highly energetic cosmic radiation trapped in the earth magnetic field as illustrated in Fig. 7-3. A second portion is received beyond the earth's magnetic field, where one is exposed to the background galactic radiation on which may be superimposed very sharp peaks of radiation due to solar flares.

The ionization produced by cosmic rays at sea level in the midlatitudes

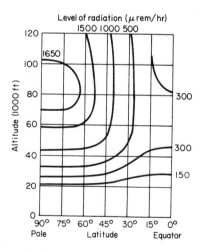

Fig. 7-2. The galactic radiation field in the earth's atmosphere, from sea level to 120,000 ft (Schaefer, 1971).

has been measured by a number of investigators and observed to vary
between 1.90 and 2.2 ion pairs/cm/sec (UNSCEAR, 1966). The dose rate
to soft tissue, owing to the ionizing component of cosmic rays at sea level,
is estimated to be 28 mrem/year.

On entering the earth's magnetic field, some of the primary particles
are deflected toward the polar regions, resulting in a somewhat lower
radiation flux at the equator. This phenomenon becomes more accentuated
with altitudes up to a few kilometers. The difference due to geomagnetic
latitude varies from 14% at sea level to 33% at 4360 m (Pertsov, 1964).

The geomagnetically trapped radiations consist mainly of protons and
electrons produced by backscatter of the primary cosmic-ray beam on the
earth's atmosphere and protons of solar origin. Because of great differences
in the mass to charge ratios of protons and electrons, the trajectories for
the two particles are widely different, giving rise to an inner radiation belt
consisting mainly of protons and an outer belt consisting mainly of electrons, as illustrated in Fig. 7-3. The proton region begins about 1000 km
above the geomagnetic equator and ends at an altitude of about 3000 km.

Definition of the geometry and density of the electron belt has been
complicated by the fact that enormous fluxes of electrons have been injected into this region of space by the explosion of nuclear and thermonuclear weapons at altitudes as high as 1275 km above sea level. Electrons
introduced in this way behave as cosmic-ray electrons trapped in the
geomagnetic field, disappearing with half-lives of several years.

The dose received by astronauts traversing the geomagnetically trapped
radiations originates from both protons and electrons, but it appears that
the latter is more important owing to production of X rays and bremsstrahlung that have far greater penetration power than the electrons that produce them. The astronauts can be exposed to dose rates of the order of
tens of rads per hour through this process.

The cosmic radiation beyond the geomagnetic field, when undisturbed

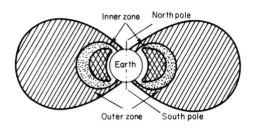

Fig. 7-3. The geomagnetically trapped corpuscular radiations.

TABLE 7-16

COMPOSITION OF THE PRIMARY COSMIC-RAY FLUX OUTSIDE THE ATMOSPHERE AT NORTHERN LATITUDES[a] (Z[b] NUMBERS ARE IN PARENTHESES)

	H (Protons) (1)	He (α Particles) (2)	C, N, or O (7)	Mg (12)	Ca (20)	Fe (26)
Particle flux[c]	4460	633	32	8.4	2.9	1.4
Absorbed dose contribution mrad/24 hr)	4	2.3	1.4	0.99	0.13	0.28
LET (keV/μ tissue)						
Minimum	0.21	0.84	10.5	30.3	84	142
Maximum	57.8	252	1200	1700	2570	3500
Absorbed dose to centrally traversed cell (rad)[d]						
Minimum	0.07	0.24	0.36	1	2.85	4.8
Maximum	20	85	420	610	870	1200

[a] From National Academy of Sciences (1967).

[b] Z numbers from 7 to 26 are group representatives.

[c] Particle intensity: particles traversing a sphere of 1 cm² cross section per hour from all directions.

[d] Dose per particle calculated for a 10 μm cell at center of track.

by exceptional solar activity, has come to be called *galactic* radiation. The origins of these particles are cloaked in mystery, but their physical properties are now well known and are given in Table 7–16.

Conversion of cosmic-ray dose estimates from rads to rems is made difficult by the wide range of energies and linear energy transfer (LET) as shown in Table 7–16. It will be seen that the LET of the incident protons varies from about 0.21 keV/μm of tissue to 57.8 keV/μm of tissue. For high Z nuclides such as iron, the LET can be as high as 3500 keV/μm. The absorbed doses for the centrally traversed cells varies from 0.07 rad for relatively low LET protons to 1200 rad for high LET iron. The biological significance of the high LET radiation remains to be evaluated (Madey, 1967).

Solar activity is capable of injecting huge quantities of high energy protons into interplanetary space, and these so-called solar flares present a potential radiation hazard to astronauts. The frequency of flare production varies with an 11-year cycle. Although optical manifestations of the flare exist for only 30 to 50 min, the increased proton flux continues to

TABLE 7-17

MAXIMUM AND MINIMUM MISSION DOSES[a]
FOR BEST AND WORST LAUNCH DATES
DURING ACTIVE PERIOD OF CYCLE 19[a]

Mission duration	Maximum dose (rad)	Minimum dose (rad)
4 years	3492	2439
3 years	3229	974
2 years	2781	526
1.5 years	2415	176
1 year	2110	15
9 months	1963	2
6 months	1963	0
3 months	1962	0
1.5 months	1492	0
1 month	1452	0
2 weeks	1452	0
1 week	1452	0

[a] From National Academy of Sciences.
[b] Surface dose inside 1 g/cm² uniform aluminum shielding.

exist in the vicinity of the earth for as much as 48 hr after the flare is observed. The total dose from a given flare may vary from a fraction of a rad to as much as 1000 rad.

The dose calculations from the solar-flare particles are very difficult owing to the wide range of energies and the complex shielding geometries presented by space capsules.

A number of authors have examined the doses that would be received by space journeys of various durations during a recent solar maximum. These data are summarized in Table 7–17 in which the doses are given for missions ranging from 1 week to 4 years. It is observed from these data that great risks may exist because of this source of radiation.

Man-Made Modifications of the Natural Radiation Environment

Using thermoluminescent dosimeters for measurement of the integrated γ-radiation doses in 100 wood frame and stucco houses, Lindeken et al.

(1971) found a 25% attenuation of the outdoor dose. Lowder and Condon (1965), in connection with a study of natural radioactivity in the granitic areas of New Hampshire and Vermont, made by portable scintillation detector measurements in 160 dwellings, found the mean indoor level to be about 70% of the corresponding outdoor dose rate. Most of the homes were of wood, which acted generally to shield the external γ radiation, but contributed little radioactivity of its own. Investigators in Sweden (Hultqvist) found that the materials of construction observed in their studies frequently add significantly to the γ-radiation dose received by the inhabitants.

In any given locality, the effect of a house on the dose received by its occupants is clearly influenced by the kinds of material used in construction. Fivefold increases in the ambient γ-radiation background have been observed in buildings constructed of granite. Hamilton (1971) measured the uranium, radium, thorium, and potassium content of building construction materials used in Great Britain and found wide variations. The radium content of conventional building materials varied by a factor of 12, with the highest values contained in gypsum made from superphosphate fertilizer-manufacturing waste products. The raw material in this process is phosphate rock, which is high in both uranium and radium. In the fertilizer production process, the uranium passes to the fertilizer and the radium remains in the waste materials from which the gypsum blocks are made. Spiers and Griffith (1956) found the dose within brick and concrete dewllings to be about double the dose from terrestrial sources in the outdoors.

It is well known that coal and oil tend to concentrate trace elements, and analysis of coals early in this century indicated the presence of heavy radioelements including radium, thorium, and uranium. However, it was not until much more recently that measurements were made of the amounts of radioactive dust emitted to the atmosphere by various types of fuel-burning power plants. These measurements have attracted considerable interest because they demonstrated that coal- and oil-burning plants in some cases discharged to the atmosphere a greater fraction of the MPC of radioactive dust than certain types of nuclear reactors (Eisenbud and Petrow, 1964). Martin *et al.* (1971) calculated that the dose rate from a 1000 MW coal-burning plant equipped with a 97.5% fly-ash removal system would, at the maximum, deliver a dose of 3.5×10^{-2} μrem/hr to the individuals inhaling the effluent from this power plant. With good dust control, a 100 megawatt equivalent (MWe) coal-burning plant would be expected to discharge about 30 mCi of mixed radium nuclides per year, but about 1 Ci could be discharged in the absence of fly-ash control equipment. As a public health hazard, this source of exposure is insignificant

but is of interest because it helps to put the subject of atmospheric radioactivity into perspective insofar as nuclear power plants are concerned.
Jaworowski *et al.* (1971) have recently shown that fossil snows contain far
less [226]Ra content than contemporary snow and that the [226]Ra content of
snow decreases inversely with distance from coal-burning power stations.

Drainage from strip-mine waste piles has been shown to elevate the α
radioactivity of river waters in Pennsylvania. Caldwell *et al.* (1970) found
the Kiskiminetas River to contain an average of 12 pCi/liter of α activity.

Areas Having Unusually High Natural Radioactivity

In recent years, there has been a growing awareness of the existence of
areas of the world in which people are exposed to unusually high levels of
natural radioactivity.

In addition to the occurrence of wells yielding drinking water having
elevated radium concentrations, described previously in this chapter, there
are two major sources of high natural radioactivity: mineral springs and
places in which people live on monazite sands or other deposits of radioactive minerals.

The Mineral Springs

It has long been known that many mineral springs contain relatively
high concentrations of radium and radon. In many places in the world
the radioactivity of the local springs has been exploited for its alleged
curative powers. Spas, such as Bad Gastein in Austria and others in South
America, Europe, Japan, and elsewhere, including some in the United
States, have commercialized the high radioactive content of local waters,
and in some places research laboratories are operated in which the physiological basis for the alleged curative effects is studied. Not only are visitors
encouraged to drink and bathe in the radioactive waters but also to sit
in "emanatoria," where they can breathe radon emanating from surrounding rock, as shown in Fig. 7-4 (Pohl-Rüling and Scheminzky, 1954).

The published values of [226]Ra in mineral waters range as high as about
100 pCi/liter (United Nations, 1958), which is approximately 1 million
times greater than the values normally reported for the public water
supplies. However, to illustrate the fact that the values reported for spring
waters are not typical of the drinking water of the region, the radium
concentration in the tap water of Bad Gastein (Muth *et al.*, 1957) is reported to be 6.2×10^{-4} pCi/ml compared to spring content of 100 pCi/ml.

Up to the present time, there are no studies which would indicate to

Fig. 7-4. Radon gallery at Bad Gastein, Austria (COSY Verlag, Salzburg).

what extent the radioactivity contained in the spring waters has been absorbed by human beings. There are, no doubt, some who have followed the practice of drinking the radioactive waters from these spas. A study of these individuals would be of interest.

MONAZITE SANDS AND OTHER RADIOACTIVE MINERAL DEPOSITS

Major anomolies in the concentrations of radioactive minerals in soil have been reported in two countries, Brazil and India.

In Brazil, the radioactive deposits are of two distinct types: the monazite sand deposits along certain beaches in the States of Espirito Santos and Rio de Janeiro and the regions of alkaline intrusives in the State of Minas Gerais (Roser *et al.*, 1964).

Monazite is a highly insoluble rare-earth mineral that occurs in beach sands together with the mineral ilmenite, which gives rise to the black sands or "aerea preta" which are known for their radioactivity by the tourists of Brazil and are much sought after for their alleged benefits to health. The external radiation levels on these black sands range up to 5 mR/hr, and people come from long distances to spend their vacations

Fig. 7-5. Beach of the "black sands" at Guarapari, State of Espirito Santos, Brazil. The dark areas contain monazite sands over which the exposure rate is as high as 5 mR/hr. The tourists travel to these beaches because of local beliefs about the beneficial effects of radioactivity.

on the black sands and in the many hotels that have been constructed to care for their needs. The most active of these Brazilian vacation towns is Guarapari, which has a stable population of about 5600 people and an annual influx of 10,000 vacationers (Fig. 7-5). Some of the major streets of Guarapari have radiation levels as high as 0.13 mR/hr, which is about 10 times the normal background. Similar radiation levels are found in the buildings of Guarapari where many of them, in parts of the village that are not built on monazite sand, are, nevertheless, elevated in radioactivity owing to the fact that beach sands were incorporated into the building materials. Roser and Cullen (1964) undertook extensive external radiation measurements throughout these areas and concluded that almost all the approximately 60,000 inhabitants of these regions were exposed to abnormally high radiation levels, but that only a small number (about 6600) were exposed to more than 0.5 rem/year. The distribution of the exposures reported by them is given in Table 7–18.

The principal radioactive constituents in monazite are from the ^{232}Th series, but there is also some uranium present and, therefore, some opportunity for ^{226}Ra uptake. However, there is very little food grown in the monazite areas, and the diets of the local inhabitants are derived principally

TABLE 7–18

SUMMARY OF POPULATION EXPOSURE
LEVELS IN STATE OF ESPIRITO SANTOS[a]

Range of radiation levels (R/year)	Population
0.09–0.13	15,000
0.13–0.17	4,000
0.17–0.22	29,000
0.22–0.35	6,000
~0.50	300
~0.95	6,000
~1.15	350

[a] Roser and Cullen (1964).

Fig. 7-6. Morro de Ferro, a radioactive mountain near Pocos de Caldas in the state of Minas Gerais, Brazil. The ambient γ-radiation levels in this area are from 0.1 to 3 mR/hr, and there is evidence of marked uptake of naturally occurring heavy radioelements by the plants. Although this area is uninhabited, it is used for grazing stock animals.

from outside sources. The exposures in the Brazilian monazite areas are, thus, due primarily to external radiation, and the internal dose is not thought to be significant (Eisenbud *et al.*, 1964; Penna Franca *et al.*, 1970).

A distinctly different source of abnormal natural radioactivity exists in the Brazilian State of Minas Gerais, near Pocos de Caldas and Araxa, an area of alkaline intrusives. The soil is generally poor in this area except in patches where the soil contains the mineral apatite, which is associated with a number of radioactive minerals containing both uranium and thorium.

A remarkable feature of this area is the Morro do Ferro, a steeply rising

Fig. 7-7. Autoradiograph of species of adiantium from the Morro do Ferro in the State of Minas Gerais, Brazil (Dr. Eduardo Penna-Franca).

hill about 300 m high where the ambient radiation levels range up to 3 mR/hr (Fig. 7-6). The flora from parts of this hill have absorbed so much radium that they can readily be autoradiographed as seen in Fig. 7-7. Penna Franca *et al.* (1970) have surveyed the dietary habits of the indigenous population of this region and have undertaken radiochemical analysis of the foods. About 1670 people live in this area, of which nearly 200 ingest radium in amounts that are 10 to 100 times greater than normal. The ratio ^{228}Ra:^{226}Ra is about 6 to 1 in the diet.

Studies have been undertaken of the exposures of rats living underground on the Morro do Ferro. Of particular interest is the dose to these rodents due to inhalation of ^{220}Rn, which was found by Drew and Eisenbud (1966) to be present in the rat burrows in concentrations up to 100 pCi/ml. Using thermoluminescent dosimeters implanted into trapped rats that were released and later recaptured, the external radiation dose to the rats

TABLE 7–19

BODY TISSUE DOSE RATES DUE TO EXTERNAL AND INTERNAL IRRADIATION FROM NATURAL SOURCES OF RADIATION IN "NORMAL REGIONS"[a]

Source of irradiation	Dose rates (mrad/year)		
	Gonad	Haversian canal	Bone marrow
External irradiation			
Cosmic rays (including neutrons)	28	28	28
Terrestrial radiation (including air)	50	50	50
Internal irradiation			
^{40}K	20	15	15
^{87}Rb	0.3	<0.3	<0.3
^{226}Ra[b]	—	0.6	0.03
^{228}Ra and decay products (equilibrium)	—	0.7	0.03
^{210}Pb[c] and decay products (50% equilibrium)	—	—	—
^{14}C	0.3	2.1	0.3
^{22}Rn (dissolved in tissue)	0.7	1.6	1.6
	0.3	0.3	0.3
Total	100	99	96
Percentage from α particles and neutrons	1.3	4.4	1.4

[a] UNSCEAR (1966).
[b] Assumes 30% retention of ^{222}Rn.
[c] Endogenous.

was estimated to be between 1.3 and 6.7 rad/year. The most astonishing finding was that the dose to basal cells of the rat bronchial epithelium was estimated to be in the range of 3000 to 30,000 rem/year. Of 14 rats trapped and sacrificed for pathological study, none was observed to show any radiation effects. This in itself means very little, since the Morro do Ferro is a relatively small area (about 3×10^5 m²), and if rats were affected by this exposure, they could be replenished rapidly from the surrounding normal areas. There are no data on population turnover rates among rats on the Morro do Ferro or in the surrounding area.

In the State of Kerala, India, the monazite deposits are more extensive than those in Brazil, and about 100,000 natives are exposed to external radiation levels that vary from 0.2 to 2.6 R/year. The epidemiological studies that might be possible with a population of this size have not as yet been carried out. In contrast to the Brazilian studies, Mistry *et al.* (1970) have reported significant uptake of ^{226}Ra from foods grown in the monazite area.

Summary of Human Exposures to Natural Ionizing Radiation

The gonadal and bone doses from the principal natural sources of ionizing radiation are summarized in Table 7–19.

Chapter 8

From Mines to Fabricated Fuel

The production of nuclear energy is presently based mainly on the fission of ^{235}U, which occurs in natural uranium to the extent of 0.7%. ^{238}U, which is the most abundant nuclide of this element, is not ordinarily fissionable but does transmute by neutron capture to ^{239}Pu, which is a fissionable material. Thus, the source materials of atomic energy may be either fissionable like ^{235}U or fertile like ^{238}U.

Thorium occurs in nature almost entirely as the isotope ^{232}Th, which is a fertile nuclide that can be transmuted to the fissionable ^{233}U. If full use is to be made of the world's potential resources in nuclear energy, thorium must ultimately be incorporated into reactor systems for the purpose of breeding ^{233}U, but comparatively little use will be made of thorium in this way for the foreseeable future.

Uranium

When interest in uranium first developed in World War II, commercially exploitable deposits were thought to be comparatively rare and to occur only in a few districts rich in the mineral pitchblende. The sources of uranium then known to exist were in the Belgian Congo, the Great Bear Lake region of Canada, and Czechoslovakia. It was also known that the carnotite deposits on the Colorado Plateau contained the elements of the

205

uranium series, and these deposits had for some years been used as a source
of radium. However, the full extent of the uranium resources of the Southwest United States was not then appreciated.

Since World War II, methods for processing low-grade ores have been
developed. An intensive worldwide search for uranium has disclosed substantial reserves, and the uranium industry in this country and abroad
has grown to major proportions (Clegg and Foley, 1958).

The present-day supply of uranium is based on relatively large reserves
of ore having a relatively low uranium content. In the United States, the
uranium content expressed as per cent U_3O_8 now varies from about 0.18
to 0.34% with an average of 0.21%. The U_3O_8 content of ores mined during
and after World War II ranged from 20 to 60%, but the total amount of
uranium available was not nearly as great as now. Based on a U_3O_8 selling
price of 8 dollars per pound, the uranium reserves in the United States at
the end of 1971 were 243,000 short tons of U_3O_8 contained in 116 million
tons of ore having an average of 0.21% U_3O_8. Exclusive of the Eastern
bloc, world reserves in January, 1970 were estimated to be 840,000 tons
of U_3O_8 (Sherman, 1971). The locations of the principal uranium-producing
areas in the United States are shown in Fig. 8-1, and the distribution of
reserves by state as of January 1, 1970 is given in Table 8-1. It is seen
that two states, New Mexico and Wyoming, hold more than 82% of the
United States total (Arthur D. Little, Inc., 1970). The principal mineral
forms in which the uranium occurs are coffinite, a basic uranium silicate

TABLE 8-1

DISTRIBUTION OF URANIUM RESERVES BY STATE[a,b]

State	Tons ore	Grade (%)	Tons U_3O_8	Percent total tons U_3O_8
New Mexico	45,357,563	0.24	108,380	44.04
Wyoming	50,884,554	0.19	95,271	38.71
Utah	2,893,717	0.31	9,064	3.68
Colorado	3,713,022	0.27	10,184	4.14
Texas	6,622,323	0.16	10,373	4.22
Others[c]	9,023,072	0.14	12,828	5.21

[a] U.S. Atomic Energy Commission (1971b).
[b] Includes reserves valued at 8 dollars per pound U_3O_8
[c] Arizona, Washington, South and North Dakota, California, Idaho, Mon-
tana, Nevada, Oklahoma, Oregon, and Alaska.

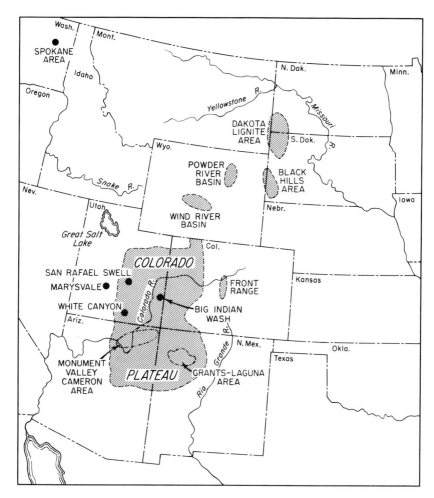

Fig. 8-1. Locations of the principal sources of uranium in the United States (U.S. Atomic Energy Commission).

usually associated with carbonaceous materials, and the uranium dioxide, *uraninite.*

About 620,000 tons of ore containing 13,000 tons of U_3O_8 were mined in the United States in 1970. This raw material is mined at more than 150 mines and is shipped to mills that produce uranium concentrate containing between 70 to 92% U_3O_8 (U.S. Atomic Energy Commission, 1970a). The output of these mills is transported to uranium refineries where the mill concentrates are converted into uranium compounds having a high degree of chemical purity. From here, the uranium goes to gaseous-diffusion plants in Ohio, Kentucky, or Tennessee for enrichment in the isotope ^{235}U. If not destined for enrichment, the uranium may be shipped as the oxide or metal to a number of privately or government-owned installations where natural uranium reactor fuel elements may be fabricated. A large fraction of the metallic natural uranium has been shipped to plutonium-production facilities at Hanford and Savannah River. The enriched uranium from the gaseous-diffusion plants is also shipped to fuel-fabrication facilities.

URANIUM MINING

Uranium mining may be carried on in a variety of ways both underground and in open pits. The size of the mines may vary from relatively small workings employing one or two men using hand tools to thoroughly mechanized enterprises employing more than 100 men.

The concentration of radon and its daughter products in the air of underground uranium mines may be excessive unless good ventilation practices are employed. As discussed in Chapter 2, it is believed that the high incidence of lung cancer among miners in Bohemia and Saxony was due to the exposure of the miners to radon. One of the tragedies of the atomic energy program in the United States is that the mining industry, and the states in which the underground uranium mines are located, did not heed the European experience. Because of a technicality in the Atomic Energy Act of 1946, the AEC did not assume responsibility for the health of the uranium miners. This left the matter to the mining states, where enforcement of recommendations of safe practice was inadequate, and many cases of lung cancer developed. The Federal government did not enter the picture insofar as enforcement is concerned until the late 1960's (Holaday et al., 1968).

Radon and its daughter products are the only significant radioactive contaminants discharged from the mines to the environment. The required ventilation rates for the mines vary from 1000 cfm to over 200,000 cfm, and the discharged air contains radon in concentrations that range from 0.5 to 20 $\mu Ci/min/1000$ ft^3 of air (Holaday, 1959). This is a relatively small amount of radon, the maximum emission rate of which is equivalent to the normal radon emissions from about 2×10^5 m^2 of earth's surface, using the estimate cited in Chapter 7.

CONCENTRATING

Mills located in Western United States concentrate the uranium by hydrometallurgical processes in which either an acid or alkali leach is employed (Clegg and Foley, 1958).

TABLE 8–2

Principal Steps in Processing Uranium Ores[a] (Feed: Uranium-Bearing Ores Containing 1.5–10 lb of U_3O_8 per Ton of Ore; Product: Uranium Oxide or Salt Concentrates Assaying 70–90% U_3O_8 by Weight)

Processing step	Methods	Notes
A. Ore preparation		
1. Size reduction	Crushing and grinding	—
2. Roasting	—	Improves uranium solubility and ore handling. Used infrequently
B. Leaching		
1. Digestion	(a) Acid leaching: Generally with sulfuric acid. Oxidant required. Conventional equipment and techniques	Normally favored for low-lime ores ($\leqq$ 12% lime content). High extraction efficiency
	(b) Alkaline leaching: Generally with a mixture of sodium carbonate and sodium bicarbonate. Conventional equipment and techniques	Normally favored for high-lime ores
2. (a) Clarification	Decantation and filtration, yielding a clear leach liquor. Flocculating agents sometimes required. Conventional equipment and techniques	Used when clear solution is required for subsequent recovery operations, as in the case of the precipitation process (see C.1.a below)
(b) Sand-slime separation	Desanding, yielding sand and slime fractions. Conventional equipment and techniques	Used in connection with the resin-in-pulp recovery process (see C.1.b below)
C. Product recovery		
1. Recovery	(a) Chemical precipitation: Conventional process permitting use of standard solid-liquid separation equipment. Reagents and conditions vary, depending primarily on whether an acid or alkaline leach has been used	Limited to ores which yield a clear leach liquor. High reagent consumption when applied to acid-leach liquors. (More recent mills using acid leach favor other recovery methods.) Product form is usually uranium oxide but may be sodium diuranate
	(b) Ion exchange: One method, using anion resins in conventional columns, requires a clear	Ion exchange used in nearly all mills built since 1950. Ion exchange has thus far been ap-

TABLE 8–2 (Continued)

Processing step	Methods	Notes
	leach liquor. A recently developed method known as the resin-in-pulp process, can be used on leach pulp. In this process, perforated baskets containing anion resin beads are moved up and down through the pulp	plied only to acid-leach liquors or pulps but application to alkaline solution is under development. Product form is ammonium diurante or uranium oxide
	(c) Solvent extraction: Organic solvents used in mixer-settler or column-type contacting equipment. Applicable to clear liquors	The application of solvent extraction techniques to uranium processing is a recent development and has not as yet been carried beyond the pilot stage. Work to date has been on acid-leach solutions and pulps
2. Final treatment	Filtration and drying of precipitate from recovery step	Final product contains 70–90 wt% uranium (as U_3O_8 or equivalent)

[a] Adopted from U.S. Atomic Energy Commission (1957b).

The mills start with ore averaging 0.21% U_3O_8, but which may vary anywhere from 0.1 to 1.0% depending on its source. The uranium is concentrated into an oxide or salt assaying 70 to 90% U_3O_8. The exact process that is used varies, depending on the nature of the ore and on the age of the mill. The principal methods are summarized in Table 8–2.

These plants are a potential source of environmental contamination because the process separates uranium from its radioactive daughter products which may then be discharged to the environment. Since the mills produce about 40 tons of uranium concentrate per day at an average concentration of 70 to 90% U_3O_8, it can be calculated that the plants must dispose of approximately 6 Ci Ra/day.

Shearer and Lee (1964) estimated that by 1964 the mills had accumulated about 34 million tons of tailings, from which it can be estimated that these piles contained about 3400 Ci of ^{226}Ra. These quantities have increased considerably since then. Cumulative uranium production, as U_3O_8, increased from 297,677 tons in 1964 to 462,836 tons by the end of 1970 (U.S. Atomic Energy Commission, 1971b), by which time the ^{226}Ra content of the tailings piles was about 5300 Ci.

The processes result in solid and liquid tailings that are usually combined and pumped to settling ponds near the mill site. About 865 gal of liquids are produced per ton of ore treated. In general, the liquid tailings are permitted to flow or seep into a nearby stream or are impounded at some distances from streams and either evaporate or seep into the ground.

The fate of the radium discharged into these streams has been the subject of a number of investigations. It had been shown (Gahr, 1959) that the ^{226}Ra content of water in the Colorado River below Grand Junction was 3×10^{-2} pCi/ml compared to 3×10^{-4} pCi/ml upstream. The San Miguel River below Uravan contained 0.086 pCi/ml compared to 0.0049 pCi/ml immediately upstream where a large mill is located.

The Animas River in southwestern Colorado serves as a public water supply for the cities of Aztec and Farmington, which have a total population of 28,000 persons, and the water is also used for irrigation. In 1955, the radium concentration below Durango, where a mill was located, was found to be 3.3×10^{-3} pCi/ml compared to 2×10^{-4} pCi/ml upstream. That considerable concentration was taking place in the stream biota was shown by the fact that plants below Durango contained 660 pCi/g compared to 6 pCi/g above Durango. Stream animals below Durango contained 360 pCi/g compared to 6 pCi/g above the mill.

More detailed surveys were conducted in 1958 and 1959 (Tsivoglou, 1959, 1960a,b). It was found that using the ICRP recommendations as a guide, consumers of untreated river water received almost 300% of the maximum permissible daily intake of radium and that the cities of Aztec and Farmington received 170 and 140% of the daily permissible intake, respectively. Durango, in which the plant was located, was somewhat less exposed because it received its water from a tributary of the Animas. Of the given total daily intakes, which comprise the intake in food as well as water, about 61% was due to the radium content of food resulting from the contaminated irrigation water.

Studies conducted by Shearer and Lee (1964) concluded that the radium was leachable from both the tailings piles and the river sediments. Thus, radium was entering the rivers not only in untreated liquid wastes but in surface runoff during rainfall as well. Steps were taken by the mill operators to correct this problem, and by 1963 the radium content of the Animas River sediments had been reduced to three times normal values, as compared to several hundred times normal several years earlier.

The tailings piles have caused additional problems because of wind-blown dust and evolution of radon. Snelling (1969, 1971) showed the external radiation levels to range from 0.5 to 0.7 mR/hr 3 ft above the piles, which covered 65 acres at one location.

Breslin and Glauberman (1970) measured the airborne dust downwind from tailings piles associated with the uranium mills and demonstrated clear relationships between the distance from the tailings piles and the concentrations of uranium and [210]Pb. Of the three tailings piles sampled, the air concentrations were well below permissible levels in two cases, but approached the upper limits recommended in 10CFR20 at a distance of about 1000 ft from the tailings piles.

The piles of tailings are now stabilized with topsoil and plantings. Radon evolution (Shearer and Sill, 1969), although higher than background, did not exceed the recommended limit for the public. However, the concentration of radon in several cities in which mills are located has been found to be elevated over normal values. A notable example is Grand Junction, Colorado, where the average concentration is about 10 pCi/liter (U.S. Public Health Service, 1969a).

The radiation exposure of Grand Junction residents has also been increased by the past practice of using mill tailings as a material of construction of homes and public buildings (Joint Committee on Atomic Energy, 1971; Dominick, 1971). The extent of exposure from this source has been under study by the EPA and the United States Public Health Service.

REFINING

The mill concentrates in this country are sent to any of several locations in which the uranium is converted to either the metal or some intermediate form, such as orange oxide (UO_3) or green salt (UF_4). The principal steps in converting the concentrates to a form that is of acceptable chemical purity are shown in Table 8–3. These processes involve potential exposure of the employees to α-emitting dust and, in the case of overseas high-grade fuels, to radon and γ radiation (Eisenbud and Quigley, 1956).

The refining operations involve the mechanical processing of dry powders of uranium compounds, and there is an opportunity to discharge uranium dust to the environment. The hastily constructed plants during World War II had insufficient control over dusts contained in their gaseous effluents, and shortly thereafter it became known that relatively large amounts of uranium were discharged to the outside atmosphere. However, uranium is so abundant in the environment that the element was undetectable above the natural background at moderate distances from the plants (Klevin et al., 1956). Present-day plants are equipped with filtration equipment that effectively removes the uranium dust, and the monetary value of uranium is such as to preclude the possibility of its being discharged to the atmosphere in significant quantities for sustained periods of time.

TABLE 8–3

PRINCIPAL STEPS IN THE REFINING AND CONVERSION OF URANIUM[a]

Feed
Miscellaneous uranium concentrates
(approximately 75% by weight U_3O_8 or equivalent

↓

Production of orange oxide (UO_3)

1. Digestion of the uranium concentrates in nitric acid
2. Solvent extraction to remove impurities and reextraction into water
3. Boil-down of the uranyl nitrate solution from (2) to a molten uranyl nitrate hexahydrate
4. Denitration of the molten salt by calcination to produce orange oxide powder

↓

Conversion to green salt (UF_4)

1. Reduction of the orange oxide to brwon oxide (UO_2) by contacting with hydrogen
2. Conversion of the brown oxide to green salt by contracting with anhydrous hydrofluoric acid

↓

Reduction to metal

1. Reduction of the green salt to massive uranium metal (derbies) by a thermite-type reaction using magnesium as the reducing agent
2. Vacuum casting of several uranium "derbies" from (1) to produce a uranium ingot

↓

Product

High purity uranium metal in ingot form

---

[a] Adopted from U. S. Atomic Energy Commission (1957b).

The kinds of liquid and solid wastes produced by the refineries depend on the type of feed that is processed. Until a few years ago, the Belgian Congo was the source of high-grade feeds that contained as much as 100 mg of radium per ton of ore. Some of the sludges from this process contained as much as 1 g of radium per ton. In recent years, however, the uranium industry has been operating with ores of much lower grade, and the uranium is usually separated near the mines, thus sparing the refineries the problem of disposing of waste products containing large amounts of radium. The limiting factor in the discharge of wastes from refineries is apt to be the chemical wastes rather than their radioactive constituents.

Isotopic Enrichment

Enrichment of the ^{235}U fraction is accomplished in the large gaseous-diffusion plants at Portsmouth, Ohio; Paducah, Kentucky; and Oak Ridge, Tennessee, where the green salt (UF$_4$) from the refineries is converted to uranium hexafluoride (UF$_6$). The UF$_6$ passes through many porous barriers in cascade, each stage of diffusion resulting in a slight enrichment in the isotope ^{235}U. The number of diffusion stages is determined by the degree of enrichment required.

Even more so than in the refineries, the economic value of the enriched uranium precludes the likelihood of widespread environmental contamination from these plants. In addition, as the uranium progresses through the diffusion plant, it becomes of increasing strategic importance and is, therefore, subject to strict accountability. The enriched uranium from the diffusion plants is destined either for assembly into weapons or into fuel elements for reactors. Large amounts of depleted uranium, which for the time being have little value, are being stored as the hexafluoride and in other forms.

Fuel-Element Manufacture

Fuel-element manufacture is currently carried on at a great many government and private facilities. Again, the relatively high cost of the uranium and the requirements for strict accountability make it unlikely that significant environmental contamination can occur from these plants, but the possibility of accidents must not be discounted entirely. Uranium chips and powders and some uranium alloys in massive form are pyrophoric. If fires and explosions occur, more than normal amounts of activity may be released. In the handling, storage, and fabrication of enriched fuels, significant amounts of contamination also may escape as the result of criticality accidents.

Thorium

Thorium (Cuthbert, 1958; Albert, 1966) is estimated to be three times more abundant in the earth's crust than uranium and may ultimately become an important source of nuclear energy as techniques are developed for converting the ^{232}Th to ^{233}U in breeder reactors. The most important known occurrences of thorium minerals are in the monazite sands of Brazil and India. Although thorium has been used for many years in the manufacture of gas mantles and at the present time has a limited application in the atomic energy industry, the production capacity for thorium is so small that there is little potential for general environmental contamination with this material or its daughter products. In time, should thorium processing be practiced on a wider scale, the daughter products might create problems more severe than those encountered in the uranium industry, since the radiotoxicity of the thorium series is believed to be greater than the uranium series.

Chapter 9

Reactors

The first nuclear reactor was demonstrated by Fermi and his associates in Chicago on December 2, 1942 less than 4 years after the discovery of nuclear fission. Under wartime pressure, reactor technology continued to develop rapidly, and only 1 year later a 3.8 thermal megawatt (MWt) research reactor began operation at Oak Ridge. Thereafter it remained in service for more than 20 years, during which time it served as the chief source of radioisotopes for research in industry in the United States and much of the Western World (Hewlett and Anderson, 1962; Tabor, 1963). Even more remarkable, the first of several reactors designed for plutonium production began operation at Hanford, Washington in 1944 at an initial power level of 250 MWt. These and additional units, at considerably higher power levels, also remained in service for more than 20 years. At present, more than 700 land-based reactors have been operated or are under construction in various parts of the world (International Atomic Energy Agency, 1970a). In addition, about 175 vessels of the American and Russian navies are powered by nuclear reactors (Joint Committee on Atomic Energy, 1970a). The United States nuclear fleet operates independently of the usual problems of fuel logistics, with submarine reactor cores now designed to last 10 years and provide propulsive power for about 400,000 miles (U.S. Atomic Energy Commission, 1971).

From 1945 until 1954 the reactor program of the United States was

216

dominated entirely by the government and was closely related to the military applications of nuclear energy. All reactors constructed until 1953 were located on government sites, but in that year a research reactor was placed in operation at North Carolina State College, the first reactor to go into operation outside an AEC facility.

Statutory changes that occurred with passage of the Atomic Energy Act of 1954 made it possible, for the first time, to disseminate a limited amount of information about reactor technology to private industry as well as to the world at large. This change coincided with the start of President Eisenhower's Atoms for Peace program, which expressed the desire of the United States to make available the civilian benefits of atomic energy on a global basis.

Great impetus to the development of a civilian reactor industry came from the First United National International Conference on the Peaceful Uses of Atomic Energy which took place in Geneva in the fall of 1955 (United Nations, 1956). At that Conference, enough formerly classified information about reactor technology was presented to enable other nations to begin development of their atomic energy resources.

The 1955 Conference was followed by the entry of many private companies into the business of designing and building reactors. It was generally recognized that it would take another decade for nuclear power to become economically viable, but meanwhile markets were developing for research reactors as well as for semiexperimental "demonstration power reactors" that were financed in part by the AEC in cooperation with the utilities and manufacturers. In parallel with the civilian program, a joint AEC-United States Navy program to develop nuclear propulsion was also proceeding rapidly, and the first nuclear power submarine, the Nautilus, was launched in 1954. The application of nuclear power to naval propulsion has been highly successful and has accelerated the development of civilian nuclear power in many ways.

In 1963, the economic viability of nuclear power was established with the decision by the New Jersey Central Power Company to build the Oyster Creek nuclear power station at Toms River. By late 1972, more than 140 nuclear power stations were in operation, in design, or under construction in the United States.

Apart from economic considerations, other factors have tended to encourage the exploitation of nuclear power. The use of electricity in the United States has been increasing at an average rate of 7% per year, necessitating that the installed generation capacity be doubled every 10 to 12 years. Moreover, during the past decade the trend has been toward larger and larger generating units to achieve economies of scale, the maximum generating unit size having increased from about 200 MW in 1950 to more than 1000 MW in 1970 (U.S. Government Office of Science and Technology, 1968). The economics of electrical power generation tends to favor large plants, so that as the plants increase in size, nuclear power becomes more desirable from an economic point of view. In addition, national concern with increasing levels of air pollution in many communities tends to favor installation of nuclear units, as does the increasing costs of fossil fuels in many parts of the country and uncertainties in regard to the long-range availability of these fuels.

The Physics and Engineering of Reactor Safety

The subject of reactor safety is an integral part of the complex reactor technology developed during the past 30 years by reactor scientists and engineers. To understand the subject fully requires knowledge of reactor physics and engineering on a level that is necessarily somewhat beyond the comprehension of anyone but a reactor specialist. However, the health physicist, physician, engineer, or health officer concerned with the influence of reactors on the environment should possess a general understanding of those aspects of design that may affect the kind and quantity of radioactive effluents discharged to the environment under normal and abnormal conditions. He should be able to evaluate the methods used to prevent uncontrolled release of radioactivity to the environment and should be capable of directing the monitoring activities that are necessary to determine if there is compliance with applicable limitations on discharges of radioactive wastes. Finally, in the event an accidental release of radioactivity does occur, the environmental specialist must be prepared to advise the measures that should be taken to minimize the consequences of the release. The discussion that follows is intended to provide the environmentalist with some of the basic aspects of reactor design that affect safety of operation. The reader who wishes to pursue the subject more comprehensively is referred to several excellent texts on the subject of reactors (Glasstone, 1955; Chastain, 1958; Dietrich and Zinn, 1958; Weinberg and Wigner, 1958; Thompson and Beckerly, 1964; Lamarsh, 1966; El-Wakil, 1962).

FUNDAMENTAL ASPECTS OF REACTOR PHYSICS

A reactor is a mass of fissionable material arranged geometrically so that nuclear fission occurs in a self-sustaining chain reaction. Contemporary reactors, with only a few exceptions, use either natural uranium or uranium in which the amount of isotope 235 has been enriched. The amount of enrichment may vary from a few tenths of 1% to more than 90%. The fuel in contemporary civilian power reactors is enriched to about 3% ^{235}U. Plutonium will be used eventually as a fuel for breeder reactors, as will be discussed later in this chapter.

One reactor operated by the Consolidated Edison Company at Indian Point, New York was fueled with a mixture of enriched uranium and thorium, the latter being utilized as fertile* material from which the fissionable isotope ^{233}U was bred. In the years to come, the supply of ^{233}U, which does not exist in nature, will become more abundant as the practice of breeding* is adopted on a more widespread scale. Because of the high specific activity of this isotope (half-life = 1.62 × 10^5 years), its use may be attended with hazards comparable with those from plutonium.

The severity of a reactor accident can be markedly influenced by chemical and physical reactions that produce heat and result in the liberation of fission products. It is thus important to avoid the use of fuels or other reactor components that can become involved in exothermic reactions in the event the core accidentally overheats. The use of metallic uranium presents the possibility that even a minute failure of the cladding surrounding the fuel will expose the metallic uranium to substances such as water and air with which it can enter into pyrophoric reactions. However, the pyrophoricity of metallic uranium can be avoided by using uranium dioxide, usually in the form of pellets enclosed in stainless-steel or zircalloy tubes. An additional advantage is that the uranium dioxide retains fission-product gases more effectively than uranium metal.

The fuel may be fabricated as rods, pins, plates, or tubes and is protected by a cladding whose function is to prevent the escape of fission products and protect the fuel from the eroding effect of the coolant. The cladding may be zirconium, stainless steel, or other special alloys. In reactors of contemporary design, the fuel is in the form of sintered UO$_2$ pellets less than 0.5 in. in diameter and about 1 in. long. These are aligned within tubes of zircalloy or stainless-steel tubes as much as 12 ft in length.† These tubes are arranged

---

* *Fissile* substances ^{235}U and ^{239}Pu can be used directly for production of nuclear power. Examples of *fertile* substances are ^{232}Th and ^{238}U, which under neutron bombardment are capable of being converted to the fissile nuclides, ^{233}U and ^{239}Pu. When the ratio of conversion to fission is greater than unity, the reactor is said to be a *breeder*.

† The UO$_2$ pellets are mounted within the tubes in such a way that a thin helium-filled gap exists between the pellet and the tube. The helium is intended to provide efficient heat transfer from the fuel to the cladding. This gap serves as a plenum within which the volatile fission products such as the halogens and noble gases accumulate.

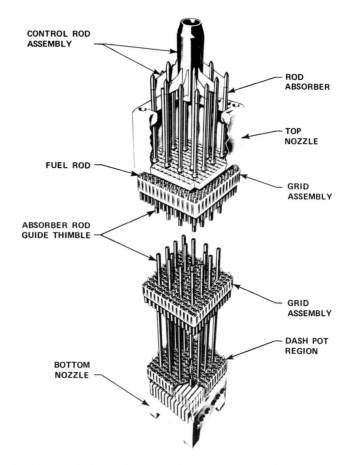

Fig. 9-1. Cutaway of a fuel and control rod assembly (Westinghouse Electric Corporation, 1971).

in bundles as shown in Fig. 9-1, within which, at selected positions, control rod guide thimbles replace the fuel pins. In a 1000 MW (electric) reactor, 193 such fuel bundles containing 39,372 fuel rods comprise the reactor core, which is about 11 ft in diameter (Westinghouse Electric Corporation, 1971).

The fission fragments produced during reactor operation vary in mass number from 72 to 160 and include over 80 isotopes produced in the frequency distribution shown in Fig. 9-2. It is seen that the yields of the mass numbers which are plotted on a logarithmetric scale range from about $10^{-5}\%$ to nearly 10% (Walton, 1961).

Fission results from the capture of a single neutron by the nucleus of a fissionable atomic species. Because more than one neutron is released in the process, a multiplication of neutrons may be achieved, affording an opportunity for additional atoms of uranium to be split which in turn will yield additional neutrons to continue the fission process.

Some of the neutrons produced by fission will escape from the reactor system and be lost. This can be minimized by surrounding the reactor core with a reflector which tends to scatter escaping neutrons back into the system. Other neutrons may be captured by nonfissionable atoms present in the materials of construction. This loss can be minimized by selecting materials of construction that have a low capture cross section* for neutrons and by eliminating high cross-section impurities by careful control over the processes by which the materials are made. Finally, some of the neutrons produced in fission will be absorbed in fissionable material and will, in turn, produce new fissions. The nuclear reactions will become self-sustaining when they are adjusted so that for every atom that fissions one fission-producing neutron remains after allowing for escape by leakage or loss by nonfission capture. At this point the reactor is said to be in a "critical" condition.

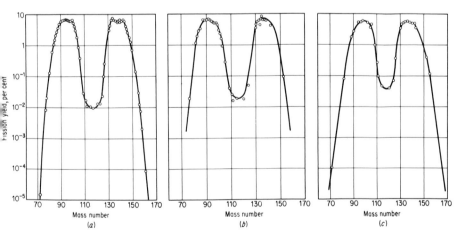

Fig. 9-2. Mass yields for slow neutron fission of (a) [235]U, (b) [233]U, and (c) [239]Pu (Steinberg and Glendenin, 1956).

* The cross section is a quantitative expression of the probability of occurrence of a given reaction between a nucleus and an incident particle. The unit of cross section is the barn, equal to 10^{-24} cm²/nucleus. The cross section of a nucleus for both fission and capture depends on the energy of the incident neutrons.

222   9. REACTORS

The neutrons produced in fission are relatively energetic, and the probability that they will produce fission can be increased by reducing their energies to the order of 1 eV. This is accomplished by distributing the fuel in discrete components between which is placed a moderator of low atomic number. Most reactors are designed for thermal neutrons and are called "thermal reactors." A fast reactor has no moderator and depends for its operation on the production of fission by fast neutrons (i.e., greater than 100 keV). Between the two extremes is an intermediate range of energies, and when a reactor core is designed to utilize neutrons in this range of energies, it is said to be an "intermediate reactor."

The state of criticality of a reactor may be expressed conveniently by k_{eff}, the effective multiplication factor, which is defined as the ratio of the number of neutrons produced by fission in any one generation to the number of neutrons produced in the preceding generation. Almost all neutrons produced will eventually be absorbed in nuclei somewhere within the reactor structure or its immediate vicinity. Absorption may occur in the fuel, moderator, materials of construction, or in impurities which may be present anywhere in the reactor configuration. A small fraction of the neutrons may escape entirely from the reactor. When criticality exists, k_{eff} equals unity, and $dn/dt = 0$, where n is the neutron flux, usually expressed as numbers of neutrons per square centimeter per second. The heat produced in a reactor, and hence its power level, is directly proportional to the neutron flux.

When k_{eff} is less than unity, the reactor is subcritical, and a chain reaction cannot be sustained. When k_{eff} is greater than unity, the reactor is supercritical, $dn/dt > 0$, and the power tends to increase with time. Unless the reactivity is reduced to ≤ 1, the power would increase to a level that would destroy the reactor. Conversely, when k_{eff} is less than unity, the power level will diminish and cannot be sustained at any given level without adding "reactivity."

Thus, a reactor is designed so that it is possible to maintain $k_{eff} = 1$ for various neutron densities corresponding to the desired power levels. When it is desired to increase the power level, k_{eff} is made slightly greater than unity ("reactivity" is added). Reactivity is defined as

$$\rho = \frac{k_{eff} - 1}{k_{eff}} = \frac{k_{ex}}{k_{eff}} \tag{9-1}$$

The reactivity of a thermal reactor is normally controlled by means of control rods containing neutron-absorbing materials such as cadmium, indium, boron, or hafnium. Insertion of the rods into the core reduces the number of neutrons available for fission, thus reducing the value of k_{eff} and causing the power level to diminish. Conversely, the rods can be withdrawn from the core in order to increase the power level.

The reactivity of a reactor is affected by a number of additional factors, including temperature and the radiation history of the core.

The effect of temperature on reactivity is one of the fundamental characteristics of a reactor and has an important influence on the ability of a reactor to self-regulate should the power be increased inadvertently. The overall effect of temperature on reactivity is the result of a number of factors. For example, the density of the coolant varies inversely with temperature. When the temperature is increased, the resulting reduction in coolant density diminishes the number of atoms per unit volume, causing the reactivity to increase because of reduced neutron capture by the coolant. In some reactors, this is more than offset by other effects that result in the moderator having a net negative effect on the temperature coefficient of the system.

Increasing the fuel temperature also reduces reactivity by increasing the probability of nonfission neutron capture. It is also possible that temperature changes can alter the geometry of the core so as to affect reactivity in one direction or the other. In general, a negative overall temperature coefficient of reactivity is desirable and serves to self-regulate the reactor. If the reverse were true and a reactor had greater reactivity on a rise in temperature, dangerous instabilities would result from the fact that any power excursion that resulted in an increase in the core temperature would cause a further increase in power and temperature because of increased reactivity and so on, until destruction of the reactor.

Reactivity is lost as the core ages owing to burnup of the uranium and because some of the accumulating fission fragments or their decay products have a high cross section for thermal neutrons and therefore increase the fraction of neutrons lost by capture. ^{135}Xe and ^{149}Sm are particularly important in this regard. The concentration of ^{135}Xe increases for several hours after reactor shutdown and may reduce reactivity to such an extent as to prevent start-up of the reactor for a day or more (Weinberg and Wigner, 1958).

Thus, because of the effects of temperature and fission-product poisons, the hot, aged core is inherently less reactive than the cold, fresh core, and the new core must have an additional measure of excessive reactivity at start-up to offset the decrease in reactivity that will occur later on. In most reactors, this is accomplished by the use of control rods, which insert neutron-absorbing compounds into the core. Another technique is the use of ^{10}B, which has a high capture cross section for thermal neutrons and which can be used to subtract reactivity when added in small amounts to the reactor core. This isotope has an advantage in that it transmutes to isotopes having a lower neutron capture cross section and, therefore, can be used as a "burnable poison," which will disappear from the reactor core somewhat in proportion to the buildup of xenon. In some water-cooled and moderated reactors, the boron is added in soluble form to the coolant.

When the control rods are removed, the value of k_{eff} increases from $k_{eff} < 1$ to $k_{eff} > 1$ and the power begins to rise. At the desired power level the rod positions are adjusted until $k_{eff} = 1$, and the steady-state critical condition is achieved.

The rate at which the power will rise or fall depends on the amount by which k_{eff} is greater or less than unity. If n neutrons are present per unit volume at the beginning of each generation and l is the generation time (or average neutron lifetime), it follows that

$$\frac{dn}{dt} = \frac{n(k_{eff} - 1)}{l} = \frac{nk_{ex}}{l} \qquad (9\text{-}2)$$

where k_{ex} is the excess multiplication factor, $k_{eff} - 1$. Integrating, we obtain $n = n_0 \exp t(k_{ex}/l)$, where n_0 is the initial neutron flux (and thus the initial power level) and n is the flux at any time t.

If we call the ratio l/k_{ex} the reactor period T so that $n = n_0 \exp t/T$, then T is equal to the time it takes to increase the power by a factor of e, and for this reason the period is sometimes called the e-folding time.

Under most conditions, l is constant, being characteristic of the core. The period is dependent solely on k_{ex}, and reactor power will rise or fall in an exponential manner, depending on whether k_{ex} is positive or negative.

The above relationships show that the reactor period is sensitive to changes in the generation time l. The average time elapsing between the production of a fission neutron and its ultimate capture is about 10^{-3} sec in a natural uranium reactor. Assuming a value of $k_{eff} = 1.005$ and a value of $l = 10^{-3}$ sec, the neutron density (power level) would increase by a factor of 150 per sec. It would be very difficult to manage a reactor if the rate of change of the power level were of this magnitude.

Fortunately, production of a small fraction of fission neutrons is delayed. The size of this fraction depends on which fissionable material is involved and varies from 0.0024 for ^{233}U to about 0.007 for ^{235}U. The delayed fraction actually controls the value of l, which is about 0.1 sec for ^{235}U. If we substitute this value in the equation for T, we find that it takes about 100 sec for the reactor power to increase by a factor of 150. This is a relatively slow and manageable rate of increase. The small fraction of delayed neutrons contributes in an important way to the manageability of thermal reactors by slowing the response of the neutron flux to changes in reactivity.

If β is the fraction of the fission neutrons that is delayed so that $1 - \beta$ is the fraction of prompt neutrons, the effective multiplication factor may be divided into two parts: $k_{\text{eff}}(1 - \beta)$ and $k_{\text{eff}}\beta$ representing the prompt neutron multiplication factor and the delayed neutron multiplication factor, respectively. When the fraction $k_{\text{eff}}(1 - \beta)$ = unity, the reactor is said to be "prompt critical," meaning that the excess reactivity is such that critically can be achieved with prompt neutrons. This condition must be avoided in practice because the period of the reactor is then determined by the mean lifetime of the prompt neutrons, and the reactor will be difficult to control.

The relationship between excess reactivity and period is shown graphically in Fig. 9-3 in which it is seen that the reactor period is independent of the prompt neutron generation time, up to reactivity values of about $0.007k_{\text{ex}}$, below which value the period is determined primarily by the generation time of the delayed neutrons. When the reactivity exceeds

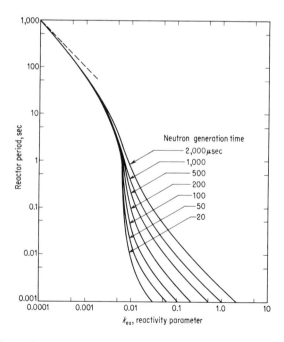

Fig. 9-3. Effects of prompt neutron generation time on reactor period. From J. W. Chastain ed. (1969). "U.S. Research Reactor Operation and Use." Addison-Wesley, Reading, Mass. Courtesy of U.S. Atomic Energy Commission.

$0.007k_{ex}$, the reactor is "prompt critical" and sensitive to differences in the generation time of the prompt neutrons. Many features of reactor design are intended to keep the reactivity well below the level required for a prompt critical condition.

Having achieved the desired neutron economy, the reactor designer must be equally attentive to the processes of heat generation and flow. Of the energy produced within the core, 80.5% is in the form of the kinetic energy of fission fragments, the range of which is of the order of 10^{-3} cm. Thus, most of the energy is converted to heat at the point of formation within the fuel (Lamarsh, 1966). Taking the β and γ radiation into consideration, about 90% of the energy is absorbed within the fuel. About 4% is absorbed within the moderator, some of the energy (mainly in the form of neutrinos) is lost to the system, and a small fraction is absorbed within the materials of construction of the reactor.

In light-water reactors, heat produced within the UO_2 fuel is transferred to the stainless-steel or zircalloy cladding across a thin film of helium that fills a gap between the fuel and cladding.

ENGINEERING ASPECTS OF REACTOR DESIGN

Reactor design is a complicated procedure in which many restrictions are placed on the physicist and engineer owing to the stringent standards of design imposed by safety considerations. This section will consider some of the engineering problems that are generally common to reactors. A discussion of the engineered safeguards designed into power reactors will be deferred until later in this chapter.

The reactor engineer is engaged in a constant battle to find materials having the desired thermal and structural properties that will not adversely affect his nuclear economy by absorbing neutrons. The thermal, nuclear, metallurgical, and structural requirements are frequently in conflict with each other as is often true in engineering (El-Wakil, 1962). It is necessary to maintain careful control over the materials of which the reactor is assembled and to take steps both in manufacture and fabrication to guard against the introduction of neutron-capturing impurities.

The designer must also use materials in which corrosion is minimal, because corrosion products in passing through the reactor become radioactive and complicate its operation. Finally, the potentially deleterious effects of intense radiation on the materials of construction are such as to place serious limitations on the kinds of materials that can be selected and the manner in which they can be fabricated.

One of the prime objectives of design is to maintain the integrity of the fuel-element cladding under all foreseeable circumstances. However, should such failure occur, there must be provision for operating safely, even though fission products diffuse from the fuel into the coolant streams.

The reactor must be equipped with sufficient instrumentation so that the operator can assure himself that the system is operating smoothly or so that he can take the necessary remedial measures in the event of some malfunction. To the extent possible, the information obtained by instruments is conveyed to electronic systems, which are designed to take corrective action automatically. To provide a reactor that will attain the desired performance characteristics and yet at all times be under complete control is the ultimate objective of the reactor designer.

An early step in ensuring the safety of a reactor operation is selection of a reactor type that has a high degree of inherent safety. Inherently unsafe reactors are not likely to get far beyond conceptual design, because neither the government nor industry would care to invest the large sums required for development and construction of a reactor that has inherently unsafe features. By the time a reactor concept is developed to the point where it is ready to be adopted for use in industry, it should be known that the kinetic features of the system are favorable. Among other things, the period of the reactor should be sufficiently long so that reactor control is facilitated, and the reactor should be stable with respect to the influences of temperature rises. The reactivity of the core under all anticipated conditions during its lifetime should be such that control is possible.

The use of materials for the core, coolant, and reactor structure should be such as to minimize the opportunities for exothermic reactions. Many of the materials that are potentially useful in reactor construction react exothermally with each other. These include metallic uranium, sodium, zirconium, and water. Some of the energy that becomes available during a serious reactor accident may be due to reactions between these materials.

The physical relationships that govern the operation of the reactors are numerous and so interrelated that the large number of parameters can only be optimized by elaborate techniques of digital computation. With the aid of large computers, a core can be designed in which the operating life is maximized, the power density variations minimized, and satisfactory control ensured.

One of the most fundamental estimates which must be made in the course of design is the temperature of the hottest channel in the core. In a well-designed reactor, power peaking, the extent to which the power produced at any given point in the reactor differs from the average power, is minimized. Nevertheless, systematic peaking in both the radial and axial directions does occur and can be predicted by calculation and experiment.

The product of the radial and axial power-peaking factors determines the amount of power that will be produced in the hottest part of the core, assuming that the peaking factors in both the radial and axial directions happen to coincide. In addition to this type of peaking, which is due to design factors, variations from the average power density will result from small manufacturing deviations in fuel-rod diameter, amount of enrichment, fuel density, the extent to which the fuel is out of round, and other similar factors. The most conservative approach would be to assume that all such deviations occur in the same direction and that they coincide spacially with respect to each other and the overall peaking factor. The product of all such factors gives a so-called "hot channel factor" which becomes a fundamental limitation in design. The amount of power produced in the hot channel and the temperature of the fuel and cladding in this channel ultimately determine the upper limitation on the ability of the reactor to produce heat.

The amount of power that can be produced by the reactor is limited by the temperature of the hot channel and the requirement that failure of the cladding be avoided. For example, in heterogeneous reactors of the water-cooled and moderated type, cladding failure might occur if the temperature of a small length of a fuel element increased to the point where heat transfer from the fuel elements to the coolant was reduced by film boiling. In a pressurized water reactor, boiling does not normally occur but may be present should the temperature rise sufficiently A moderate amount of boiling is not deleterious and may, in fact, increase the amount of heat transfer from the fuel. However, if the boiling, instead of being nucleate, occurs as a film along the surface of the fuel element, heat transfer is greatly reduced, and the temperature of the fuel and its cladding may rise dangerously. The reactor is designed so that the temperature at which departure from nucleate boiling (DNB) occurs is not approached. The safety factor in this regard is the DNB ratio, which is the temperature at which DNB takes place, to the design temperature of the hot channel. This ratio is another fundamental parameter of reactor safety.

The intense radiations within a reactor can affect reactor materials by producing changes in such physical properties as thermoconductivity, resistivity, hardness, and elasticity. In addition, dimensional changes in some materials may occur, and in graphite these changes may be associated with the storage of relatively large amounts of energy.

The interaction of neutrons with the crystalline form of graphite displaces carbon atoms from their normal positions in the molecular lattice. Vacancies in the normal crystalline structure occur, and atoms of carbon appear interstitially. This has a number of effects on the physical properties of the graphite, including dimensional growth which may occur linearly to the extent of about 3%. The ultimate effect of continued irradiation is breaking of the crystalline structure with the carbon appearing as carbon black after 10^{21} neutrons/cm^2 (Harper, 1961; Wittels, 1966–1967).

If the crystalline structure has not been destroyed, the original molecular form can be restored by an annealing process, which permits the interstitial atoms to diffuse to vacant positions in the crystalline lattice. The disarray of carbon atoms represents stored energy which is released during the annealing process. As much as 500 cal/g of graphite may be released, enough to raise the temperature of the graphite to a dangerous level. This was the cause of the destruction of the Windscale reactor, as we will see in Chapter 16.

Whether or not the changes produced by a given radioactive environment are deleterious to the materials can best be answered by subjecting test specimens to conditions simulating those expected in the reactor.

The fuel, cladding, brazing materials, and various pieces of hardware that may be included in fuel-element design are particularly important. For example, irradiation may affect the rate at which fission gases such as xenon and krypton are released from the fuel into the gas-filled gap that in some designs is present between the fuel and the clad. This affects heat transfer from the fuel to the clad and thus affects the rate at which heat can be removed from the core by the pressurized water coolant. The effect of irradiation on the integrity of the cladding must also be studied. If the physical properties are changed so that the cladding erodes or otherwise becomes unable to withstand the rigors of the thermal and radioactive environment within the core, fission products may be released from the fuel. These and other similar questions can best be answered by actual irradiation of test specimens under simulated in-pile conditions.

To provide the radioactive and thermal environment for such tests is one of the principal functions of reactors like the Materials Testing Reactor and the Engineering Testing Reactor located at the National Reactor Testing Station in Idaho. At these and other reactors, test specimens can be inserted into the reactor in such a way that not only can one simulate the neutron and γ environment that is desired, but other factors such as the temperature, pressure, and rate of coolant flow can be simulated as well. In order to simulate the physical conditions that may eventually be encountered in practice, it may be necessary to subject the test specimens to irradiation for considerable lengths of time, perhaps as long as 2 or 3 years. Studies of these kinds are among the many steps that must be taken in order to be certain that the materials that go into a reactor will perform efficiently and safely.

During every phase of reactor design the engineer and physicist must examine each decision for its possible effect on the safety of the system. Decisions regarding materials, dimensions, equipment, operating temperatures, and other aspects of design must be carefully reviewed.

Types of Reactors

Reactors can be constructed to serve primarily as sources of radiation or heat. Included among those constructed as radiation sources are: (1) "production" reactors in which products of neutron irradiation such as plutonium, other transuranic elements, or ^{60}Co are produced; (2) research reactors such as those located on university campuses and other research centers; and (3) industrial-type test reactors that are used to study the effects of radioactivity on materials of construction and equipment components. The reactors used as sources of heat are used primarily as sources of power for electric generators but may also be used to generate steam for space or process heat. Production and test reactors will not be included in this discussion.

Research Reactors

A large number of research reactors are now in operation on university campuses, in industrial laboratories, and in medical research institutions. Research reactors function basically as sources of neutrons or other radiations used to activate samples, study the structure of matter, or irradiate living things. At the end of 1970, there were 72 operating research reactors in the United States (USAEC, 1971a), and in May, 1970 the International Atomic Energy Agency reported (IAEA, 1970a) that 367 research reactors were in operation in 47 countries.

Research reactors may be fueled with either natural or enriched uranium, but a natural-uranium reactor is inherently large and, for this reason, would only be located at a major research center where the large size of the reactor would be advantageous in permitting access by large numbers of investigators. When moderated by graphite, these reactors are cubes about 25 ft to each edge. Only two such reactors have been built in the United States, at Brookhaven and Oak Ridge National Laboratories, and both of these reactors were shut down after about 2 decades of useful life. The elaborate nature of the experimental equipment utilized at a reactor of this kind is shown in Fig. 9-4. Both the Oak Ridge and Brookhaven reactors were air cooled.

Natural-uranium research reactors can be moderated with heavy water,

Fig. 9-4. The Brookhaven National Laboratory air-cooled, graphite-moderated reactor. This photograph illustrates the large size of this reactor and the large number of experiments that can be accommodated (Brookhaven National Laboratory).

which reduces the size of the reactor compared with a graphite-moderated reactor, but which entails greater expense for the heavy water. The Brookhaven medical research reactor and the MIT research reactor (Fig. 9-5) are of this type.

Research reactors employing enriched fuel vary in power from a few milliwatts in the case of the Argonaut to several megawatts in the case of

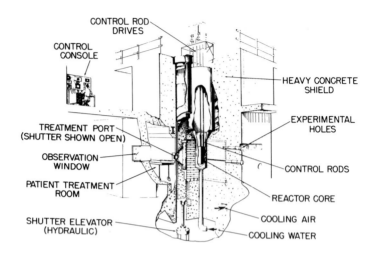

Fig. 9-5. The heavy-water-moderated medical research reactor at Brookhaven National Laboratory (Brookhaven National Laboratory).

water-cooled-and-moderated reactors. The Argonaut is a relatively inexpensive reactor which ranges in power from a few milliwatts to about 5 W and is useful for a variety of training exercises and simple physics experiments. It consists of uranium dioxide distributed throughout a polyethylene core for which no cooling is necessary (Chastain, 1958).

One of the more popular reactors during the late 1950's and early 1960's was the so-called "pool reactor" which ranges in power up to about 5 MW. As shown in Fig. 9-6, the fuel elements are assembled in an open-top tank under about 25 ft of water. The water acts as a transparent shield through which the core is accessible from a bridge over the top of the reactor. At power levels over 1 MW, special attention must be given to the problem of removing the activation products ^{41}A, ^{24}Na, and ^{17}N, and because of this factor, it becomes necessary to enclose the pool completely at levels higher than 5 MW. In these "tank" reactors, less water is used than in the case of pool reactors, and more shielding is therefore required.

Another very popular type of research reactor is the Triga (Fig. 9-7), which is a pool-type reactor moderated by zirconium hydride homogeneously combined with 20% enriched uranium. A distinctive characteristic of this reactor is the highly negative temperature coefficient of reactivity, which makes it possible to pulse the reactor to peak power levels a thousand times greater than the average level. The reactor can be pulsed by rapid withdrawal of a control rod which causes a transient for several thousandths of a second, which is terminated by the inherent nuclear characteristics of the reactor. Thus, a Triga reactor capable of operating at 250 KW under continuous conditions can be pulsed to 250,000 KW for a brief period.

Research reactors are utilized either by providing beams of neutrons that irradiate specimens external to the reactor or by providing tubes that make it possible to place specimens within the reactor. Many experiments require thermal neutrons, a requirement that necessitates special attention to the core and moderator geometries so as to provide columns of thermal neutrons accessible for research.

Some of the larger reactors may have pneumatically operated "rabbits" which pass a specimen quickly through the core for very brief periods of irradiation. This provides a quick means of getting specimens in and out of the reactor for studies of short-lived radionuclides.

In pool reactors the specimens to be irradiated are frequently placed in boxes having the same dimensions of the fuel elements which can then be placed in the core in place of the fuel element.

Fig. 9-6. Pool-type research reactor at Union Carbide's Nuclear Center, Sterling Forest, New York showing the way in which the water-moderator-coolant serves as a transparent shield for operators on the bridge above the reactor (AMF Atomics).

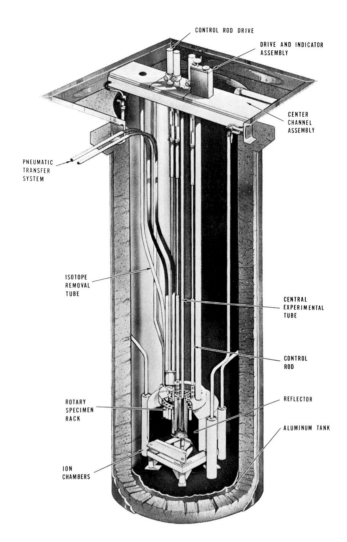

Fig. 9-7. Schematic cutaway diagram of a Triga research reactor (Gulf General Atomic).

POWER REACTORS

Power reactors can be divided into two major categories depending on the energy of the neutrons that produce fission. When they utilize neutrons having energies in excess of 0.1 MeV, the reactors are called fast, whereas in thermal reactors the neutrons are moderated to energies of a few hundredths of an electron volt. There has been a limited amount of experience with reactors that utilize neutrons in the intermediate range of energies.

Among the advantages of the fast reactor is that no moderator is required, and hence it is possible to design cores that are small in comparison with thermal reactors. In addition, the capture cross sections of most substances for fast neutrons are relatively low, permitting a wider choice in the selection of construction materials. The fission products also have low capture cross sections for fast neutrons, and greater burnup of fissionable material is, therefore, possible because certain fission products do not act as poisons in fast reactors. Finally, because the nonfission capture cross sections of fast neutrons by the fertile material are relatively high, the fast reactor is inherently favorable for breeding fissionable ^{239}Pu from ^{238}U.

One of the disadvantages of the fast reactor is the fact that the selection of a coolant is restricted to those having no moderating effect. Liquid sodium, which is frequently used for cooling fast reactors, has obvious disadvantages because of its chemical reactivity with air and water, but this is offset by excellent heat transfer properties and the fact that high pressures are not required when sodium is used as a coolant. The small size of the fast reactor reduces the surface areas available for heat transfer and, thus, further limits the choice of coolants to those having high conductivity and specific heat.

Because of the favorable neutron economy referred to earlier, it is possible to design fast reactors with relatively small amounts of excess reactivity, but an extensive core meltdown might result in a configuration having a higher reactivity than the original configuration. Should sufficient melting occur, it is possible that the molten fuel could rearrange itself in such a way as to produce a critical mass. Although this can probably be avoided by proper design, this factor has led to a great deal of conservation in regard to the use of fast reactors for civilian power (Okrent, 1965).

Light-Water Reactors

With but few exceptions, the power reactors constructed by United States power companies are of the so-called light-water type. Prior to about 1963, when the light-water reactors emerged as economically viable and operationally practical, a number of other systems were demonstrated, some in the laboratory and some as power generators. However, the homogeneous types of reactors, the organic moderated reactors, and others of the period gave way to the light-water reactors, which found favor in the United States Navy, setting the basis for its development as an electrical power producer.

This class of reactors includes two types, the boiling-water reactors

(BWR) and the pressurized-water reactors (PWR) (Bright, 1971). The fuel in both is usually slightly enriched uranium in the form of oxide pellets contained in stainless-steel or zircalloy tubes. Water is used for both coolant and moderator.

Uranium oxide has the desirable characteristic of trapping fission products within its crystal structure, thereby greatly reducing the probability of escape of fission products. Use of this material also avoids the hazards due to the pyrophoric nature of metallic uranium, which can react exothermically with both water and air.

The PWR (Fig. 9-8a) includes a primary system within which the reactor core is enclosed in a several-inch-thick pressure vessel, usually made of stainless steel (Fig. 9-9). Water is pumped through the core where it absorbs heat produced by the nuclear reactions. Since the entire primary system is maintained under pressure, the water does not boil but passes through a heat exchanger which serves as a steam generator for water in a secondary loop. The steam thus produced is pumped through a steam drier and then to a turbogenerator from which the steam tailings are condensed and returned to the boiler. Thus, in the PWR the steam used to drive the turbogenerator does not pass through the reactor but receives its heat via the heat exchanger.

In the boiling-water reactor (Fig. 9-8b) the water passing through the reactor core is allowed to boil, and the steam goes directly to the turbogenerator from which the steam tailings are condensed and returned to the reactor.

The thermal efficiency, i.e., conversion of heat energy into electric energy, is 31–33% for light-water reactors. Therefore, if the power capacity of a nuclear plant is 3000 MW thermal (MWt), the electrical generating capacity will be approximately 1000 MW electrical (MWe).

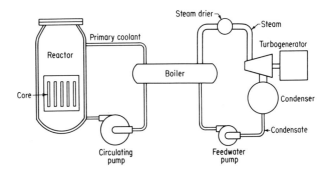

Fig. 9-8a. Schematic flow diagram of pressurized water reactor.

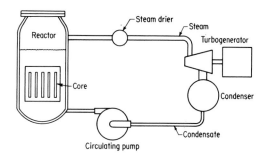

Fig. 9-8b. Schematic flow diagram of boiling-water reactor.

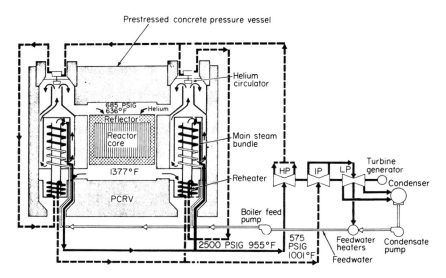

Fig. 9-8c. Schematic flow diagram for large, high temperature gas reactor (Gulf General Atomic).

Gas-Cooled Reactors

Although the British (Kaplan, 1971; Goodjohn and Fortescue, 1971) have placed a good deal of emphasis on the development of gas-cooled reactors and have had a number of such reactors in operation since the mid-1950's, this system has only begun to receive attention in the United States. One high-temperature gas-cooled reactor (HTGR) built by the Philadelphia Electric Company at Peach Bottom has been in operation since 1967, and the Public Service Company of Colorado is constructing a second unit of this type at Fort St. Vrain.

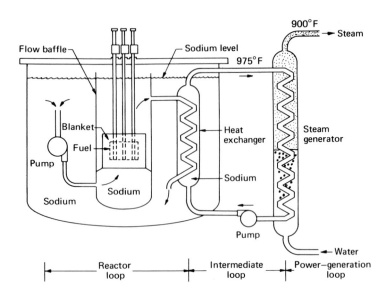

Fig. 9-8d. Schematic diagram of the liquid metal fast-breeder reactor.

In the HTGR, the reactor core is constructed from hexagonally shaped graphite fuel elements within which the fuel is contained as rods of uranium dicarbide and thorium carbide. The latter would serve as a fertile material to breed ^{233}U, which is fissionable. As shown in Fig. 9-8c, the reactor is cooled by helium maintained under pressure. One characteristic of the HTGR is that it operates at higher temperatures and pressures, is, therefore, more efficient thermodynamically, and discharges less waste heat to the environment.

Liquid Metal-Cooled Fast Breeder Reactors

Light-water reactors, which are currently predominant in the nuclear power generation field, are inherently inefficient and convert only 1 to 2% of the potentially available energy of uranium into heat. In contrast, a fast-breeder reactor can economically use up to about 75% of the energy contained in uranium, thereby achieving efficiencies about 40 times greater than light-water reactors (Seaborg and Bloom, 1970). The exponential growth of the demand for electrical energy and limitations in the known uranium reserves require that as a matter of world policy the breeder reactor be developed to a practical state.

Of the various alternative ways of designing a breeding reactor, the

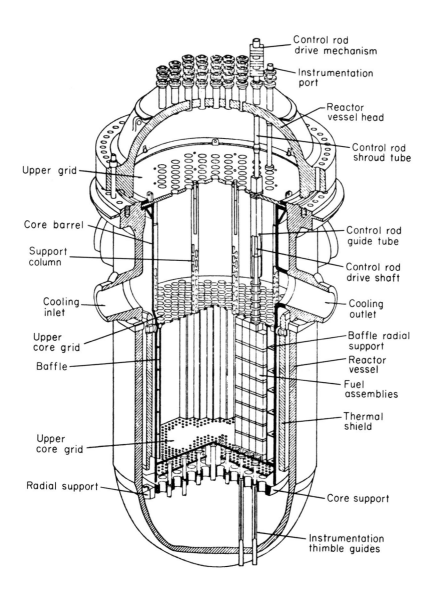

Fig. 9-9. Cutaway view of the pressurized water reactor (PWR) No. II at the Indian Point Nuclear Power Plant, New York showing the arrangement of the fuel assemblies, control rods, pressure vessel, and other components (International Atomic Energy Agency, 1968).

so-called liquid metal-cooled fast-breeder reactor (LMFBR) has received the most attention in the United States. The LMFBR will not only utilize uranium more efficiently in its own operation, but it will add to the efficiency of contemporary light-water or high temperature gas-cooled reactors by making it possible to use the plutonium produced in these power generators.

The LMFBR is illustrated schematically in Fig. 9-8d. It is cooled with liquid sodium in loop A, from which the reactor-produced heat is passed to an intermediate loop B, which, in turn, produces steam in loop C, which drives a conventional turbogenerator. The intermediate loop is necessitated by the fact that sodium becomes highly radioactive in the course of passing through the reactor, and the ultimate consequences of a possible sodium–water reaction in the event of failure of the barriers between loops B and A or C would be somewhat mitigated by the fact that the radioactive sodium cannot come in contact with the water of loop C. The high boiling point of the sodium is an attractive feature that enables the reactor to operate at high temperature and low pressures.

The advantage of the LMFBR is that it actually produces more fissionable material than it uses. A conventional 1000 MWe light-water during a 30-year lifetime will require 3900 metric tons of natural uranium and will produce 3870 metric tons of depleted uranium and 5300 kg of plutonium. A similar sized LMFBR would use 100 metric tons of depleted uranium and 2300 kg of plutonium, but in addition to the power generated, it would produce 50 metric tons of depleted uranium and 7700 kg of plutonium. The reactor economy of the future may thus be based on the use of plutonium, a very toxic radioelement.

The use of plutonium, the opportunities for sodium–water reactions, and the remotely possible risk of an accidental change in core configuration that could result in prompt criticality are factors that have made some observers uncomfortable about the safety of the LMFBR. However, it is important to recognize that this program is in an early stage of development, and there is ample time in which to prove the safety of the system that is scheduled for demonstration in the early 1980's (USAEC, 1971e; Baker *et al.*, 1970; Keilholtz and Battle, 1969).

Inventory of Radioactivity in Light-Water Reactors

The radioactive substances that accumulate in reactors are primarily those produced within the reactor core by the fission process and only secondarily the activation products formed when impurities in the coolant pass through the core. The water used to cool and moderate the reactor accumulates traces of corrosion products and other impurities that undergo

neutron bombardment in passing through the core, thus yielding radionuclides of such elements as chromium, cobalt, manganese, and iron. In
practice, the fission product inventory is very much larger than the inventory of corrosion products, but the nature of reactor operation is such that
the corrosion products may be present in relatively greater quantities in
the aqueous wastes.

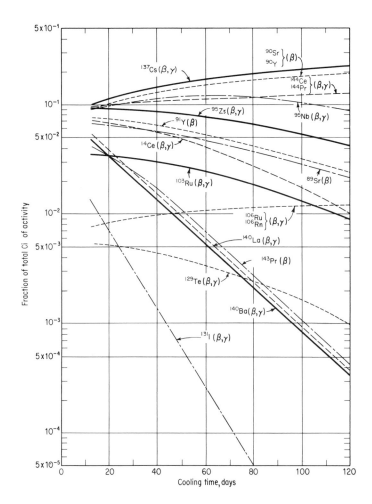

Fig. 9-10. The principal radionuclides in fission products at various times after
reactor shutdown. It is assumed that the reactor has operated for an extended period
and that an approximate equilibrium has been attained prior to shutdown (Glasstone,
1955).

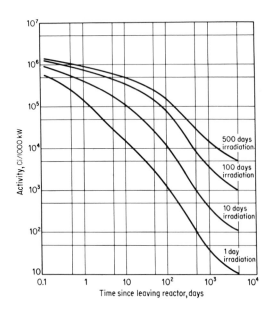

Fig. 9-11. Fission-product inventory in a reactor core after various periods of irradiation and shutdown (Parker and Healy, 1956).

Figure 9-2 gives the yields of the various fission fragments for three types of fissionable material. The inventory of radionuclides in any given core is a complicated function of the operating history of the reactor and the power level. The rate of accumulation of each fission product can be calculated during any period of reactor operation at any given power level. Correspondingly, the decay of the various radionuclides can be calculated during periods of shutdown or reduced power. If a reactor operates sufficiently long so that equilibrium with the short-lived nuclides has taken place, the distribution of the radionuclides as a function of time after reactor shutdown is given in Fig. 9-10. The total amount of radioactivity as a function of time after shutdown for various radiation histories is given in Fig. 9-11 in which it is seen that a 1 MW reactor will accumulate about 1 MCi of radioactivity after 500 days at full power. The principal radionuclides contained in a reactor core that has been shut down for 1 day subsequent to 2 years of continuous operation are given in Table 9-1.

The data of Fig. 9-11 are obtained from the following equation proposed by Way and Wigner (1948):

$$\text{Ci} = 1.4P \left[(t - T_0)^{-0.2} - t^{-0.2} \right] \tag{9-3}$$

where

Ci = Curies at any time t (in days) after shutdown
P = power levels (in watts)
T_0 = duration of continuous reactor operation prior to shutdown (in days)

RADIOACTIVE WASTES UNDER NORMAL OPERATION

Although they accumulate enormous inventories of fission products, most reactors discharge relatively insignificant amounts of radioactivity to the environment (IAEA, 1966; Logsdon and Chissler, 1970). In view of the special importance of light-water nuclear power reactors, the sources of radioactivity from them will be discussed in detail after we have reviewed briefly the sources of environmental radioactivity from single-pass, air-cooled and water-cooled reactors.

TABLE 9-1

INVENTORY OF SELECTED RADIONUCLIDES
1 DAY AFTER 2 YEARS OF REACTOR OPERA-
TION[a]

Selected isotopes	Half-life	Activity in fuel (kCi/ MWt)
^{3}H	12.3 years	0.0043
^{85}Kr	10.7 years	0.25
^{89}Sr	51 days	24
^{90}Sr	28.9 years	1.8
^{90}Y	64 hr	1.8
^{91}Y	58.8 days	32
^{99}Mo	66.6 hr	40
^{131}I	8.06 days	28
^{133}Xe	5.3 days	54
^{134}Cs	2.06 years	0.61
^{132}Te	78 hr	34
^{133}I	20.8 hr	22
^{136}Cs	13 days	0.74
^{137}Cs	30.2 years	2.4
^{140}Ba	13 days	46
^{140}La	40.2 hr	49
^{144}Ce	284.4 days	35

[a] National Academy of Engineering (1972).

Single-Pass, Air-Cooled Reactors

Air-cooled reactors discharge radioactive atmospheric effluents from several sources. Neutron activation of normal atmospheric argon produces ^{41}Ar, and other induced activities may result from neutron capture by elements present in the ordinary dust carried by the cooling air. In addition, fission products may be released through defective cladding of the fuel.

Isotope ^{41}Ar has a half-life of 1.8 hr, and the rapid decay of this isotope and its chemical inertness tend to simplify the problem of control. When the air-cooled research reactor at Brookhaven National Laboratory was in operation, as much as 750 Ci of ^{41}Ar was discharged each hour into the atmosphere, but exposure beyond the perimeter of the plant site was readily controlled by utilizing meteorological dilution from a 400 ft stack. When atmospheric conditions were sufficiently unfavorable, the power level of the reactor was reduced.

Emission of radioactive dust from air-cooled reactors can be avoided by filtering the supply of air, thereby eliminating particles that become radioactive by neutron capture when passing through the reactor core. Filtration of the effluent gases is of value in arresting radioactive particles which may have flaked off the interior of the reactor structure as well as particles from the release of fission products through ruptures which may occur from time to time in the protective cladding of the fuel elements.

Single-Pass, Water-Cooled Reactors

The only reactors of this type in the United States were the original plutonium production reactors at Hanford, which were fueled with natural uranium, graphite moderated, and cooled by treated Columbia River water. These units were operated for about 20 years during and after World War II.

Radioactivity owing to neutron activation of impurities in the water was minimized by treating the water to remove suspended solids and dissolved impurities. However, traces of impurities remained or were added by contact of the water with the aluminum cladding surrounding the fuel elements. Sometimes the cladding developed pinhole failures through which fission products entered the cooling water. The distribution of radionuclides in the reactor effluents of the Hanford reactor is shown in Fig. 9-12. Because many of these isotopes were short lived, the radioactivity of the discharged coolant could be diminished by 50 to 70% by holding the effluent in tanks for 1 to 3 hr before the water was returned to the river. The radioactivity diminished further as the effluent entered the river and decayed to less than 10% of its original value by the time it moved 35 miles downstream (Foster, 1959).

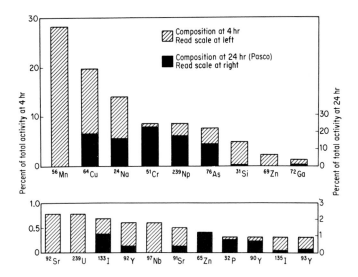

Fig. 9-12. Isotopic composition of the water effluents from the Hanford reactors at 4 and 24 hr after irradiation (Foster, 1959).

Light-Water Power Reactors

The radioactive substances generated by a water-cooled-and-moderated reactor are the induced radionuclides produced by neutron bombardment of substances contained in the primary coolant or the fission products that enter the primary coolant. The latter may sometimes be due to uranium present as a contaminant on the surface of the reactor parts or to defective fuel elements. These radionuclides exist in gaseous form and dissolved solids and, to a lesser extent, as suspended solids. One of the objectives of good reactor design is to provide a chemical-processing system which removes these radionuclides and concentrates them into a form satisfactory for ultimate disposal (Blomeke and Harrington, 1968).

The induced activities that occur in the primary coolant will depend on the materials of construction of the core, pressure vessel, pumps, piping, and other components in contact with the water. The induced activities will result to a lesser extent from impurities ordinarily present in water, because they are removed by water treatment before the coolant is introduced into the reactor.

When stainless steel is used as a material of construction in pressurized-water reactors, some corrosion takes place, and a number of induced activities may be found in the coolant, including ^{60}Co, ^{59}Fe, ^{51}Cr, ^{54}Mn, and ^{55}Fe. These isotopes have moderately long half-lives and ultimately present a waste-disposal problem in addition to being an operational problem because of their γ activity.

Traces of "tramp" uranium will sometimes contaminate the surfaces of fuel elements or other reactor-core components, and this contamination may undergo fission following startup of the reactor (Beaver, 1961). Small, but nevertheless annoying, concentrations of fission products may be encountered as a result of this contamination. Every effort should be made to decontaminate the surfaces of reactor components of tramp uranium and other contaminants that might result in the formation of radionuclides. Except for the adventitious presence of uranium on the exterior surface of the core, the presence of fission products in coolant water may be taken to signify failure of cladding.

In-pile tests of uranium oxide fuel have shown that the various isotopes diffuse into the coolant at different rates, depending on their relative mobility (USAEC, 1960a). Table 9-2 summarizes the escape-rate coefficients measured in pressurized-water reactor fuel. Elements such as cesium, iodine, xenon, krypton, rubidium, and bromine have escape-rate coefficients more than 1000 times greater than strontium and barium.

Liquid Wastes. The fission and activation products that diffuse from the fuel into the coolant can be removed continuously by means of a purification system that is usually designed to have sufficient capacity to permit continuation of reactor operation in the event of failure of 1% of the fuel rods.

Tritium is discharged by light-water reactors in quantities that are relatively copious compared to other radionuclides. Fortunately, tritium is a pure β emitter of very low energy, which is usually introduced to the environment in the form of water. It does not concentrate significantly in biological systems and has a relatively rapid turnover rate. Thus, the maximum permissible concentration (MPC) in drinking water is relatively

TABLE 9-2

ESCAPE-RATE COEFFICIENTS[a]

Elements	Escape-rate coefficient (sec^{-1})
Cs, I, Xe, Kr, Rb, Br	1.3×10^{-8}
Sr, Ba	1.0×10^{-11}
Zr, Ce, and rare earths	1.6×10^{-12}
Te	1.0×10^{-9}
Mo	2.0×10^{-9}

[a] U.S. Atomic Energy Commission (1960a).

TABLE 9–3

THE ANNUAL AVERAGE RADIONUCLIDE
COMPOSITION RELATIVE TO THE CON-
CENTRATION OF ^{137}Cs FROM A TYPICAL
PRESSURIZED WATER REACTOR

Nuclide	Relative concentration[a]
^{3}H	5.3×10^2
^{51}Cr	$<1.7 \times 10^{-2}$
^{54}Mn	2.8×10^{-1}
^{55}Fe	8.1×10^{-2}
^{59}Fe	$<2.4 \times 10^{-3}$
^{58}Co	2.1×10^{-2}
^{60}Co	1.1×10^{-1}
^{65}Zn	$<3.1 \times 10^{-3}$
^{90}Sr	6.6×10^{-4}
^{91}Y	2.6×10^{-3}
^{95}Zr-Nb	$<2.8 \times 10^{-2}$
^{103}Ru	$<3.1 \times 10^{-2}$
^{106}Ru	$<8.6 \times 10^{-2}$
^{131}I	$<5.2 \times 10^{-3}$
^{134}Cs	4.8×10^{-1}
^{137}Cs	1.0×10^0
^{144}Ce	$<3.8 \times 10^{-2}$

[a] ^{137}Cs = 1.0.

high, and the large amount of tritium discharged from reactors does not imply a correspondingly high public health risk (Jacobs, 1968).

The tritium originates in two ways. Albenesius (1959) first demonstrated that tritium is produced in fission at a rate of about 1 triton/10,000 fissions. Tritium is also produced by spallation following neutron irradiation of ^{10}B, and this is a major source of tritium in reactors that use ^{10}B as a burnable poison. The use of boron in this way is confined primarily to pressurized-water reactors which for this reason tend to discharge more tritium than boiling-water reactors.

Differences in the diffusion rates of the various radionuclides from intact fuel cladding cause the coolant radioactivity in most reactors to be relatively depleted in less labile fission products such as ^{90}Sr or ^{89}Sr. However, in the event of fuel failure, direct exposure of the uranium oxide fuel to the coolant can cause these nuclides to be more abundant.

The exact composition of the liquid wastes from light-water reactors will vary from reactor to reactor depending on the materials of construction and the condition of the fuel. Table 9-3 lists the radionuclides in the primary coolant of a pressurized water reactor. The effect of the relatively low diffusivity of the strontium nuclides is manifested by their low concentration relative to ^{137}Cs.

In light-water reactor purification systems, the coolant is first filtered to remove suspended radionuclides. The water then passes through cation- and anion-exchange resins. If necessary, the coolant can be passed through a gas stripper in which the water is percolated over plates across which a countercurrent stream of steam is passed, serving to remove dissolved gases such as air, fission gases, and hydrogen. The combined decontamination factor of the purification system including the filters and ion-exchange units is between 10^{-2} and 10^{-3} (Bethel et al., 1959).

Apart from the need to purify the coolant, there are many ways in which radioactive liquid wastes are produced. Leaks of coolant from valves, flanges, and pumps will ultimately result in the contamination of sump water. Components, which are removed for repair, must first be decontaminated, and this will result in contaminated water, as will the operation of washing casks, sluicing resin beds, laundering contaminated clothes, and washing contaminated laboratory ware. In addition, it may be expected that the cooling pools for spent fuel may in time become contaminated as a result of failures in the fuel element cladding. Provision must be made for containment of these waste liquids and their treatment and ultimate disposal.

The system for treating such wastes at a PWR is shown in Fig. 9-13 in which low-level contaminated water is accumulated for periodic processing. The wastes are accumulated in storage tanks and then are passed through a waste-gas stripper. Steam generated in the stripper scrubs incoming water free of gas as it passes vertically downward through sections of porcelain saddles. The stripped gases, after passing through a condenser, are routed to the waste-gas system. The effluent from the waste-gas stripper is passed to an evaporator from which the vapors are condensed and passed to a demineralizer having cation and mixed-bed resins. The sludge from the evaporator bottom is passed to storage tanks from which it goes to drumming stations. The effluent from the demineralizer passes to storage tanks where it can be sampled. If it is of satisfactory quality, it can be discharged via the condenser discharge canal (Fig. 9-14) or it may enter the purification system for reuse in the primary system. In the event the quality is not satisfactory, it may be returned for reprocessing.

Gaseous Wastes. In a PWR, the waste fission-product gases that are stripped from liquid wastes can be passed through a condenser which removes the condensible portion, passing the remainder to holdup tanks and hence to catalytic hydrogen recombiners which remove the radiolytic

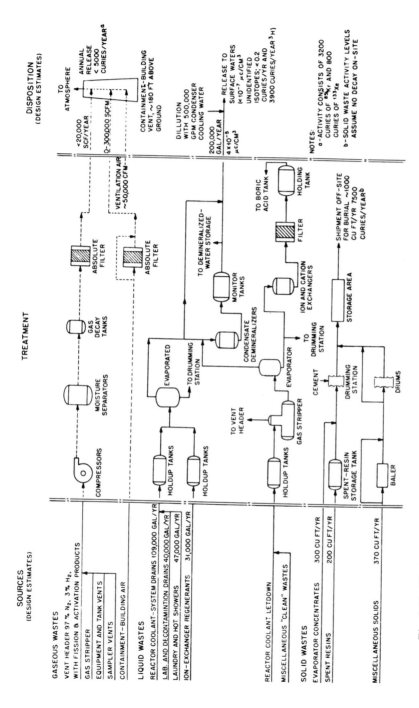

Fig. 9-13. Waste management flow diagram for a large pressurized water reactor (Blomeke and Harrington, 1968).

Fig. 9-14. Condenser discharge for a pressurized water reactor. Low-level liquid wastes can be released into this tunnel through which condenser coolant is flowing at a rate of 300,000 gpm (gal/min). The pipe at the left edge of the canal is a sampling manifold. Sampling intake pipes can also be seen (Consolidated Edison Company).

hydrogen. The residual gases can then be pumped to holdup tanks to permit decay, following which the gases can be passed through an absolute filter and to a stack for discharge to the atmosphere. After 3 months the radio-iodine has decayed, and the remaining radioactivity is due mainly to ^{85}Kr, which can be released to the atmosphere or, in special circumstances, removed for off-site disposal.

Other sources of PWR gaseous wastes are leaks in the primary system. Small quantities of primary coolant can leak directly to containment and be vented to the atmosphere from time to time when the containment building is purged to permit access by personnel. Small leaks can also develop in the heat exchanger, causing quantities of primary coolant to pass to the secondary loop. When this happens, radioactive gases can pass to the atmosphere without treatment via the boiler blowdown.

In the BWR, the fission and radioactive noble gases boil off with the steam, pass through the turbine, and then go into the condenser from which they are removed by the air ejector. Leakage of air into the condenser occurs because it is normally operated under vacuum. Small quantities of air leak into it and become mixed with the gaseous fission and activation products.

The gaseous wastes from many BWR's are held for 30 min to permit decay of the short-lived nuclides before being discharged to the atmosphere, but means are available by which longer holdup can be provided. The gases can be stored on activated charcoal to allow for decay of all but ^{85}Kr which,

TABLE 9-4

COMPOSITION OF GASEOUS WASTES FROM
DRESDEN I[a]

Gaseous waste	Rate of emission (μCi/sec)
Nuclide	
^{3}H	9×10^{-3}
Noble gases	
^{85m}Kr	391
^{85}Kr	0.12
^{87}Kr	978
^{88}Kr	771
^{133}Xe	483
^{135}Xe	1200
^{138}Xe	3270
Particulates and radioiodine	
^{58}Co	26×10^{-6}
^{60}Co	25×10^{-6}
^{89}Sr	970×10^{-6}
^{90}Sr	5×10^{-6}
^{137}Cs	35×10^{-6}
^{140}Ba	430×10^{-6}
^{131}I	920×10^{-6}

[a] Kahn *et al.* (1971b).

as in the PWR, can be released to the atmosphere under controlled conditions or removed for offsite storage or disposal. The ^{85}Kr from such a system would be contained in 500–1200 standard cubic feet of gas per year (Kent *et al.*, 1971). The radioactive gaseous wastes include mainly short-lived noble gases, as shown in Table 9-4. Stigall *et al.* (1971) have concluded that the inherent barriers to ^{131}I transport through a BWR system function with a high degree of efficiency and that ^{131}I is not a significant nuclide in the gaseous releases from this type of reactor.

RADIATION EXPOSURE OF THE NEARBY PUBLIC

The release of liquid wastes from civilian power reactors has resulted in insignificant exposure of the general population. Table 9–5 summarizes the

TABLE 9-5

RELEASES OF RADIOACTIVITY IN LIQUID EFFLUENTS FROM U.S. POWER REACTORS (1970)[a]

| Facility | Mixed fission and corrosion products | | | Tritium | |
	Released (Ci)	Concentration limit (10^{-7} μCi/ ml)	Percent of limit[b]	Released (Ci)	Percent of MPC
Oyster Creek	18.5	1	71	22	0.0028
Dresden I	8.2	1	25	5	0.0005
Nine Mile Point	28	1	21	20	0.0015
Dresden II	13	1	15	31	0.0012
Humboldt Bay	2.4	1	15	<7	<0.0014
San Onofre	7.6	1	12	4800	0.26
Ginna	10	3	5	110	0.005
Big Rock Point	4.7	15	3.1	54	0.018
Conn. Yankee	6.7	3.6	2.8	7400	0.38
Saxton	0.012	1	0.33	10	0.01
Indian Point I	7.8	70	0.28	410	0.03
Peach Bottom	0.006	1	0.15	<50	<0.05
Yankee Rowe	0.034	1	0.14	1500	0.21
La Crosse	6.4	300	0.067	20	0.002

[a] Data from AEC.
[b] In most cases the facility operators chose to use 10^{-7} μCi/ml as the permissible concentration limit. This is the AEC limit for unidentified mixed fission and corrosion. With isotopic analysis, the limits could have been raised substantially, with a corresponding reduction in the "percent of limit."

liquid wastes discharged to the environment by the 13 United States civilian power reactors that operated during 1970. In all cases the quantities released are a very small percentage of the amounts permitted under AEC rules. Moreover, because of dilution that takes place in the general environment, the actual dose to the general population is very much less than these percentages would indicate. In the case of the Yankee Atomic Power station, a PWR that has been in operation since August, 1960 and which was surveyed about 10 years later, the EPA (Kahn *et al.*, 1971a) found only minimal increases in the radiation background. The dose rate at the nearest point of habitation (0.4 km from the plant) was about 5 mR/year. At 1 km, the dose rate was estimated to be 2.5 ± 2.5 mR/year. A similar conclusion was reported following a study of the Dresden I reactor in Illinois by the same agency in which it was found that "exposure to the surrounding population through consumption of food and water from radionuclides released at Dresden was not measurable" (Kahn *et al.*, 1971b). This was a particularly significant conclusion in view of the fact that this BWR had been in operation for 6 years and a considerable amount of fuel cladding development had been undertaken in the course of which severe fuel cladding damage was known to have occurred.

A study has also been made of the environs of the PWR (Unit No. I) at Indian Point on the east bank of the Hudson River (Lentsch *et al.*, 1972). During much of its operating history, this reactor had a leak from the primary system into the secondary (steam) system with the result that low-level contamination found its way directly to the river via boiler blowdown. During the period of this study, this one reactor released a significant fraction of the total amount of radioactivity discharged to receiving waters by all reactors in the United States. Nevertheless, the radioactivity from this reactor could only be detected in the Hudson River by highly sensitive measuring techniques, and it was estimated that the maximum dose to humans would be 0.05 mrem/year, assuming a person consumed 11 kg/year of fish caught near the outfall.

Despite the fact that large-scale testing of nuclear weapons had been terminated in 1962 (see Chapter 15), the radioactivity of the Hudson River was dominated through 1970 by radionuclides introduced from fallout, as can be seen from Table 9–6, which compares the annual discharges from Indian Point I to fallout from weapons testing in past years. The presence of ^{137}Cs from such fallout is so predominant that the reactor contribution could only be detected using the isotope [134]Cs as a tag. This nuclide is formed in reactors as a product of neutron capture by [133]Cs, the daughter of the fission product [133]Xe. The isotope [134]Cs is not normally present in nuclear weapons fallout.

TABLE 9–6

COMPARISON OF DISCHARGES OF THE PRINCIPAL RADIONUCLIDES FROM INDIAN POINT I
AND FALLOUT FROM WEAPONS TESTS (MEASURED IN CURIES)[a]

	^{90}Sr	^{54}Mn	^{137}Cs	^{3}H (tritium)
Annual discharge from Indian Point I (1968)	0.008	5.8	2.3	810
Fallout from weapons				
a. On Hudson River watershed (35,000 km²)	825	1236	1320	205,000
b. On Hudson River surface	3.7	5.5	5.9	920
c. On mixing zone of river, 16 km above and below plant	0.58	0.58	0.93	144

[a] For purposes of comparison, the year of heaviest fallout (1963) is compared to the year of maximum reactor discharge (1968). These data have been assembled from several sources.

The pressurized water reactors of the United States Navy have discharged relatively small amounts of radioactivity into the various harbors in which they have operated and been serviced. As of the end of 1969, when there were 87 nuclear-powered submarines and 4 nuclear-powered surface ships in operation, the total annual discharges averaged about 0.2 Ci/year in 13 harbors in the United States and a number of harbors overseas (Mangeno and Miles, 1970).

The atmospheric releases of radioactive rare gases from operating reactors from 1967 through 1970 are given in Table 9–7 in which it is seen that for all reactor types except the BWR's the emissions were less than 1% of the limits specified. One BWR (Humboldt Bay) discharged as much as 57% of the AEC permissible limit, but at no time were the limits exceeded. The relatively high releases at Humboldt Bay were due originally to defective stainless-steel fuel cladding that has since been replaced by zircalloy. However, exposed pieces of UO_2 were evidently dispersed within the reactor at the time of failure, resulting in relatively high release rates even after the defective fuel was replaced. This has also been true, but to a lesser extent, at Dresden I and Big Rock Point.

In the PWR and other closed-cycle reactors, the fission and activation gases normally remain within the primary coolant long enough to permit decay of all but the longer lived gases, resulting in the relatively small atmospheric releases noted. Moreover, a provision is usually made for radioactive gases that are removed from the coolant in the course of purification to be stored in tanks to permit all but the [85]Kr to decay. However, it is possible for dissolved gases in the PWR primary system to pass via boiler leaks directly to the secondary system, and when this happens, shorter-lived gases may pass to the atmosphere by means of the boiler breakdown. However, in one plant in which this is known to have occurred, the atmospheric releases were, nevertheless, lower than 0.01% of the AEC permissible amount.

The principal radionuclides released in the gaseous emissions from a BWR in which the gases are stored for 30 minutes prior to release were given in Table 9–4. Two long-lived nuclides are present, 10.7-year [85]Kr and 12-year tritium, but only in insignificant amounts. The remainder is mainly radioactive noble gases with half-lives ranging from minutes to a few days.

Studies at Dresden I (Stigall *et al.*, 1971) have shown that when the primary coolant contained 2200 pCi ^{131}I/ml (equivalent to 0.42 Ci ^{131}I in the total primary system), the ^{131}I was being discharged via the stack gases at a rate of 840 pCi/sec or about 725 μCi/day. Compared to this the discharge rate for noble gases was about 1000 Ci/day, and the ratio of the

TABLE 9–7

RELEASES OF RADIOACTIVITY IN GASEOUS EFFLUENTS FROM U.S. POWER REACTORS (1970)[a]

Facility	Noble and activation gases			Halogens and particulates		
	Released (Ci)	Permissible (Ci)	Percentage of permissible	Released (Ci)	Permissible (Ci)	Percentage of permissible
Oyster Creek	110,000	9.4×10^6	1.2	0.32	130	0.25
Dresden I	910,000	1.8×10^7	5.2	3.3	75	4.3
Nine Mile Point	9,500	2.7×10^7	0.037	<.06	63	<0.1
Dresden III	250,000	2.2×10^7	1.1	1.6	110	1.4
Humboldt Bay	540,000	1.6×10^6	34	0.35	5.6	6.2
San Onofre	4,200	5.7×10^5	0.75	<0.0001	0.8	<0.001
Ginna	10,000	3.6×10^5	2.8	0.05	1.7	3
Big Rock Point	280,000	3.0×10^7	0.88	0.13	37	0.35
Conn. Yankee	700	2.8×10^5	0.24	0.0015	0.2	0.7
Saxton	2,200	3.7×10^3	59	0.15	10	1.5
Indian Point I	1,800	5.3×10^6	0.03	0.075	7.6	1
Peach Bottom	5.7	1.9×10^5	0.003	<0.0006	0.1	<0.6
Yankee Rowe	17	6.6×10^3	0.26	Not measured		—
La Crosse	950	3.2×10^5	0.3	<0.063	1.6	<4

[a] Data from USAEC.

discharge rates of ^{131}I to noble gases was about 7.2×10^{-8}. [131]I could not be detected in the environs of the Dresden I reactor (Kahn et al., 1971b). [131]I is thus released at a rate between six and seven orders of magnitude lower than the noble gases, and a few nuclides such as [58]Co, ^{60}Co, ^{89}Sr, ^{90}Sr, [137]Ce, and [140]Ba have been reported as traces in the boiling-water reactor stack effluents. The condensation process within the BWR seems to provide an effective distillation barrier that prevents all but the noble gases from passing to the stack in significant amounts.

Various techniques are available by which one can reduce the amounts of gaseous waste discharged from boiling-water reactors. If one recombines the radiolytic hydrogen and oxygen that find their way into the condenser off-gas system, the waste gas volume is reduced appreciably with a correspondingly greater holdup time as the smaller volume of gases passes through the delay system. One can also store the off-gases on activated charcoal, as has been done in the KRB reactor in Germany. In this way, one can hold the off-gases long enough to permit all but the [85]Kr to decay, and if one wishes to, this nuclide can be stored in tanks for off-site storage. It is also possible to remove krypton by cryogenic or chemical means, but the practical feasibility of this has not yet been demonstrated (see Chapter 12).

Reactor manufacturers have proposed systems in which the liquid wastes can be so thoroughly purified that the aqueous product from the decontamination system can be returned via storage tanks for reuse in the primary system. The radioactive products, except tritium, can be removed from the aqueous wastes in a manner which permits packaging for off-site disposal. The tritium, which cannot be separated by any practical means, is returned with the treated waste water to tanks where it can be stored awaiting reuse in the reactor.

Thus far we have discussed the releases from light-water reactors under what may be described as normal conditions. Malfunctions can of course develop, not all of which are abnormal any more than it is abnormal for an incandescent bulb to burn out or an electrical appliance switch to fail. It is considered normal for occasional fuel elements to fail or for small boiler leaks to develop, and there are other mishaps that are anticipated in design and are normal to the extent that there is a high probability of occurrence during the lifetime of the reactor. The reactor is designed accordingly to deal with such malfunctions in such a way that the radioactive releases can be handled in a routine manner.

Apart from the normal malfunctions, one must also consider the possibility of more severe accidents of types that can conceivably occur with a probability that is inversely related to its severity but is not expected to occur during the lifetime of the reactor. Although the reactor designer spends a good deal of effort and expense to reduce the probability of occurrences of such accidents to the minimum, the design must also include means by which the consequences of a serious accident can be minimized and the spread of radioactivity avoided should the accident occur. The only noteworthy accident among the nuclear power plants that have operated in the United States to date was that which occurred on October 5, 1966 at the Enrico Fermi Power Plant, a sodium-cooled fast reactor, as the result of a core obstruction that caused partial melting of 2 out of 103 fuel subassemblies. Some radioactive gas escaped within the building, but no release to the environs was detected outside the building (see Chapter 16). Except for this one case, no incidents have occurred in the operation of the civilian power reactors other than those of a routine nature.

Types of Reactor Mishaps

The kinds of accidents that must be considered depend on the type of reactor being designed. The sequence of events by which one can analyze the effects of a reactor accident is summarized in Fig. 9-15, which illustrates three important types of accidents: excessive reactivity, loss of coolant, and coolant-flow stoppage. Although various combinations of the blocks can be assembled to describe many accidents resulting from malfunctions of the reactor, not all the blocks will apply in any one mishap. For example, the potential for excess reactivity can be limited by design factors to such an extent that the probability that a nuclear excursion might destroy a well-designed thermal reactor is not regarded as "credible" in current analyses of reactor hazards.

The more common types of accidents that are considered include the effects of stuck control rods, the effect of loss of pumping power for coolant, and the sudden addition of cold water to the coolant system. Each of these contingencies must be studied and its effect on the permanence of the reactor evaluated.

There have been a great many studies of the ultimate consequences of the "maximum credible reactor accident." Some of these studies were made early in the history of the World War II atomic energy program and led to great conservatism in the sites selected for the location of the early reactors. However, it must be remembered that when the first reactors were designed their performance could only be predicted from theoretical considerations and, time was required for operating experience that would demonstrate the validity of the underlying safety principles.

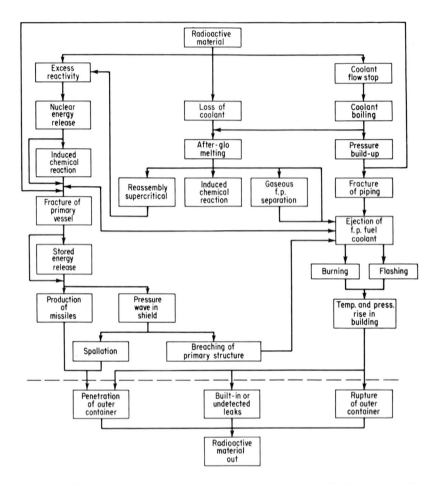

Fig. 9-15. The various factors involved in accidental releases of fission products from reactors. Not all factors will be involved in any one type of accident (Brittan and Heap, 1958).

The negative temperature coefficient of reactivity is a very important underlying characteristic of reactors designed to produce power, but this characteristic took years to fully understand. It was also clear that the opportunities for destruction of large reactors owing to the pyrophoric nature of certain materials could be prevented simply by avoiding the use of such materials. The use of uranium oxide in place of uranium metal was a major step forward in narrowing the kinds of accidents that could take place.

The early studies (Parker and Healy, 1956; Marley and Fry, 1956; USAEC, 1957a) did, however, make it clear that the enormous inventory of radioactive fission products in a reactor core could cause catastrophic consequences if they were disseminated to the environment in a mishap. The most comprehensive early study was that undertaken by the AEC in 1957, "Theoretical Possibilities and Consequences of Major Accidents in Large Nuclear Power Plants" (USAEC, 1957a). This study, widely known as WASH-740 because this was the official report number, undertook to analyze the consequences that would result from destruction and volatilization of the core of a 500 MWt reactor that had been operating for 180 days. The fission product inventory would then be 4×10^8 Ci measured 24 hr after an accident. The reactor was assumed to be located about 3 miles from a major city.

The study was performed at Brookhaven National Laboratory when power reactor technology was in its infancy and the first commercially owned nuclear generating station was not yet in operation. By defining the enormous potential consequences of dispersing a reactor core of this size, the report served a useful purpose by pinpointing the kind of research that would be needed in order to determine which radionuclides would become aerosolized in the event of an accident and the manner in which the releases could be contained. In the 1957 study, there were essentially two boundary conditions: (1) The partially volatilized reactor core is fully contained, and, therefore, essentially no contamination of the environs occurs; and (2) 50% of the core is assumed volatilized without containment, resulting in restrictions in the use of 150,000 square miles of land on which almost 4 million people live. For the latter case, the cost estimates ranged up to 7.2 billion dollars.

As a result of extensive laboratory and pilot scale investigation, and also as a result of studies of accidental core melt-downs that have occurred, it is now known that the assumptions made in the 1957 report, though properly prudent at the time in the absence of more realistic information, were unnecessarily pessimistic (Cottrell *et al.*, 1965; Nuclear Safety Information Center, 1968; and DiNunno *et al.*, 1962). Whereas WASH-740 assumed that 50% of the entire core was volatilized and escaped to the environment, experience has shown that it is the volatile fission products that are limiting. Based on the works of Parker and Creek (1958), it is now assumed that in the event of fuel failure in light-water reactors 100% of the noble gases, 50% of the halogens, and 1% of the remaining radionuclides in the fission product inventory will be released. It is also known that the radioiodines, which are the limiting radionuclides for purposes of hazard calculation, tend to plate out on surfaces within the building in which the release occurs

(Kaplan, 1964; Keilholtz and Barton, 1965; Walker, 1970) and that escape of ^{131}I to the general atmosphere would be limited to about one-half of that which is released from the core initially, or 25% of the original radioiodine inventory of this nuclide.

Thus, the essential conclusion of all studies conducted to date is that the energy released in a light-water power reactor accident is limited by the energy contained in the steam. To a lesser extent, one must be concerned with energy released from metal–water reactions, zircalloy and water in particular. The limiting radiological risk is due to the radioiodines. An important consideration is the fact that a number of organic compounds of radioiodine are formed within containment. Existing techniques for radioiodine removal are relatively ineffective for the organic compounds (Keilholtz and Barton, 1965).

Comparatively little attention has been given to the risk from nuclides other than the radioiodines in the event of a reactor accident. Raabe (1970) examined the full spectrum of radionuclides from the point of view of the quantities available for release and their behavior when inhaled. He concluded that for a core that has operated for 1 year the second most hazardous nuclide (after ^{131}I) would be [144]Ce/[144]Pr, which would deliver a dose to the lung equivalent to about 11% of the thyroid dose. Other nuclides deliver doses to their respective critical organs equivalent to less than 3% of the thyroid dose. According to Ergen (1963), the second class of nuclides, following the radioiodines in importance, would be the radioactive noble gases, which would deliver a body dose equal to about 4% of the dose to the thyroid from radioiodine exposure.

It becomes quickly apparent that certain engineered safeguards should be provided to prevent dispersion of the radioiodines and other radionuclides to the general environment. It is essential to design the reactor systems in such a way that the probabilities of mechanical failures that can result in core failure are vanishingly small. However, after having designed the reactor in such a way that failure should not occur, it is then necessary to assume that core destruction does in fact take place and that exposure to nearby populations must be controlled by engineered safeguards.

It is commonly accepted that the maximum credible mishap that could result in massive release of fission products would be the sudden structural failure of the primary coolant system. This is the "loss-of-coolant accident."

A fundamental question is whether massive failure of the piping or other components of the primary system is in fact credible (USAEC, 1970b). There have been no such accidents in 30 years of experience with nuclear systems, and, more important, massive failures are unknown in high-pressure central steam boilers which provide considerably more experience

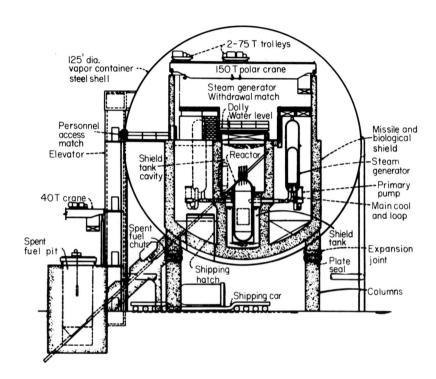

Fig. 9-16. The spherical steel containment for the PWR at Rowe, Massachusetts.

than is available in the nuclear industry. A study of some 500 boiler steam drums, designed for pressures over 600 psi, representing 4000 boiler years of operating experience, showed no failures of the steam drums themselves. Failures did occur in other parts of the high-pressure steam system, but they were not the massive type of failure that would cause a sudden release of coolant from a water reactor (Miller, 1966).

Should such a failure occur the coolant would flash to steam, causing the core to become subcritical. The fuel temperature would rapidly rise because of radioactive decay of fission products, and without a heat sink, fuel cladding failure could occur and possibly some melting of the fuel. To prevent the core from overheating, an emergency core cooling system could be provided which would flood the core and keep the temperature below that at which the cladding would fail.

An essential additional safeguard is provision of a containment building

designed to confine the steam and any entrained radioactive substances.
The containment structure must be designed to withstand a variety of
mechanical stresses, including the release of steam from the reactor system
and the impact of missiles that could conceivably be produced by such an
explosion (Gwaltney, 1969). Containment structures have become more
complex since the reactors have become larger and have been built closer
to centers of population. The first containment vessels were simple spheres,
about 125 ft in diameter, and fabricated from 1 in. welded steel plate (Fig.
9-16). Subsequent containment vessels have been constructed of concrete,
with linings fabricated from steel plate.

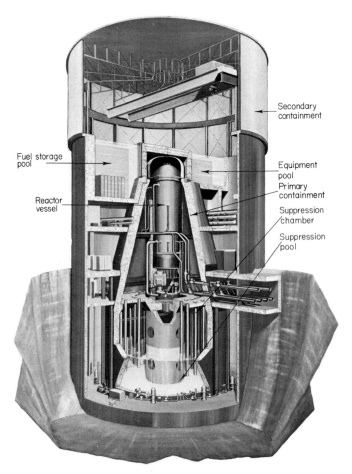

Fig. 9-17. BWR conical concrete pressure suppression containment (General Electric
Company).

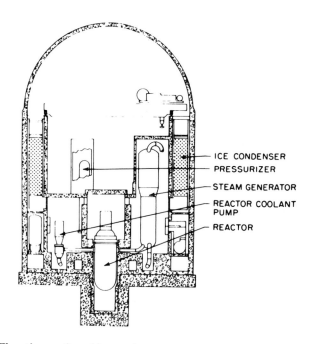

Fig. 9-18. Elevation section of ice-condenser containment system (Weems *et al.*, 1970).

In the place of a containment structure, means can be provided to rapidly condense the steam so that the pressure is greatly reduced, making it possible to use a less massive containment building. In one such system (Fig. 9-17) designed by General Electric Co., the steam would be vented into a water-filled pool that serves as a heat sink. In Fig. 9-18 is shown another type of vapor suppression system, an ice condenser designed by Westinghouse, in which the steam is condensed by venting through an ice-filled structure (Weems *et al.*, 1970). It is said that if the ice is in the form of cubes containing sodium thiosulfate, the system is additionally useful as an effective means of removing radioiodine in inorganic form. The system does not efficiently remove radioiodine in the organic form (Soldano and Ward, 1971).

Additional protection is provided by a system of sprays that wash the radioiodine and other fission products from the containment atmosphere into a sump. Addition of sodium thiosulfate to the spray adds to the efficiency of the radioiodine scavenging process (American Nuclear Society, 1971; Parsly, 1971). Alternatively, or concurrently, the atmosphere within containment can be recirculated through activated charcoal to remove the radioiodines.

Finally, the containment structure itself must be designed so that leakage is minimized. It is conventionally assumed that the containment atmosphere leaks at a rate of 0.01% per day, but techniques are available by means of which this can be reduced even further, and methods are available for testing the leak rates (Zapp, 1969).

Analysis of the so-called loss-of-coolant accident and the design of safeguards intended to mitigate the effects of such an accident occupy much of the vast amounts of material that are required to be submitted to the AEC by an applicant for a power reactor license. During the past decade the vast amount of technical information derived from laboratory study, pilot scale experiments, and theoretical analysis has been adapted to computerized methods of predicting the pressure and temperature transients involved in the various mishaps that might occur, and the results of analysis are then frequently tested in the laboratory.

One method of mitigating the consequences of a severe reactor accident would be to place the power plant underground. The reactor containment vessel would thus be placed in a rock chamber that could afford additional physical protection. Whether or not this is a feasible approach depends on local geological conditions and the economies of excavating the large chambers necessary to house the reactor containment vessel, fuel storage pools, and other components of the nuclear portion of a power plant. Rogers (1971) has estimated the additional costs of placing a plant underground at 10 million dollars for a 1000 MWe reactor.

Criteria for the Selection of Reactor Sites

Selection of a site for a nuclear power plant, or any power plant for that matter, is a complex procedure in which many constraints are imposed. These include economic factors, availability of rights of way for transmission lines, availability of a natural source of cooling water (or the feasibility of installing cooling ponds or towers), the geological and seismological characteristics of the region, the distance to the load center, and ecological or public health implications of atmospheric and liquid effluents from the plant. Until 1969, nuclear power plants were regulated exclusively by the AEC, whose authority was limited to control over the radiological hazards, and who ascertained whether the plant could be built without jeopardizing the public health and safety insofar as radiological hazards are concerned. All other environmental questions were not covered by the AEC, presumably because the Atomic Energy Act did not give them, or anyone else at that time, the required legal authority to regulate other environmental aspects such as the effects of waste heat. However, with the passage of the National Environmental Policy Act of 1969 the picture changed drastically, and any Federal agency with the authority to issue a license of any kind is now required to examine all environmental aspects of the proposed construction. In the course of this examination, a comprehensive analysis must be submitted by the applicant to the licensing agency and be reviewed by other agencies of the government concerned with the various relevant environmental effects. In this chapter we will outline only the radiological part of the reactor licensing process.

The AEC has the authority to issue licenses for nuclear power plants, but the Environmental Protection Agency can be expected to have a more and more influential role in the establishment of the standards of radioactive environmental pollution. Whether a given reactor design is suitable for a given site is evaluated under the rules promulgated in Parts 50 and 100 of Title 10 of the Code of Federal Regulations. An application to construct and operate a power reactor involves procedures that are highly complex from technical, legal, and administrative points of view. The sheer mass of highly technical reports developed in the course of a typical licensing procedure is shown in Fig. 9–19.

The first formal application to the Commission must be accompanied by (1) a Preliminary Safety Analysis Report (PSAR) and (2) the Environmental Impact Report.

The PSAR is a comprehensive document of several volumes that includes a complete description and safety assessment of the site, the hydrology, geology, meteorology, and ecology of which must be adequately documented. The analysis must include a preliminary design of the reactor with an analysis and evaluation of the design and performance of structures, systems, and components of the facility. A description of the quality assurance program to be applied in design, fabrication, construction, and testing must also be included as well as a preliminary plan for the applicant's proposed organization and personnel training program. To the extent that full technical information is not available, the PSAR must include an identification of the research and development programs that must be completed before the design can be finalized. Finally, the PSAR must include a preliminary plan for dealing with emergencies.

The PSAR is reviewed by the AEC staff and by a statutory committee of experts known as the Advisory Committee on Reactor Safeguards (ACRS). The staff and ACRS review is accompanied by many conferences with the applicant and a massive amount of highly technical and voluminous correspondence. If the reviews are favorable, and depending on the

Fig. 9-19. Mass of technical reports and hearing transcripts accumulated in the course of application for an AEC power reactor operating license. Shown are the Preliminary Safety Analysis Report, Final Safety Analysis Report, Environmental Reports, and transcripts of AEC construction license hearings (Consolidated Edison Company).

desire of the public to participate, the Commission can elect to hold a public hearing by an Atomic Safety and Licensing Board (ASLB). In recent years, public intervention at these hearings has been intensive and has delayed the licensing procedures extensively.

If the ASLB rules in favor of the application, then a license to construct the reactor is normally issued by the AEC.

Several years later, as the reactor is nearing completion, essentially the same procedure is followed for the operating license. The applicant submits a Final Safety Analysis Report (FSAR), which goes through the same reviews leading up to public hearings and a final decision as to whether the license will be issued. By late 1972, the licensing procedures had become so unwieldy that modifications were badly needed that would streamline the procedures without sacrificing their thoroughness.

The detailed reactor siting criteria used by the AEC are contained in Part 100 of Title 10 of the Code of Federal Regulations. The rationale for these criteria is given in technical information document TID-14844, "Calculation of Distance Factors for Power and Test Reactor Sites" by DiNunno *et al.* (1962), the purpose of which is to provide a uniform approach to site evaluation.

According to Part 100, three zones surround the reactor.

1. "Exclusion area" means that area surrounding the reactor in which the reactor licensee has the authority to determine all activities including exclusion or removal of personnel and property from the area. This area may be traversed by a highway, railroad, or waterway, provided these are not so close to the facility as to interfere with normal operations of the facility and appropriate and effective arrangements are made to control traffic on the highway, railroad, or waterway, in case of emergency, to protect the public health and safety. Residence within the exclusion area shall normally be prohibited. In any event, residents shall be subject to ready removal in case of necessity. Activities unrelated to operation of the reactor may be permitted in an exclusion area under appropriate limitations, provided that no significant hazards to the public health and safety will result.

2. "Low population zone" means the area immediately surrounding the exclusion area which contains residents, the total number and density of which are such that there is a reasonable probability that appropriate protective measures could be taken in their behalf in the event of a serious accident. These guides do not specify a permissible population density or total population within this zone because the situation may vary from case to case. Whether a specific number of people can, for example, be evacuated from a specific area, or instructed to take shelter, on a timely basis will depend on many factors such as location, number and size of highways, scope and extent of advance planning, and actual distribution tion of residents within the area.

3. "Population center distance" means the distance from the reactor to the nearest boundary of a densely populated center containing more than about 25,000 residents.

Using these definitions, methods are proposed (DiNunno *et al.*, 1962), for calculating the size of the required exclusion area, low population zone, and population center distance for a contemplated reactor design. The distances should be selected so that the following distances will exist following a maximum credible accident.

a. An exclusion area of such size that an individual located at any point on its boundary for 2 hr immediately following onset of the postulated fission product release would not receive a total radiation dose to the whole body in excess of 25 rem or a total radiation dose in excess of 300 rem to thyroid from iodine exposure.

b. A low population zone of such size that an individual located at any point on its outer boundary who is exposed to the radioactive cloud resulting from the postulated fission product release (during the entire period of its passage) would not receive a total radiation dose to the whole body in excess of 25 rem or a total radiation dose in excess of 300 rem to the thyroid from iodine exposure.

c. A population center distance of at least one and one-third times the distance from the reactor to the outer boundary of the low population zone. In applying this guide, due consideration should be given to the population distribution within the population center.

The Commission was careful to note that use of the 25 rem dose for whole-body exposure and 300 rem for thyroid exposure is not to be construed, generally, as permissible doses to the public in the event of accidents. "Rather, this 25 rem whole body value and the 300 rem thyroid value have been set forth in these guides as reference values, which can be used in the evaluation of reactor sites with respect to potential reactor accidents of exceedingly low probability of occurrence, and low risk of public exposure to radiation."

Document TID-14844 defines the basic ground rules for estimating the amount of radioiodine available for leakage to the environment and the emergency doses to the surrounding population that should serve as the design criteria. Ever since publication of these criteria, reactor designers have been attempting to devise safeguards, such as those previously described, that can be used as a substitute for distance. Ideally, one would hope that the safeguards would in time be regarded as sufficiently reliable to permit reactors to be built in populated zones.

An example of the progress that has been made is shown in Table 9–8 in which are summarized the calculated off-site doses under a variety of circumstances following loss of coolant in a large PWR. Whereas for design purposes 10CFR100 suggests a thyroid dose limit of 300 rem in the low population zone following a loss-of-coolant accident, the calculated thyroid dose would be less than 1 rem if a reasonable combination of safeguards are in operation.

The assumption on which Table 9–8 is based is that the radioiodine released to the environment is that which is contained in the gap between the fuel and the cladding. The more pessimistic assumption of TID-14844,

TABLE 9–8

Summary of Offsite Exposure Calculations for Loss of Coolant Accident[a]

Dose	2-hr Exposure at 520 m (minimum exclusion radius)	Total exposure at 1100 m (minimum low population zone radius)
I. Thyroid dose (based on zero to 5% of airborne as CH_3I)		
Containment leakage terminated in 1 min by isolation valve seal water system	0.7 rem	0.36 rem
Gap release[b]—continuous leakage with 2 spray pumps and 5 fan filters operating	0.8–1.45 rem	0.42–0.68 rem
Gap release[b]—continuous leakage with one spray pump and 3 fan filters operating	1.7–2.7 rem	0.85–1.4 rem
Gap release[b]—continuous leakage with 5 fan filters operating	8.8 rem	4.8 rem
Gap release[b]—continuous leakage with 2 spray pumps	0.95–6.7 rem	0.55–13.9 rem
10CFR100 suggested limit	300 rem	300 rem
II. Whole-body dose		
Containment leakage terminated in one minute by isolation valve seal water system	<1 mrem	<1 mrem
Gap release—continuous leakage	18 mrem	68 mrem
TID-14844 release—continuous leakage	3.8 rem	4.9 rem
10CFR100 suggested limit	25 rem	25 rem

[a] USAEC Docket 50, Exhibit B.
[b] TID-14844 initial iodine leakage inventory of 25% of core equilibrium quantity will result in thyroid dose 10 times value shown.

that 25% of the core inventory of radioiodine is released to the containment structure, would result in dosages approximately 10 times those shown. Despite the severity of a loss-of-coolant accident, the whole-body dose to nearby inhabitants would be a few millirem, well within the limits for normal reactor operation.

The Consequences of a Nuclear Attack on a Power Reactor

The question frequently arises as to what additional risks would be involved if a nuclear power plant were the target of nuclear attack. This question has been analyzed by Chester and Chester (1970) who concluded that the fission product inventory in a power reactor would not cause significantly increased casualities unless the fission products were deposited in sufficient quantity outside the depopulated target zone. They calculated that failure of the pressure vessel would require a blast over-pressure of 200 psi sec, which would result from a 100 kiloton weapon detonated less than 200 ft from the reactor or a 10 megaton weapon detonated less than 2000 ft away. A 10 megaton weapon detonated about 15 miles from the reactor would be expected to rupture the containment structure and the primary coolant lines but not the pressure vessel. The reactor site would thus be within a large devastated area (see Chapter 13), and although it is hard to argue that the additional radioactivity from these circumstances would not be a significant additional problem, it probably can be accepted that a nuclear attack so saturates the target area with problems that the additional radioactivity from the reactor would simply contribute to what is already an "overkill." When a 1000 MWe reactor has been operating for about 3 years, the radioactivity of the core is about equal to the radioactivity released by a 1 megaton thermonuclear weapon. Using information that was available to them, Chester and Chester calculated that for a 50% probability of a weapon landing within the distance required to rupture the pressure vessel, it would be necessary to fire about 20 one-megaton missiles or about 75 one-hundred-kiloton missiles at the reactor.

Nuclear Rockets

The development of a nuclear rocket is a major objective of programs in space technology and atomic energy in the United States, but its future is uncertain for lack of appropriations for research and development. In the opinion of many students of the subject, the advantages of nuclear energy over chemical power for lengthy manned missions are such that the nuclear rocket is essential for the successful exploration of deep space.

The first development efforts began in early 1957 (Joint Committee on Atomic Energy, 1962), and in 1959 ground tests began in Nevada. The fuel zone of the reactor that has been under development consists of a thick-walled cylinder of graphite loaded with ^{235}U, filled with heavy water, and surrounded by a graphite reflector (Fig. 9-20). Rocket power would

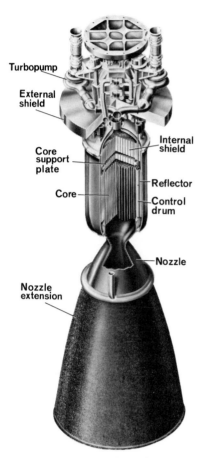

Turbopump

External
shield

Core
support
plate

Core

Internal
shield

Reflector

Control
drum

Nozzle

Nozzle
extension

Fig. 9-20. Cutaway view of the NERVA flight engine mock-up (National Aeronautics and Space Administration).

be provided by heating a propellant such as hydrogen in the reactor core and ejecting the gas through jet nozzles.

The nuclear rocket program is known as Project NERVA and has the objective of ground testing a full scale (75,000 lb thrust, 825 sec specific impulse) engine by the late 1970's, with a flight readiness by 1978 (JCAE, 1970a). Whether these goals will be achieved remains to be seen in view of the spasmodic nature of support for the program.

Chapter 10

Miscellaneous Sources of Exposure to Radioactive Substances

A wide variety of radioactive substances are used in industry, medicine, and research. In addition, there is a possibility of exposure to radioactive materials in the course of transportation. This discussion will be divided into three sections: (1) radioactive substances that occur naturally, such as radium, uranium, and thorium; (2) the radioactive by-products of the fission process; and (3) transportation.

Radium, Uranium, and Thorium

Of the naturally occurring radioactive materials, it is ^{226}Ra that is of special interest, owing to its unique history, the many applications that were found for it, and the mischief it caused by its misuse early in this century. Radium was discovered by the Curies on December 28, 1898, but they did not produce the first 100 mg until nearly 4 years later. It is reported that about 1300 Ci of radium were sold in the United States between 1912 and 1961, when the use of artificially prepared radionuclides replaced radium to such an extent that new production had ceased (Stevens, 1963; U.S. Public Health Service, 1971). Since the vast majority of the radium-poisoning cases discussed in Chapter 2 were caused by material produced prior to 1925, the approximately 60 known deaths from radium dial painting evidently occurred at a time when the radium industry had extracted only about 200 g of this substance. This is all the more astonishing when one realizes that much of the 200 g was used in radiation therapy and did not find its way into luminizing compounds.

Of the 1300 g known to have been sold in the United States, only about 480 g were accounted for in 1971. The balance, 1300 to 1700 g, may have been used largely for luminous compounds, static eliminators, or other ionization sources (U.S. Public Health Service, 1971). It is possible that some is stored, perhaps unknowingly, in safe deposit boxes or attics by survivors of early radiologists (Villforth, 1964; Villforth et al., 1969).

From the time of its discovery and continuing up to the present time, radium has been a source of many problems. The tragic misuses in luminous dial painting and quack medicine have already been reported. A continuing problem is that when used for medical purposes the radium is fabricated into small sealed capsules within which pressure develops, primarily from production of radiolytic hydrogen and oxygen from the water of crystallization and secondarily from evolution of helium and radon. This has resulted in occasional ruptures of sources, necessitating expensive building decontamination. In one case, rupture of a single 50 mg radium sulfate capsule used for instrument calibration caused such extensive contamination that abandonment of a 250,000 dollar building was necessary (Gallaghar et al., 1957).

Another problem associated with radium sources, as well as sources using artifically produced γ emitters such as ^{137}Cs, is that they are of necessity small in size and are frequently lost in hospitals or in industry where they are used for industrial radiography. Every large city has had instances in which frantic searches of hospital incinerators, plumbing systems, municipal sewage treatment plants, or land fills were undertaken to find lost radium. Villforth et al. (1969) have analyzed 415 mishaps involving radium up to 1969. Sixty-five percent of the incidents involved loss of radium, mostly from medical facilities.

The use of radium dial wristwatches and other timepieces results in general irradiation of the population to a slight degree. It is estimated that 3 million radium-bearing timepieces are sold in the United States annually (U.S. Public Health Service, 1971). An additional 30,000 miscellaneous devices containing radium are estimated to be sold each year, including static eliminators, fire alarms, electron tubes, gauges, and educational products. In Europe, a number of studies have shown the per capita gonadal dose to range from 0.5 to 3.3 mrem/year (Robinson, 1968). The radium content of men's wristwatches has been reported to range from about 0.01 to 0.36 μg (Seelentag and Schmier, 1963). The International Atomic Energy Agency (1967a) recommended that the radium content of timepieces be limited to 0.15 μCi. It was estimated by that organization that the annual dose to the critical organs of an individual wearing such a wristwatch would be 4800 mrem to the skin of the wrist, 110 mrem to the lens of the eye, 30 mrem to the blood-forming tissue, and 10 mrem to the gonads.

Radium is gradually being replaced, for most purposes, by less expensive, artificially produced radionuclides. Although the price of radium remains high, artificially controlled at about 20,000 dollars per gram, there is a dwindling market for it, and ultimate disposition of radium to avoid environmental contamination will continue to present the public health officer with a continuing problem for some time to come (Pettigrew et al., 1971).

Natural uranium and ^{232}Th are used in certain consumer products but do not add significantly to the radiation exposure of the general public. Uranium has been a popular coloring agent in ceramic glazes over the years, producing colors that range from orange-red to lemon yellow. The dose to the hands in contact with these glazes ranges from 0.5 to 20 mrad/hr for glazes produced prior to 1944, but there is evidence that the dose from ceramics produced since 1944 may be less, by a factor of about 5 (Menczer, 1965).

Thorium has been used in incandescent gas mantles ever since they were invented by Wellsbach in the last century, and thorium oxides are used as glazes as well as for refractory furnace linings (Albert, 1966).

The Fission By-Products and Transuranic Nuclides

Production of reactor-produced "by-product materials" in the form of formerly unavailable radioisotopes has found wide application in civilian life. In general, these nuclides, which have come to be known loosely as "isotopes," may be employed for a wide variety of uses, depending on their chemical and physical properties.

The total quantities of the various radionuclides produced each year, the fraction of each that decays, the fraction that remains in use, and the amounts that go to waste disposal sites, are difficult to estimate. During the first 11 months of 1970, more than 1 million Ci were distributed by Oak Ridge National Laboratory, but far greater quantities must have been distributed by other producers, the records of which are not readily available. However, it is comforting that after 25 years of widespread use the system of regulation has been adequate in controlling the manner of use of these radioactive substances, at least to the extent that no known cases of injury or significant exposure to the general public has been reported.

The use of radioisotopes as tracers is particularly widespread. A few radioactive isotopes that exist in nature among the heavy elements have been used as tracers, and early in this century it was shown by von Hevesey and Paneth that radium D (^{210}Pb) could be used as an indicator of lead in studies of the solubilities of lead compounds (von Hevesey, 1966). Subsequently, with the aid of particle accelerators such as the cyclotron and the Van de Graffe generator, it became possible to produce a number of radioisotopes that do not exist in nature. These include ^{11}C, ^{32}P, ^{24}Na, ^{131}I, elements which are important in many biological processes. The value of the artifically produced radionuclides as tracers was recognized in the early 1930's and exploited at a number of laboratories. However, the amounts then available were relatively small, and the instrumentation by which their use could be exploited was nonexistent, except in the very few laboratories that were then at the forefront of the nuclear sciences.

The situation changed drastically with the discovery of fission. In addition to the copious amounts of a wide variety of radionuclides that became available as by-products of the fission process, the intense neutron fluxes available in reactors made it possible to produce isotopes conveniently by neutron irradiation. The Manhattan District was quick to recognize the potential civilian value of these isotopes and took steps early in 1946 to make these by-products available to science and industry. When the Atomic Energy Commission (AEC) succeeded the Manhattan District in 1947, the utilization of isotopes for peaceful purposes became one of the major program objectives. The AEC regulations that pertain to the use of the by-product materials are contained in Part 30 of Title 10 of the Code of Federal Regulations (CFR).

The value of radioisotopes as tracers is too well known to require elaboration. Radioisotopes are now one of the standard tools of the research worker, and equipment such as scintillation counters, geiger counters, and scalers are part of the normal laboratory scenery in university, government, and industrial laboratories everywhere. The variety of radionuclides matches the variety of uses to which they are put for tracer work.

The sensitivity of the instrumentation used in laboratory tracer studies is such that in many cases the isotopes can be used in such small quantities that their use does not involve appreciable risks and thus does not require AEC licenses. The exempted quantities of some of the more useful radionuclides are given in Table 10–1.

TABLE 10-1

QUANTITIES OF CERTAIN RADIONUCLIDES THAT ARE EXEMPT
FROM LICENSING UNDER AEC REGULATIONS[a]

Radionuclide	μCi	Radionuclide	μCi
^{140}Ba	10	^{55}Fe	100
^{109}Cd	10	^{59}Fe	10
^{45}Ca	10	^{85}Kr	100
^{14}C	100	^{87}Kr	10
^{144}Ce	1	^{54}Mn	10
^{137}Cs	10	^{59}Ni	100
^{36}Cl	10	^{32}P	10
^{51}Cr	1000	^{210}Po	0.1
^{60}Co	1	^{24}Na	10
^{64}Cu	100	^{89}Sr	1
^{198}Au	100	^{90}Sr	0.1
^{3}H	1000	^{65}Zn	10
^{131}I	1		

[a] See Schedule B of Paragraph 30.71 of 10CFR30 for full list and recent changes. The values in this table were valid in early 1972.

MEDICAL USES

There are many clinical procedures in which radionuclides are used for either diagnostic or therapeutic reasons. It is estimated that during 1966 radionuclides were administered to more than 1.5 million patients. More than 300,000 doses of ^{131}I-labeled NaI were administered for the measurement of thyroid uptake (U.S. Public Health Service, 1970).

By the end of 1966, the AEC and the states authorized by the AEC to regulate the use of radioisotopes had issued 5028 licenses for medical use. The principal nuclide used in medicine is ^{131}I, followed by ^{32}P, ^{60}Co, ^{198}Au, and ^{59}Fe. In addition to the nuclides administered internally, about 157,000 patients are annually receiving treatment from ^{60}Co and ^{137}Cs teletherapy units.

It is apparent that radionuclides are used clinically on a massive scale, and there is evidence that those uses will continue to increase. The Federal government has estimated that the sales of radiopharmaceuticals is increasing at a rate of 25% per year (USAEC, 1970a), and that the whole-body population dose from these procedures will be about 3.3×10^{6} man-rem by 1980 (Minx et al., 1971).

When used as sources of γ radiation, radioisotopes are usually prepared in forms that lend themselves to convenient handling without the opportunity for environmental contamination under normal circumstances. Thus, ^{60}Co in the form of sealed γ-ray sources has been used extensively for the irradiation of cancerous tissue and also as a source of γ radiation for industrial radiography.

Although millions of curies of radionuclides are produced and sold each year, the largest portion is in the form of sealed sources used in radiography and radiotherapy. The market for sealed γ-ray sources, mainly ^{60}Co and ^{137}Cs, may increase even further in the future as applications for megacurie sources of γ radiation increase in the food processing and chemical industries. Several applications for megacurie sources have already developed in which γ radiation is used to produce plastic and concrete polymers and to accelerate a number of chemical reactions. Use of γ radiation for food sterilization has been demonstrated to be feasible, but assurances have not yet been fully provided that deleterious trace chemicals are not formed in the process (World Health Organization, 1970). Although these kinds of applications will require millions of curies of γ-emitting sources, there is ample evidence, based on past experience, that the sources can be used safely without causing environmental contamination.

RADIONUCLIDES AS SOURCES OF POWER

An interesting application for sealed sources of radioactive substances is thermoelectric conversion of the heat of radioactive decay. With the advent of miniaturized, transistorized electronic circuits, there has developed the need for small power sources in the range of 1 to 100 W that can serve data-gathering and communications circuits for prolonged periods of time without maintenance of any kind. An obvious requirement for such power sources is the program of space exploration in which it is important to maintain communications with satellites, space probes, and moon-based scientific equipment. In addition to the space applications, a number of opportunities exist for isotopic power units in such devices as weather stations located in inaccessible regions, marine buoys, oceanographic equipment, and other situations in which it is necessary to gather information and telemeter data from places where electric power is not available and where other types of packaged power are impractical because of servicing difficulties.

Radioactive power units are also being studied as sources of power for mechanical hearts (Cole *et al.*, 1970) and artificial cardiac pacemakers (Anonymous, 1969–1970).

In order for an isotope to be useful as a source of thermoelectricity, its half-life must be sufficiently long to offer the prospect of a reasonably long

TABLE 10–2

TYPICAL RADIONUCLIDE HEAT SOURCES[a]

Nuclide	Mode of decay	Half-life	Fuel form	Density (g/cm^3)	Thermal power (W/cm^3)	Fuel required (Ci/W)
^{210}Po	α	138 days	Po	9.3	1320	31.2
^{242}Cm	α	162 days	Cm_2O_3	11.75	1169	27.2
^{238}Pu	α	86.4 years	PuC	12.5	6.9	30.3
^{144}Ce	β	285 days	CeO_2	6.4	12.5	128
^{147}Pm	β	2.6 years	Pm_2O_3	6.6	1.1	2700
^{137}Cs	β	33 years	CsCl	3.9	1.27	320
^{90}Sr	β	28 years	$SrTiO_3$	4.8	0.54	153

[a] Adopted from Harvey and Morse (1961).

period of service and yet sufficiently short so that its specific activity is not too low. An additional requirement is that the particle emitted by the isotope be sufficiently energetic to yield a reasonable amount of power per curie. Seven isotopes having a desirable combination of properties are shown in Table 10–2. It is seen that the thermal power yield per curie ranges from 27.2 Ci/W for ^{242}Cm to 2700 Ci/W for ^{147}Pm. Conversion of heat to electric power is accomplished by a variety of thermoelectric techniques with foreseeable conversion efficiencies up to 10% (Anonymous, 1969).

The isotopic power generators were originally conceived as Satellite Nuclear Auxiliary Power (SNAP) units, but several terrestrial applications have since developed. In general, ^{90}Sr has been popular as a heat source for land-based devices, and ^{238}Pu and ^{210}Po have been used in the space program (Morse, 1963).

Figure 10-1 illustrates the construction of one of the earliest and most successful of the terrestrial units. It was a 5 W source, powered by 17,500 Ci of ^{90}Sr, and designed to supply power for a weather station in an isolated location in the Canadian Arctic (Fig. 10-2). In the design shown, the source is contained in a $\frac{1}{2}$ in. thick cylinder of Hastelloy C, which, in turn, is surrounded by a $1\frac{1}{4}$ in. tungsten encasement surrounded by a $\frac{1}{4}$ in. Hastelloy shell. When α emitters are used in this way, one must design for a buildup of internal pressure due to the formation of helium. If cerium oxide is used, still another source of pressure is the release of oxygen in copious amounts (Harvey and Morse, 1961).

A number of other ^{90}Sr units have been built to provide power for buoys,

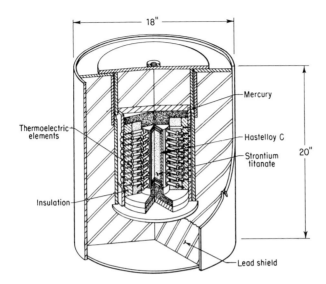

Fig. 10-1. Construction of a 5 W isotopic power generator utilizing 17,500 Ci of ^{90}Sr in the form of strontium titanate pellets. An unattended weather station utilizing such a unit was installed on Axel Heiberg Island 700 miles from the North Pole in June, 1961. It has a range of up to 1200 miles and was designed for a minimum of 2 years of operation without servicing (Martin Marietta Corporation).

Fig. 10-2. Installing the ^{90}Sr powered weather station on Axel Heiberg Island (Martin Marietta Corporation).

lighthouses, and weather stations (Shor *et al.*, 1971), and the AEC has separated several megacuries of ^{90}Sr from fission product wastes to provide a stockpile for future ^{90}Sr generators (USAEC, 1968f).

Auxiliary power from isotopic sources has become commonplace in the space field with ^{238}Pu the nuclide of choice (Prosser, 1970) for missions of 1 to 6 years duration, and ^{242}Cm and ^{210}Po are used for periods of 3 to 5 months. During the Apollo 12 moon landing, 44,500 Ci of ^{238}PuO$_2$, designed to deliver 63 We (watt electric), was left on the moon to power five scientific experiments for 1 year (Fig. 10-3).

The power levels of space power systems is increasing exponentially. The first unit, launched in 1961 to provide power for the Transit satellite (Fig. 10-4), produced 5 W of electric power and was still operating 9 years later (USAEC, 1971f). Future plans include devices that will produce

Fig. 10-3. Apollo 12 astronaut deploying SNAP 27 73 W generator powered by ^{238}Pu (National Aeronautics and Space Administration).

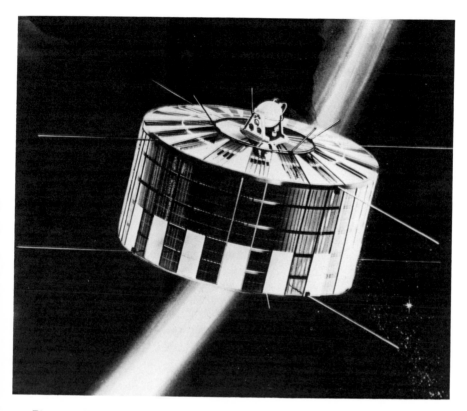

Fig. 10-4. The TRANSIT satellite, powered by isotopic power from ^{238}Pu. The SNAP unit is the white sphere atop the drum-shaped satellite (Martin Marietta Corporation).

several kilowatts by the Brayton cycle in the late 1970's (USAEC, 1968f). Higher power requirements for space exploration can best be met by reactors of special design, as was discussed in Chapter 9.

The potential environmental hazards of the isotopic power generators are different, depending on whether they are to be used in terrestrial or extraterrestrial applications. It is apparent that insofar as their terrestrial and marine applications are concerned very substantial quantities of isotopes are involved in these applications, but the risk from their use can be eliminated by massive design of the containment and by restricting the locations in which the units will be used to ones which are so isolated as to preclude violation of the containment by unauthorized human contact with the device. It is conceivable that very substantial amounts of radio-isotopes could be utilized in this way and provide a beneficial solution to the problem of disposing of fission products produced by power reactors.

Hazards Control for Satellite Nuclear Auxiliary Power Devices

The environmental risks may be considered in five stages (Branch and Connor, 1961): (1) prelaunch handling of the device; (2) a launch-pad mishap, as, for example, the explosion of the rocket; (3) a mishap during ascent of the rocket between the launch pad and the approach to orbit; (4) failure to orbit; and (5) reentry into the atmosphere after successful orbit has been achieved.

Prelaunch Handling. Prelaunch handling of the isotopic power units will involve procedures adopted elsewhere in the atomic energy industry, such as in the handling of spent fuel and teletherapy units. Since the nature of the mission requires extremely rugged construction, there should be little concern over the possibility that the radionuclides might escape into the environment as the result of mishandling during the period when the power unit is being transported to the launch pad.

Launch-Pad Aborts. During the launching procedure, a fuel-tank rupture or malfunction in the guidance system might result in a total destruction of the vehicle. The danger of rupture or melting of the capsule containing the radioactive isotopes is avoided by incorporating the pellets of radioactive compounds in blocks of metal such as molybdenum or inconel.

Controlled Flight Phase. After the vehicle leaves the launch pad it is subject to the control of the range safety officer for several thousand miles. Launches from the United States are made either at Cape Kennedy in Florida or Port Arguello in California, both locations having been chosen because long stretches of open water are available in the down-range direction. Any deviation from the desired trajectory during the period of controlled flight would result in destruction of the vehicle by the range safety officer. If the capsules for the isotopic power sources can be designed to resist the effects of a ground explosion and fire, they presumably can survive the explosion aloft which will be less violent. The sources would fall into the ocean and sink to the bottom. If they are designed to resist impact against the water and the corrosive action of salt-water immersion, the radioisotopes will decay at the bottom of the sea. It is conceivable that the capsule might be dredged up in some manner, but this would not be likely in view of the large areas involved.

Near-Orbital Failure. As the launch velocity increases, the down-range impact distance likewise increases until, as shown in Fig. 10-5, an orbital velocity of 25,600 fps is reached. Between the time the vehicle leaves

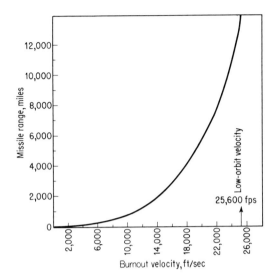

Fig. 10-5. Distance of impact from launch pad versus velocity of the vehicle at burnout. Orbit will be achieved if the velocity reaches 25,600 fps. If this velocity is not attained, impact will occur at down-range distances that can be calculated from the known burnout velocity (Branch and Connor, 1961).

control of the range safety officer and the moment it achieves a satisfactory orbit, there is a period when it is beyond the earth's atmosphere traveling at speeds close to orbital velocity. Until this point, an object of the design of the isotopic power unit is to maintain its integrity so as to prevent dispersion of the radioisotopes. However, if the unit failed to achieve its orbit just short of the required escape velocity, it might fall onto land under conditions that would be a hazard to nearby inhabitants. It would not be possible to predict exactly where the landing would occur, and this would lead to an undesirable situation in which substantial amounts of radioactive substances might be deposited in some unknown location.

Three distinct types of behavior can be described following reentry of an isotopic power unit into the earth's atmosphere: (1) The unit could be designed to withstand reentry heating and impact on the earth's surface, thereby retaining the integrity of the encapsulated source of radioactivity; (2) the unit could be designed to be dispersed as a fume into the upper atmosphere by reentry heating; or (3) because of failure of design, only partial burn-up is achieved or the capsule fails on impact at the earth's surface.

The first alternative was originally excluded because radionuclides that emit penetrating radiation, such as ^{90}Sr (bremsstrahlung) and ^{137}Cs (γ), were being considered, and it would be hazardous to allow impact of the sources in situations where people might be exposed unknowingly to the radiation. Moreover, it first seemed that one could be more certain of designing a source that would disperse as a fume than a source that would retain its integrity if, for example, it impacted on a granite surface. Thus, the first SNAP units (about 1955) were designed subject to conflicting requirements. One was required to maintain the integrity of the isotopic power capsule under such severe conditions as launch-pad failure, but it was likewise required that the capsule should disintegrate and disperse the radionuclide on reentry into the earth's atmosphere. The first capsules were designed so that if they reentered at speeds above 24,000 fps and remained in the upper atmosphere for at least 300 sec, they would burn up completely before reaching 100,000 ft in altitude (Branch and Connor, 1961).

At this point in its trajectory, the fate of the rocket is very sensitive to small changes in velocity. If the velocity is less than 24,000 fps, the rocket will descend in distances within 10,000 miles from the launching point, well within the expanse of open water on both the Atlantic and Pacific missile ranges. At some point above 24,000 fps but less than the orbital velocity of 25,600 fps, the rocket would pass beyond open water and might fall on land. Hence, burn-up is desirable for rockets that do not quite achieve their orbits.

Within this narrow range of velocities the burn-up behavior is sensitive to the angle of reentry into the atmosphere and the geometry of the reentering vehicle. Burn-up behavior can only be determined experimentally to a limited extent; for the most part the behavior is calculated from complex equations having both a theoretical and an empirical basis. An example of the calculated fate of a SNAP device that did not quite achieve its orbit is shown in Fig. 10-6 (Hagis et al., 1961). We will see in Chapter 16 that a SNAP device that reentered the atmosphere abortively in 1967 behaved as was predicted by the models that had been developed previously.

When it became apparent that the SNAP units would be fueled with α emitters such as ^{238}Pu or ^{210}Po, the possibility of total burn-up became less attractive because of uncertainties as to the hazards of inhaled α-emitting particulates. At the same time, by excluding nuclides emitting penetrating radiation, the risks associated with an intact capsule were very much reduced. When field and laboratory tests demonstrated that the capsules could survive impact and retain their integrity long enough for the nuclide

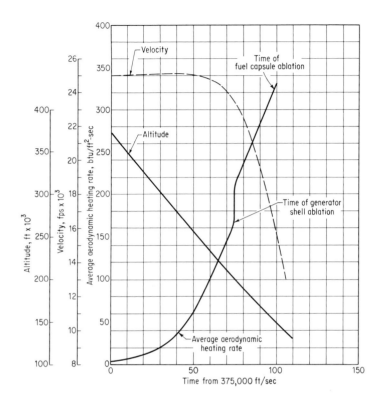

Fig. 10-6. Calculated behavior of an isotopic power unit during reentry into the atmosphere following a hypothetical near-orbital failure (Hagis *et al.*, 1961).

to decay, it was decided that the capsules should be designed to withstand both reentry heating and impact (Bustard *et al.*, 1970). A ^{238}Pu generator was aboard the Apollo 13 moon mission which failed in April, 1970. The spacecraft, with the generator in a transport cask, reentered the earth's atmosphere abortively after orbiting the moon. The SNAP unit survived reentry and is believed to be resting in the South Pacific at a depth of 20,000 ft (USAEC, 1971a).

The Artificial Heart and Pacemaker

^{238}Pu is being considered as a source of heat (Mullins and Leary, 1969–1970) for thermoelectric devices designed to supply power for an artificial heart, but their application is by no means near at hand. An artificial heart would require a power supply in the range of 1 to 7 W, which would necessitate heat rejection at a level of about 50 W. The nuclide of choice for this application is ^{238}Pu, although impurities such as ^{236}Pu would result in appreciable irradiation to the user. As presently visualized, the power supply would be implanted in the abdominal cavity, which would receive a dose estimated to be of the order of 25 mrem/hr, in part from γ radiation but also from α, n reactions within the fuel, as well as spontaneous fission. ^{147}Pm and ^{171}Tm are also being considered for this application.

Considerably closer to practicality is the use of isotopic power for cardiac pacemakers. The principle of the artificial pacemaker has been well demonstrated, and by 1970 about 10,000 persons with heart block were being assisted by implantation of electrical pacemakers powered by chemical batteries. These devices stimulate the heart contractions at a nominal rate of about 70 times per minute. The power requirement of this device is about 160 μW, which can be supplied by a ^{238}Pu source that would weigh only 100 g and would last for 10 years (Anonymous, 1969–1970).

Transportation of Radioactive Substances

The rules and regulations that govern transportation of radioactive material are complex owing to the wide variety of shipments to which they are applicable (Conlon and Pettigrew, 1971; Cornish and Simens, 1971). In the United States, the Department of Transportation (DOT) has primary responsibility for regulating the shipment of most radioactive substances, the AEC role being limited to specifying the criteria for packaging and shipping fissile and large amounts of radioactive material (Grella, 1971). The DOT is also the "Competent National Authority" for the purpose of dealing with the International Atomic Energy Agency, which has developed regulations intended for international application (Swindell, 1971; IAEA, 1970c).

Under DOT regulations, a radioactive material is defined as one that has a specific activity in excess of 0.002 μCi/g of material [Title 49, Code of Federal Regulations, Part 173 (49CFR173)]. This is a very conservative definition, as can be seen from the fact that this specific activity is only about 2.5 times that of elemental potassium. The regulations divide radioactive shipments into seven "transport groups," depending on the relative hazard of the nuclides in shipment. ^{239}Pu and ^{226}Ra are examples of nuclides included in transport group I, and certain forms of tritium are in the least hazardous category, group VII (49CFR170). The designated transport groups for some of the more important nuclides are shown in Table 10–3 (U.S. Public Health Service, 1971).

The packaging and labeling requirements are further determined by the amounts of radioactivity involved. The shipments are divided as shown in

TABLE 10-3

Transport Groups of the More Important Nuclides[a]

Radionuclide	Group	Radionuclide	Group
^{241}Am	I	^{238}Pu	I
^{140}Ba	III	^{239}Pu	I
^{14}C	IV	^{226}Ra	I
^{45}Ca	IV	^{228}Ra	I
^{144}Ce	III	^{222}Rn	II
^{60}Co	III	^{106}Ru	III
^{51}Cr	IV	^{35}S	IV
^{137}Cs	III	^{122}Sb	IV
^{55}Fe	IV	^{89}Sr	III
^{59}Fe	IV	^{90}Sr	II
^{125}I	III	Tritium	IV
^{131}I	III	Tritium activated	VII
^{85}Kr	III	^{228}Th	I
^{85}Kr (uncompressed)	VI	^{232}Th	III
Mixed fission		^{238}U	III
products	II	Uranium enriched	III
^{99}Mo	IV	Uranium depleted	III
^{22}Na	III	^{65}Zn	IV
^{24}Na	IV	^{95}Zr	III
^{237}Np	I		
^{32}P	IV		

[a] U.S. Public Health Service (1971).

Table 10-4 into types A and B, depending on the transport group and quantity (in curies).

Tests designed to assure the suitability of packaging techniques under normal operations include protection from heat and cold (tests cover the range −40 to 130°F), vibration, wetting by a 30-min water spray, free-fall impact on a hard surface, and various other stresses. It must also be demonstrated that the shipping package can retain its integrity after a free fall of 30 ft to an unyielding surface, or a fall of 40 in., landing on the upraised tip of a 6 in. steel bar. The thermal test consists of exposing the package to a temperature of 1475°F for 30 min with no allowance for artificial cooling until at least 3 hr after conclusion of the test period. A water immersion test requires that the package be submerged for not less than 8 hr under at least 3 ft of water.

The applicable regulations are highly detailed as to methods of packaging and labeling and should be consulted directly for guidance as to the current requirements.

TABLE 10-4

TYPE SHIPMENT ACCORDING TO TRANSPORT GROUP[a]

Transport group	Type A quantity (Ci)	Type B quantity (Ci)
I	0.001	20
II	0.05	20
III	3	200
IV	20	200
V	20	5,000
VI and VII	1,000	50,000
Special form	20	5,000

[a] U.S. Public Health Service (1971).

TRANSPORTATION ACCIDENT EXPERIENCE

Hundreds of thousands of shipments of radioactive materials have been made since 1949, and through 1970 a total of 160 shipments have been involved in accidents (McCluggage, 1971). The types of accidents have been classified by the AEC in accordance with a system originally proposed by Morgan *et al.* (1961) in which six classes of accidents are included as follows.

Class I Radiation Release. The vehicle has been involved in an accident or package damage is suspected. The shipment is delayed or stopped. No radioactive material is released and there is actually no loss of integrity to the package.

Class II Radiation Release. The package integrity is breached. However, there is no release of radioactive materials.

Class III Radiation Release. Radioactive material is released from the package but is confined to the vehicle.

Class IV Radiation Release. Radioactive material is released to the ground or traffic way with no runoff or aerial dispersal.

Class V Radiation Release. Radioactive material is released, resulting in aerial dispersal.

Class IV Radiation Release. Radioactive material is released and enters a watercourse, either directly or after spilling to the ground or traffic way.

Note that in Classes I–III there is no release of radioactive material beyond the vehicle in which the shipment is made. McCluggage has summarized the types of accidents experienced in transporting radioactive materials from 1949 through 1970. A total of 26 accidents have been classified as Class IV, V, or VI during this 20-year period. In only four of the accidents (Classes V and VI) was radioactive material dispersed as an aerosol or discharged into a waterway. It is significant that none of these accidents involved type B shipments, and there were no recorded instances of overexposure of the public. A total of nine claims have been made to insurance companies for property damage due to accidents involving the transport of nuclear material (Cummings, 1971). The average loss was about 1700 dollars per accident, and the most costly was 3519 dollars. All things being taken into consideration, the regulations that govern the packaging and transportation of radioactive materials seem to be highly effective in protecting the public during the past two decades.

Chapter 11

Fuel Reprocessing

When a reactor core has reached the end of its useful life, only a small percentage of the ^{235}U will have been consumed in fission and an additional small fraction of the ^{238}U will have been transmuted to ^{239}Pu and other transuranic elements. The core with its inventory of fission products and long-lived α-emitting heavy radioelements is then removed from the reactor, stored under water in the fuel storage pools associated with the reactor, and then transported (Fig. 11-1) to a fuel-reprocessing plant in which the spent fuel is chemically treated to: (1) convert the fission products into a form suitable for long-term storage (see Chapter 12); and (2) recover the remaining ^{235}U and the transuranic elements. The presence of large amounts of fission products in the irradiated fuel greatly complicates the processing procedures and makes it necessary to adopt elaborate measures to protect the operating personnel and to avoid environmental contamination.

Description of the Fuel Reprocessing Industry in the U.S.

Until recently, there were only four centers of atomic energy development in this country with facilities for reprocessing reactor cores, and all were government owned. Two of the plants, at Hanford and Oak Ridge, were built during World War II; the other two were added since that time at Savannah River, South Carolina and at the National Reactor Testing

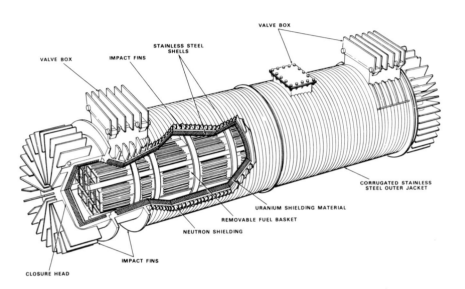

VALVE BOX

STAINLESS STEEL
SHELLS

VALVE BOX

IMPACT FINS

CORRUGATED STAINLESS
STEEL OUTER JACKET

URANIUM SHIELDING MATERIAL

REMOVABLE FUEL BASKET

NEUTRON SHIELDING

IMPACT FINS

CLOSURE HEAD

Fig. 11-1. General Electric IF 300 spent fuel cask, designed for transportation of all types of light-water-moderated fuels. During normal operation, the cask is filled with water, which provides heat transmission to the walls by natural circulation. The fins are intended for impact protection. The outer surface of the cask is cooled by air blowers not shown (General Electric Company).

Station at Idaho Falls, Idaho. However, these fuel-reprocessing plants no longer can serve the needs of the expanding nuclear energy industry. The first privately owned nuclear fuel-reprocessing plant, Nuclear Fuel Services Inc. (NFS), went into operation in 1966 in Cattaraugus County, New York with a daily processing capacity of 1 metric ton (tonne) of low-enriched uranium oxide fuel. Other privately owned plants are being constructed by General Electric Co. in Morris County, Illinois and the Allied Chemical Corporation at Barnwell, South Carolina (Unger *et al.*, 1971; USAEC, 1970a).

The national requirement for fuel-reprocessing capacity will increase rapidly in the 1970's. In 1970 only 70 tonnes of uranium were discharged from power reactors for reprocessing, but this is expected to increase to 1300 tonnes in 1975 and about 3000 tones in 1980 (USAEC, 1970a).

Economic considerations rule against construction of more than relatively few fuel-reprocessing plants, but the existing and projected plants will undoubtedly be increased in capacity as the need arises. NFS is expanding its capacity to 3 tonnes/day, and plants expected to be operating by the mid-1970's will have processing capacities as high as 5 tonnes/day.

When the fuel is removed from a reactor, it is stored for 90 days or longer to permit decay of the shorter radionuclides and is then placed into casks for shipment (Fig. 11-1) to reprocessing centers.

The fuel-reprocessing methods vary, depending on the materials from which the fuel is fabricated, but all fuel-processing plants use some form of the Purex process, a solvent extraction system using tributyl phosphate (TBP) diluted with kerosene (Etherington, 1958; Stoller and Richards, 1961). The fuel elements are prepared for reprocessing by being sheared into small lengths, after which the cladding is removed chemically. The fuel is then dissolved in nitric acid, following which a series of successive TBP extraction and stripping steps, under controlled chemical conditions, results in separation of the original solution into the required fractions, transuranic elements, uranium, plutonium, and fission products. The waste products are in the latter fraction and are produced in highly radioactive form. After being concentrated by evaporation, the wastes may be stored in underground tanks until the decay heat has subsided, following which the wastes are prepared for ultimate disposal as will be discussed in Chapter 12.

Zirconium cladding may be removed by dissolution in ammonium fluoride solutions in a process that produces about 4600 gal/ton of fuel.

Stainless steel, type 304, which is a popular cladding for uranium oxide fuels, is removed by sulfuric acid, producing about 3000 gal of solution/ton of fuel. The solutions that result from removal of the cladding contain of the order of 10^{-4} of the radioactivity in the fuel (Bruce, 1960). In the Purex process, about 1200 gal of acidic wastes are produced initially per tonne of fuel processed, but the volume may be reduced to about 55 gal/ tonne by evaporation followed by sodium hydroxide neutralization.

The intensely radioactive Purex concentrates cause radiolytic decomposition of H_2O and $NaNO_3$, which produces HNO_3 and has resulted in corrosion of the liners of the waste storage tanks. It is estimated that the Hanford liners are corroding at a rate of 10^{-4} to 10^{-5} in./year (Bruce, 1960; Schneider et al., 1971). We will see in Chapter 12 that storage of high level wastes in liquid form is regarded only as an interim procedure in the United States and that after a period of years the wastes are converted to a solid form, which facilitates long-term storage.

The potential danger of environmental contamination arising from the operation of fuel-reprocessing plants is due to the enormous amounts of radioactivity in spent fuel, as was discussed in Chapter 10. The inventories of the principal nuclides vary with the reactor design and fuel irradiation

TABLE 11-1

Composition of Typical BWR and PWR
Spent Fuels at 30,000 MW Days/
Metric Ton Uranium after 150
Days Cooling[a]

Spent fuel	Composition (per ton)
Plutonium	10 to 12 kg
Gross β,γ	5×10^6 Ci
^{90}Sr	98,000 Ci
^{85}Kr	8,400 Ci
^3H	375 Ci
^{131}I	2.2 Ci
^{129}I	0.03 Ci

[a] Runion (1970).

history, but the values shown in Table 11-1 may be taken as representative. The projected inventories for the year 2000, based on an annual requirement to process 3000 tons of fuel, can be estimated from Table 11-1 and indicate that the annual quantity of ^{90}Sr to be produced will be 3000 × 98,000 = approximately 300 MCi/year. Since the average life of ^{90}Sr is 40 years, if we assume no further growth in the nuclear industry, a sustained ^{90}Sr production of 300 MCi/year would eventually result in an accumulation of 40 × 300 MCi = 1.2×10^4 MCi ^{90}Sr. Continued growth of the nuclear industry beyond the year 2000 is likely, with even greater accumulations of ^{90}Sr as a prospect. This illustration could be repeated for other long-lived reactor-produced radionuclides.

Thus, the large fission-product inventories, together with the very nature of the fuel-reprocessing methods, present opportunities for major environmental contamination unless strict procedures are followed to avoid release of radioactive substances to the vicinity of the plant. Fortunately, the potential hazards of these operations were anticipated even before the first chemical-reprocessing plants were built during World War II, and techniques have been used from the very start that have proved effective in safeguarding the operation of these plants.

As have other segments of the atomic energy industry, the fuel-reprocessing plants have established excellent records of safety during the three decades since the first operations began. In order to protect the employees from the intense γ radioactivity of the materials being processed, it has been necessary to learn how to undertake the separation processes by remote control and remote maintenance. This has resulted in the development of a whole new technology in which the chemical processing and maintenance of equipment are carried on behind many feet of concrete. Irradiation exposures to the employees who operate these plants are minimal (Kelly and Wenstrand, 1971). Similarly, by developing careful techniques for confining the liquids and treating the gaseous effluents of the process, it has proved possible to operate these plants without environmental contamination occurring to a hazardous degree. However, it should be observed that only five such plants have operated in the United States to date and that four of the five installations were built as part of large government-owned complexes of reactors and production facilities. The cost of controlling the effluents from those plants was not a limiting factor in determining the courses of action that have been taken. When the first privately owned fuel-reprocessing plant began operation in New York State in 1966, it was able to draw on nearly 25 years of prior experience in the government-owned plants.

Sources of Radioactive Emissions

As in other components of the nuclear industry, the opportunities for environmental contamination can be divided into those that occur in normal plant operation and those that result from accidents so severe as to overwhelm the defenses against uncontrolled releases. The nature of the process is such that chemical explosions are conceivable, and since fissionable materials are processed, it is possible for critical masses to be assembled accidentally. However, as in the case of reactors, the potential hazards have been offset by conservative process design to minimize the probability of a serious accident. As in the case of reactor design, however, having achieved this objective, it is nevertheless assumed that an accident will happen, and compensating safeguards are provided to minimize releases to the environment. The safety of commercial fuel-processing plants is regulated by the AEC by a system as rigorous as that used to regulate reactor design and operation.

The most serious accident that can occur in a fuel-reprocessing plant would be an explosion that would result in dispersal of fission products, plutonium, or perhaps other transuranic radioactive elements. The explosion could occur because a critical mass had been inadvertently assembled from rupture of a pressurized vessel or pipe, or from a chemical fire or explosion. The latter opportunities are minimized by using solvents such as tributyl phosphate and kerosene that have low volatility and explosibility.

Accidental criticality is prevented by limiting the amount of fissionable material being processed. The geometrical design of process vessels and storage containers is carefully controlled and is a basic method by which the opportunity for assembling a critical mass is prevented. In some situations, neutron-absorbing components are placed in tanks in the form of rings or parallel plates (Unger *et al.*, 1971; McBride *et al.*, 1965).

Dispersal of airborne material in the event of an accident is controlled by constructing the process building as a series of shells having gradations of negative atmospheric pressure. The physical design of the processing cells is such that they can withstand the blast effects from the maximum credible explosion (Unger *et al.*, 1971).

The reprocessing process results in enormous quantities of fission products, the management of which will be discussed in Chapter 12. In this chapter, we will limit our discussion to low-level wastes that are discharged to the atmosphere or nearby surface waters.

The primary sources of gaseous wastes are the fuel-element chopping and dissolution processes (Logsdon *et al.*, 1970). Owing to the relatively long storage time before the fuel is processed, most of the ^{131}I (half-life 8.1 days) has decayed, but enough remains to require special treatment of the gases. The iodine can be removed chemically with caustic scrubbers or other means, such as by reactions with mercury or silver.

A potential long-range problem exists because of ^{129}I, a little-known nuclide with a half-life of 1.7×10^7 years. Because of its long half-life, this nuclide has received little attention. However, it is produced in fission with a yield of 1% and is, therefore, present in reprocessing wastes in relatively large quantities. Because of its long half-life, it will accumulate in the environment, become part of the iodine pool, and deliver a thyroid dose to the general population that could increase in proportion to the rate of nuclear power production. For purposes of estimating future dose projections, Russell and Hahn (1971) calculate that the world-wide inventory of ^{129}I will be 2 MCi, and if the removal efficiency of iodine treatment processes is 99%, about 20 kCi of ^{129}I will be distributed in the environment. It does not seem likely that this quantity could cause significant exposure on a world-wide basis, but Russell and Hahn have noted that this nuclide is accumulating in the immediate vicinity of a fuel-reprocessing plant. More information is needed before the long-range significance of this nuclide can be evaluated.

Other radioactive gaseous releases from a nuclear fuel-reprocessing plant include ^{85}Kr and ^{3}H. The gaseous release of ^{3}H has been small compared to the releases in liquid form. The stack releases of ^{85}Kr are substantial but have not been a source of significant exposure in the vicinity of the proces-

TABLE 11-2

RADIONUCLIDES DISCHARGED FROM NUCLEAR FUEL SERVICES
REPROCESSING PLANT, MAY–OCTOBER, 1969[a]

Nuclide	Curies discharged	Nuclide	Curies discharged
^{3}H	1.7×10^3	^{95}Zr	4.6×10^{-3}
^{106}Ru	5.2×10^1	^{60}Co	2.0×10^{-1}
^{137}Cs	8.0	^{125}Sb	5.9×10^{-1}
^{134}Cs	2.0	^{54}Mn	2.7×10^{-3}
^{90}Sr	8.3	Pu (α)	1.8×10^{-3}
^{144}Ce	1.6×10^{-1}	U (α)	3.2×10^{-2}
^{147}Pm	9.2×10^{-2}	Gross α	4.8×10^{-2}
		Gross β	8.2×10^1

[a] Magno et al. (1970a).

sing plants. According to Slansky et al. (1969), 9200 ± 2300 Ci ^{85}Kr are
released with about 43,000 ft^3 of waste gases for each ton of uranium
reprocessed. The 10.5-year half-life of ^{85}Kr and its almost total chemical
inertness are causing this nuclide to accumulate in the atmosphere. The
concentrations of ^{85}Kr in future years and the dose that will be delivered
to the world's population have been projected well into the 21st century
by Coleman and Liberace (1966). These and other findings as to the potential population dose from ^{85}Kr will be discussed in more detail in Chapter
13, in which it will be seen that provision may be necessary in the fuel-

TABLE 11-3

A SUMMARY OF ESTIMATED YEARLY DIETARY
INTAKES AND EXPOSURE TIMES OF POPULATION
NEAR NUCLEAR FUEL-REPROCESSING PLANT[a]

Media	Typical		Maximum
Milk	310	liter	380 liter
Deer	2	kg	50 kg
Fish	5.5	kg	40 kg
Drinking water	440	liter	730 liter
Air	8736	hr	8736 hr
River bank	0	hr	200 hr

[a] Shleien (1970b).

TABLE 11-4

DOSE COMMITMENT ACCRUED IN 1968 FROM ALL MAN-MADE SOURCES IN THE VICINITY OF THE NUCLEAR FUEL SERVICES PROCESSING PLANT[a]

Radio-nuclide	Critical organ	Media	Dose commitment[b] (mrem) Typical individual	Maximum individual
[90]Sr[c]	Bone	Diet[d]	77	90
		Deer	1.9	227
		Fish	16	215[e]
[137]Cs	Whole body	Diet	1.7	4.5
		Deer	1.6	200[f]
		Fish	0.2	8.6
[134]Cs	Whole body	Deer	0.4	48
[60]Co	G.I. tract	Deer	<0.01	0.2
		Fish	0.05	0.5
[106]Ru-Rh	G.I. tract	Fish	0.9	16.4
[3]H	Whole body	Diet	<0.1	0.3
		Deer	0.1	0.3
		Fish	<0.01	0.3
		Drinking water	<0.17	1.0
		Air (Inhalation and skin absorption)	<0.05	0.1
[85]Kr	Whole body	Submersion in air	0.05	0.3

[a] Shleien (1970b).

[b] Dose per 50 years or per lifetime.

[c] For ^{90}Sr, ~7% of the dose commitment to bone delivered is delivered during the first year after intake. Diet; 5.4 mrem (typical) and 6.3 mrem (maximum). Deer; 0.1 mrem (typical) and 16 mrem (maximum). Fish; 1.1 mrem (typical) and 15 mrem (maximum). For other radionuclides >90% of the dose delivered in the first year after intake.

[d] Diet intake indirectly determined through use of radionuclide value reported in milk.

[e] Represents whole fish. The majority of ^{90}Sr concentrates in bone which is usually considered inedible. Therefore, the dose commitment may represent a high estimate.

[f] From sample caught on site.

reprocessing plants of the future for cryogenic or chemical means of removing [85]Kr from the gaseous wastes.

The liquid wastes from the fuel-reprocessing plants vary enormously in radioactivity. The most radioactive solutions are those from the first extraction cycle, in which the aqueous raffinate may contain 99.9% of the fission products in the dissolved fuel. These intensely radioactive solutions produce radiolytic hydrogen and large amounts of heat for which removal processes must be provided. The management of high-level wastes will be the subject of Chapter 12.

Experience with Commercially Operated Plants

Magno *et al.* (1970a) and Logsdon and Hickey (1971) have studied the liquid-waste effluents from the nuclear fuel-reprocessing plant in New York State. The plant is designed to process 1 tonne of uranium per day. The radionuclides discharged from the plant are listed in Table 11–2, in which it is seen that tritium is the predominant nuclide followed by [106]Ru, ^{137}Cs, ^{90}Sr, and [144]Ce. Compared to the releases from nuclear power reactors, these discharges are relatively large and require correspondingly greater caution in siting a reprocessing plant, particularly with respect to sources of human food and water.

The dose to man from the discharges from this plant has been investigated by Shleien (1970b). Estimates were made of the dietary habits of the local population and the annual intakes of the "typical" and "maximum" individual (Table 11–3). Estimates were also made of the time spent at the river bank and in the open air. Samples of food, air, and water were then analyzed, and the dose commitments from 1 year of habitation were calculated. These calculated doses are given in Table 11–4. It is seen that the dose from ^{90}Sr is controlling for both the typical and maximum case, but these doses are approached by that from ^{137}Cs in the case of the individual with a high rate of deer-meat consumption. It was estimated that the "maximum" individual is represented by perhaps 48 hunters near the site during the deer season and less than 12 at other times. A slightly larger number of fishermen were involved. More significant is the fact that the data of Table 11–4 are from all sources, weapons fallout, and the plant. Shleien concluded on the basis of these data that the dose commitment "accrued in 1968 to a hypothetical typical individual is not thought to differ significantly from that for the average adult population in the rest of New York State and is attributable essentially to fallout."

Chapter 12

Management of Radioactive Wastes

Previous chapters have considered the kinds of wastes produced by the major components of the atomic energy industry and the factors that influence the behavior of wastes when they are discharged into the environment. In this chapter the more general implications of radioactive waste disposal will be discussed, and, in particular, an attempt will be made to appraise the total magnitude of the waste-disposal problem now and in the future. The specific techniques by which radioactive waste products are managed will be mentioned only in passing. For a more detailed discussion of the subject, the reader is referred to the four Proceedings of the United Nations Conferences on the Peaceful Uses of Atomic Energy (United Nations, 1956, 1958, 1965, and 1972); the various symposia held by the International Atomic Energy Agency; the transcript of the hearings on Industrial Radioactive Waste Disposal (Joint Committee on Atomic Energy, 1959c) conducted by the Joint Committee on Atomic Energy of the Congress of the United States; and texts by Mawson (1965), Glueckauf (1961), and Collins (1960).

The radioactive waste products from laboratory, medical, and industrial sources are characterized as either *low-level* wastes, which can be discharged to the environment under controlled conditions without violating applicable regulations, or *high-level* wastes, which are too radioactive to be discharged directly into the environment and, therefore, must be treated for storage under controlled conditions. Intermediate wastes may be defined as those requiring dilution or some degree of decontamination before being discharged to the environment (Belter, 1965).

The previous reviews have shown that mining, milling, refining, and fuel fabricating are not likely to produce widespread radioactive contamination, although we found in Chapter 8 that wastes from uranium mills of the Southwest United States have caused significant low-level radium contamination in nearby areas. It will undoubtedly be necessary to deal with problems of this kind from time to time; however, the waste-disposal problems between the mines and the refineries are similar to those characteristic of chemical industries, and significant environmental contamination can be avoided with common-sense procedures.

In the case of the fuel fabrication plants, the extent of environmental contamination by enriched uranium is limited by economic considerations. The value of enriched uranium is so great as to make it inconceivable that widespread environmental contamination could be produced with this substance. In future years, it is possible that plutonium and other α emitters that are more toxic than uranium will be utilized for reactors, and fuel fabrication operations may then be required to exercise additional care in the release of these materials to the environment because of their greater toxicity.

The methods for limiting the wastes from reactors was discussed in Chapter 9 and will not be further reviewed here.

Low-Level Waste Management

Low-level solid wastes, consisting of paper, wood, biological materials, scrap metal, and building materials, can be effectively controlled, although it is sometimes not a simple matter to define quantitatively the extent of radioactive contamination of a heterogeneous mix of solid materials. When scrap materials are gathered together from an area in which unsealed radioactive materials are processed or otherwise used, whether or not an item is found to be "radioactive" is apt to depend on the techniques by which one monitors the material. It can literally be like searching for a luminous wristwatch in a haystack. For economic reasons, the practice most often adopted is to assume that all scrap originating in a "radiation area" is contaminated and that it should be handled accordingly. Such wastes usually lend themselves to baling, following which they can be wrapped, labeled, and shipped to a disposal area. If the waste is combustible, it is possible to incinerate, provided the AEC regulations for gaseous wastes are not exceeded. Incineration has been shown to be a practical method for reducing the volume of combustible wastes from fuel fabrication plants (Glauberman and Loysen, 1964). Studies of a number of commercial incinerators showed that natural and enriched uranium was effectively retained in the ash and that atmospheric releases were insignificant even without the use of dust control equipment. However, incineration is not often practiced except in hospitals (Wollan *et al.*, 1971) or biological laboratories where incineration of wastes is frequently desirable for sanitary reasons.

Disposal of low-level solid wastes such as these is almost entirely by near-surface land burial, similar to the sanitary landfills used for municipal refuse (Fenimore, 1964; IAEA, 1967b). Until 1963 private industry arranged to send such wastes to AEC sites (Fig. 12–1), but in that year AEC withdrew from offering that service and began to license private companies for the operation of waste burial grounds. It is difficult to estimate the number of curies of wastes that repose in this way, but the total amount must be small, particularly since many of the nuclides are short lived. The volume of the low-level solid wastes in 1970 was estimated to be a

Fig. 12-1. The solid waste disposal area at Oak Ridge National Laboratory. This location is a major burial site in the Eastern United States. It uses about 5 acres/year (U.S. Atomic Energy Commission).

little less than 1 million ft³ but is expected to increase sixfold by 1980 (USAEC, 1970a). In 1970 there were five approved burial sites in the United States located at Richland, Washington; Beatty, Nevada; Sheffield, Illinois; Morehead, Kentucky; and West Valley, New York. In addition, 23 commercial companies were licensed to receive low-level wastes for shipment to these burial sites.

The Atlantic and Pacific Oceans and the Gulf of Mexico were used beginning in 1946 for disposal of packaged low-level wastes originating mostly from research and development facilities such as Brookhaven, where 55 gal 18 gauge steel drums were used to contain mixtures of low-level wastes and cement, which was permitted to harden before disposal into the Atlantic Ocean. The practice of sea dumping was essentially discontinued by 1970, during which year only 36 containers were dumped containing 3 Ci of miscellaneous radioactive waste. In 1962, 4087 containers containing a total of 478 Ci were handled in this way. The total ocean deposits of the United States of packaged wastes between 1946 and 1970 totaled 94,600 Ci, but most of this has long since decayed since much of the radioactivity was from the short-lived nuclides used in tracer applications and clinical practice. The practice of ocean dumping has been discontinued in the United States mainly because land burial is more economical but also because of mounting pressure against sea disposal of any kind.

For many years, the once-through cooling systems of the plutonium production reactors on the Columbia River at Richland, Washington were a major source of short- and intermediate-lived nuclides that eventually found their way into the ocean (Foster, 1959). In December, 1959, the *daily* rate of passage of the principal nuclides to the Pacific Ocean at Vancouver was 39 Ci of ^{32}P, 2000 Ci of ^{51}Cr, 27 Ci of ^{65}Zn, and 200 Ci of ^{239}Np.

Other countries have continued to practice sea disposal to some extent. The European Nuclear Energy Agency carried out two large-scale disposal operations off the coasts of Portugal and the United Kingdom in the late 1960's in which about 30,000 Ci of various nuclides were deposited in about 5000 m of water (Belter, 1971). We will see later that the British discharge large quantities of intermediate-level waste into the Irish Sea.

In 1961, the AEC undertook studies (USAEC, 1961) at two locations off the California coast where low-level wastes had been deposited in 55 gal steel drums for a number of years. An extensive sampling program at the ocean bottom was unable to detect any radioactivity above the normal background. These disposal sites were in use since 1946 in one case and 1953 in the other. At the older of the two sites, some 14,000 Ci in 22,000 packages had been dumped (Pneumo Dynamics Corporation, 1961; Brown *et al.*, 1962).

It is of interest that in 1958 a panel of marine scientists of the National Academy of Sciences considered the problem of radioactive waste disposal into the Atlantic and Gulf coastal waters and concluded (National Academy of Sciences—National Research Council, 1959) that relatively large quantities of radioactive wastes could be deposited safely in shallow coastal waters. Twenty-eight possible locations were selected that could be used for this purpose without limiting the areas for other uses. The total quantity of radioactivity that could be deposited in *any one* disposal area in any one year was estimated to be about 250 Ci of ^{90}Sr or the biological equivalent of other isotopes. Compared to actual practice, this was a rather liberal recommendation. The calculated quantities of selected radioisotopes equivalent ecologically to 250 Ci of ^{90}Sr are given in Table 12–1. These data are presented to emphasize the relatively large quantities of some of the more common radionuclides that could be dumped safely in coastal waters, according to the NAS Committee.

TABLE 12–1

QUANTITIES OF SELECTED RADIOISOTOPES EQUIVALENT TO 250 Ci OF ^{90}Sr, SHOWING THE INITIAL QUANTITIES THAT WILL DECAY TO 250 EQUIVALENT Ci ALLOWING 1 MONTH AND 1 YEAR CONTAINMENT[a,b]

	Curies		
Isotope	No containment	1 Month containment	1 Year containment
^{24}Na	5.0×10^7	10^{24}	10^{183}
^{32}P	15.5	68.6	1.1×10^9
^{35}S	3.1×10^6	3.9×10^6	5.6×10^7
^{42}K	3.1×10^6	10^{14}	10^{226}
^{45}Ca	1.6×10^5	1.8×10^5	7.5×10^5
^{59}Fe	1.2×10^3	1.9×10^3	3.3×10^5
^{60}Co	6.2×10^3	6.3×10^3	7.0×10^3
^{64}Cu	5.0×10^4	10^{21}	10^{201}
^{65}Zn	1.4×10^4	1.5×10^4	3.8×10^4
^{90}Sr	250	250	250
^{131}I	9.3×10^2	1.2×10^4	10^{16}
^{137}Cs	9.3×10^4	9.3×10^4	9.3×10^4

[a] NAS-NRC (1959).
[b] Equivalence based upon ratios of permissible sea-water concentrations.

The National Academy of Sciences—National Research Council (1962) has also examined the practicality of disposal of low-level wastes into Pacific coastal waters and selected 40 sites along the western coast of North America where relatively large quantities of packaged radioactive wastes could be dumped safely.

Management of High-Level Wastes

Until comparatively recently, most of the high-level wastes generated in the United States came from the large AEC facilities for the production of plutonium for military purposes. However, these programs were greatly curtailed in the late 1960's, and to an increasing extent the high-level waste products will originate in the future from reprocessing spent fuel from nuclear power stations. The annual production of high-level liquid wastes through the year 2000 is summarized in Table 12–2 in which it is estimated that the quantity of spent fuel to be processed will increase from about 94 metric tons in 1970 to 15,000 metric tons (tonnes)/year by the end of the century. The volume of high-level liquid wastes generated by fuel reprocessing will increase from 17,000 gal/year in 1970 to 4.6 million gal/year by the year 2000 (National Academy of Engineering, 1972), and the accumulated liquid wastes by 2000 A.D., if no further processing is undertaken, will be about 60 million gal. However, it has been AEC policy since 1969 to require that high-level liquid wastes must be converted into solid form for ultimate storage. This will be discussed further in a later section of this chapter.

STORAGE IN TANKS OR IN THE GROUND

In the United States until recently, high-level wastes have been stored in elaborate tank farms. The AEC production and research facilities at Hanford, Savannah River, and Idaho have nearly 200 underground tanks in which are stored about 80 million gal of radioactive waste liquids and sludges, ranging from 10^{-3} to 1 Ci/ml (National Academy of Engineering, 1972). Despite their rugged construction (Fig. 12-2), estimates of their life range from only 15 to 40 years (Schneider, 1969), necessitating that reserve tanks be maintained so that the contents of leaking tanks can be pumped into new tanks if necessary. This practice is obviously suitable only as an interim measure.

In exceptional cases where sufficient land has been available and there was a thorough knowledge of the geology and hydrology of the area, the ground has been used for disposal of liquid radioactive wastes. This practice

TABLE 12-2

PROJECTED FUEL-PROCESSING AND WASTE MANAGEMENT REQUIREMENTS FOR THE CIVILIAN NUCLEAR POWER PROGRAM[a]

	Calendar year			
	1970	1980	1990	2000
Installed capacity, MW(e)[b]	14,000	153,000	368,000	735,000
Electricity generated, 10^9 kWhr/year	71	1,000	2,410	4,420
Spent fuel shipping				
Number of casks shipped annually	30	1,200	6,800	9,500
Number of loaded casks in transit	1	14	60	85
Spent fuel processed, metric tons/year	94	3,500	13,500	15,000
Volume of high-level liquid waste generated[b,c]				
Annually, 10^6 gal/year	0.017	0.97	2.69	4.60
Accumulated, 10^6 gal	0.017	4.40	23.8	60.1
Volume of high-level waste, if solidified[c,d]				
Annually, 10^3 ft^3/year	0.17	9.73	26.9	46.0
Accumulated, 10^3 ft^3	0.17	44.0	238	601
Solidified waste shipping				
Number of casks shipped annually	0	3	172	477
Number of loaded casks in transit	0	1	4	10
Significant radioisotopes in waste				
Total accumulated weight, metric tons	1.8	450	2,400	6,200
Total accumulated β activity, MCi	210	18,900	85,000	209,000
Total heat-generation rate, MW	0.9	80	340	810
^{90}Sr generated annually, MCi	4.0	230	560	770
^{90}Sr accumulated, MCi	4.0	960	4,600	10,000
^{137}Cs generated annually, MCi	5.6	320	880	1,500

TABLE 12-2 (continued)

	Calendar year			
	1970	1980	1990	2000
^{137}Cs accumulated, MCi	5.6	1,300	6,500	15,600
^{129}I generated annually, Ci	2.0	110	440	670
^{129}I accumulated, Ci	2.0	480	2,700	7,600
^{85}Kr generated annually, MCi	0.6	33	90	150
^{85}Kr accumulated, MCi	0.6	124	570	1,200
^{3}H generated annually, MCi	0.04	2.1	6.2	12
^{3}H accumulated, MCi	0.04	7.3	36	90
^{238}Pu generated annually, MCi	0.0007	0.041	0.2	0.6
^{238}Pu accumulated, MCi	0.0007	1.20	8.3	31
^{239}Pu generated annually, MCi	0.00009	0.005	0.05	0.2
^{239}Pu accumulated, MCi	0.00009	0.02	0.24	1.3
^{240}Pu generated annually, MCi	0.00012	0.007	0.06	0.21
^{240}Pu accumulated, MCi	0.00012	0.04	0.4	1.9
^{241}Am generated annually, MCi	0.009	0.5	4.4	15
^{241}Am accumulated, MCi	0.009	2.3	23	120
^{243}AM generated annually, MCi	0.00021	0.01	0.1	0.5
^{243}Am accumulated, MCi	0.00021	0.23	1.5	5.2
^{244}Cm generated annually, MCi	0.13	7.4	18	23
^{244}Cm accumulated, MCi	0.13	30	140	260
Volume of cladding hulls generated				
Annually, 10^3 ft^3	0.3	8	40	90
Accumulated, 10^3 ft^3	0.3	40	320	1,030

[a] From National Academy of Engineering (1972) after Oak Ridge National Laboratory (1970).

[b] Based on an average fuel exposure of 33,000 MWd/ton, and a delay of 2 years between power generation and fuel processing.

[c] Assumes wastes concentrated to 100 gal/10,000 MWd (thermal).

[d] Assumes 1 ft^3 of solidified waste/10,000 MWd (thermal).

Fig. 12-2. Steel-lined concrete tanks for the storage of high-level wastes under construction at the AEC's Savannah River Plant. The tanks are 85 ft high and rest on steel saucers. The tank designs are complicated by the corrosive nature of the wastes and the high temperatures due to decay heat (U.S. Atomic Energy Commission).

was followed on a very large scale at Hanford, where in excess of 2.5 million Ci of fission products contained in several million gallons of liquid were discharged to cribs and trenches (Linderoth and Pearce, 1959; Pearce *et al.*, 1960). It should be noted that the Hanford Works occupies an area exceeding 500 square miles and that the resources in real estate, equipment, and human talent have been possibly unique. Similar disposal techniques have been practiced at Oak Ridge, Tennessee and Chalk River, Canada (Mawson, 1965).

Figure 12-3 illustrates the movement of radioactivity below one Hanford crib site containing 40 million gal of predominately acid waste containing 753,000 Ci of gross β activity. Ruthenium shows the greatest mobility, with strontium being the intermediate and cesium being the most stable. This distribution may be considered applicable only to the Hanford-type soils and acid or neutral wastes. In alkaline wastes, the strontium is believed to move more rapidly than cesium.

The horizontal movement of the radioactive wastes deposited in the ground at Hanford has also been studied during the past 25 years by monitoring the radioactivity of more than 200 wells surrounding the disposal area (Essig, 1971). From 60 to 90% of the observed radioactivity

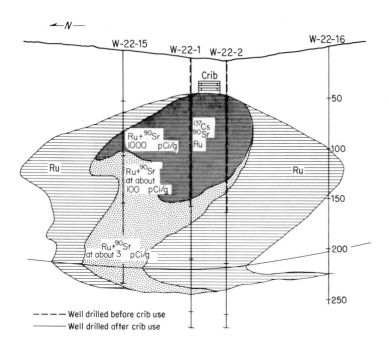

Fig. 12-3. Migration of radionuclides below a Hanford crib containing 40 million gal of acid wastes (Pearce *et al.*, 1960).

has been found to be ¹⁰⁶Ru, with the remainder due to ⁹⁹Tc, and ⁶⁰Co. The more toxic radionuclides such as ⁹⁰Sr and ¹³⁷Cs have been tightly held by the soil minerals and detected in the ground water only in the immediate vicinity of some disposal sites. Figure 12-4 shows that by 1970 the plume extended for about 10 miles. The limit of the plume in this figure is arbitrarily defined by the 0.2 pCi/ml isopleth, which is about 2% of the NCRP maximum permissible drinking water concentration for the general population. Between 1967 and 1970, the plume seems to have receded, reflecting either a change in ground water movement or, possibly, the past history of disposal.

The British have for some years followed the practice of discharging fission products from a fuel-reprocessing plant at Windscale into the Irish Sea (Dunster, 1969; Howells, 1966). Experimental releases were first made in 1952 when about 10,000 Ci of radioactive effluent were discharged over a period of about 6 months. The activity of the contaminated water ranged from 0.01 to 0.1 μCi/ml. Based on studies of the ecological behavior of the radioactivity released during this and subsequent experiments, the quantities released were gradually increased, so that during the period 1955–1965 the annual releases ranged from 3742 Ci/month to 7659 Ci/month. The composition of the effluent (Table 12–3) is seen to vary from month to month, presumably owing to changes in processing conditions. During most of this 10-year period, the principal nuclide was [106]Ru, which has a half-life of 1 year.

By studying the dietary habbits of the nearby populations in relation to radioecological factors, it was possible to derive working limits of permissible contamination of silt, sand, seaweed, and fish. The early studies identified contamination of local fish and a species of seaweed used in the preparation of laverbread as the critical pathway for human exposure (Dunster, 1958). Having defined the working limits of contamination of these vectors of human exposure, it was possible to calculate the permissible rates of release of the radioactive aqueous wastes. It was found that the [106]Ru content of seaweed was limiting and that the permissible discharge rate of [106]Ru was 7000 Ci/month. If it were not for the seaweed–

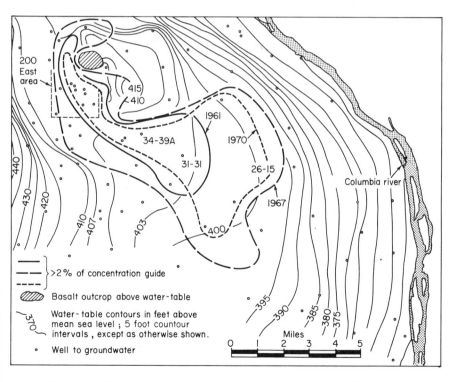

Fig. 12-4. Changes in [106]Ru groundwater plume with time (Essig, 1971).

TABLE 12-3

THE COMPOSITION OF WINDSCALE WASTES DISCHARGE RATES TO SEA[a]

Radio-nuclide	Discharge rate, Ci/month								
	1957	1958	1959	1960	1961	1962	1963	1964	1965
^{106}Ru	2218	3522	2956	3302	2095	1916	2781	1924	1752
^{103}Ru	300	492	746	964	265	153	800	100	150
^{90}Sr	137	210	129	43	41	85	46	81	97
^{89}Sr	248	72	170	82	114	42	14	16	14
^{144}Ce	215	497	583	74	180	200	116	267	288
^{91}Y and rare earths	300	567	506	83	201	125	90	90	73
^{137}Cs	310	516	165	76	91	92	31	111	97
^{95}Zr	59	210	415	196	140	78	47	1797	1479
^{95}Nb	535	510	845	523	658	356	272	1735	2803
Total β	5366	6846	7659	6461	3981	3742	4020	5055	4560
Total α	4.8	5.2	5.6	6.8	11.1	15.5	19.0	23.5	33.8

[a] Howells (1966).

laverbread pathway, fish would have been limiting, in which case the permissible release would have been over 1 million Ci/month. This was one of the earliest and most practical applications of the "critical pathways" technique, which will be discussed in Chapter 17.

SOLIDIFICATION OF HIGH-LEVEL WASTES

The AEC has for many years sponsored research and development of methods of solidifying high-level liquid wastes. The principal advantages of solidification are that the volume is reduced enormously and that the waste can be converted to a form that lends itself to more complete long-term control. Four processes have been developed: pot calcination, spray solidification, phosphate glass solidification, and fluidized bed calcination (Schneider, 1970; Schneider et al., 1971). All four processes have some characteristics in common in that the liquid wastes are heated to from 400° to 1200°C, volatilizing the water and nitrates (Schneider, 1969).

In pot calcination, a batch process that was developed at Oak Ridge National Laboratory, the liquid waste is heated and calcined to a solid state in a stainless-steel vessel that serves as the final container for the solidified waste.

Spray solidification is a continuous process developed by Battelle Northwest Laboratory. The liquid waste is pneumatically atomized at the top of a heated tower in which the atomized waste is dried and calcined as it settles into a heated vessel that melts the powdered material (800°–1200°C) and then allows the melt to pass to a pot in which it cools to a solid form. The process can be varied, depending on the chemical nature of the wastes, and the final product may be in the form of orthophosphates or semiglassy materials.

Phosphate glass solidification is also a continuous process, developed at Brookhaven National Laboratory. The liquid wastes with chemical additives are first concentrated by evaporation to a thick, aqueous phosphate slurry which is then fed to a heater in which total volatilization of water, nitrates, and other volatiles is accomplished. A molten glass is produced that is potted and cooled for permanent storage.

In fluidized bed calcination, a continuous process developed at Argonne National Laboratory, the waste is atomized into a fluidized bed of heated granular particles which become coated with the atomized waste. The particles are continuously removed and transported to storage tanks from the wastes can be taken for packaging into a suitable form.

All four processes have been demonstrated at AEC facilities, but the first commercial application will be at the Midwest Fuel Recovery Plant in Illinois, where beginning in 1972 the fluidized bed calcination process will be used to convert the high-level liquid wastes from the 1 tonne/day, fuel-reprocessing plant (McElroy *et al.*, 1971; Blasewitz *et al.*, 1971).

Having solidified the high-level waste, the question remains as to its ultimate disposition. The longer-lived constituents of the estimated national inventory of radioactive wastes by the year 2000 are listed in Table 12–2, from which it is apparent that storage arrangements must be made on a time scale without precedent in human experience. Whereas ^{90}Sr and ^{137}Ce will essentially disappear in the first thousand years, about 0.5 MCi of ^{239}Pu will remain after 30,000 years and about 250 MCi after 300,000 years. The 7600 Ci of ^{129}I, having a half-life of 16 million years, remain unchanged. For all practical purposes, the wastes must be safeguarded in perpetuity.

This problem has been actively considered for many years by a committee of the National Academy of Sciences—National Research Council, which first recommended in 1957 that the wastes be buried in natural salt deposits (NAS-NRC, 1957a). Salt beds have a number of advantageous characteristics, including shielding properties approximately equal to concrete, natural plasticity that will effectively seal the waste containers, and the highest capacity to dissipate heat of any type of rock. Based on the recommendations of this report, a 19-month test of this disposal concept

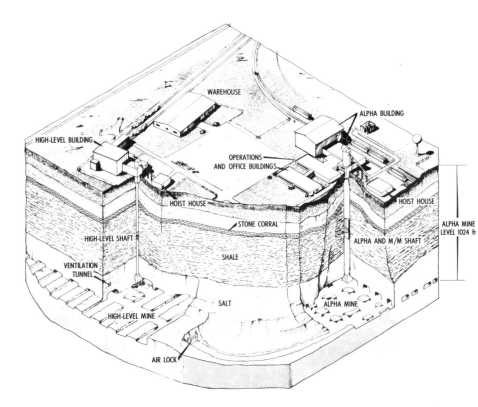

Fig. 12-5. Artist's conception of the proposed Federal repository for high- and low-level radioactive waste near Lyons, Kansas. The facility, served by a railroad spur, would handle commercial wastes from nuclear power plants through the year 2000, storing them 1000 ft underground in bedded salt formation (Oak Ridge National Laboratory).

was undertaken (Bradshaw *et al.*, 1970) in which 4 MCi contained in 21 capsules were implanted in an inactive salt mine at Lyons, Kansas. The success of this experiment resulted in a recommendation (NAS-NRC, 1970) by the same committee that the Lyons site be prepared for a long-term demonstration facility adequate to store the solidified high-level wastes from United States power reactors until the end of the century (Fig. 12-5).

Surprisingly, as can be seen from Table 12–2, the total volume of high-level wastes when solidified will not be very great. By the year 2000, these wastes will be generated in solid form at a rate of about 46,000 ft³/year, and it is estimated that by that time the total national accumulation of high-level solid wastes will be about 600,000 ft³. The heat produced by the radioactive decay of the accumulated wastes will be about 810 MW! The production of heat from high-level wastes is shown in Fig. 12-6 in which it is seen that strontium and cesium account for most of the heat production after 5 years. However, it is seen that removal of these isotopes before several years of aging would have little effect on the amount of heat generated (USAEC, 1968a).

A number of questions remain to be investigated before general acceptance of permanent storage in salt beds will be achieved. In particular, corrosion of the cannisters, together with possible flooding of the beds, has caused some skepticism as to the long-term feasibility of such storage. The need for a system of recovery of the containers in the event of some unforeseen development has been suggested and deserves serious consideration (Meyer, 1971).

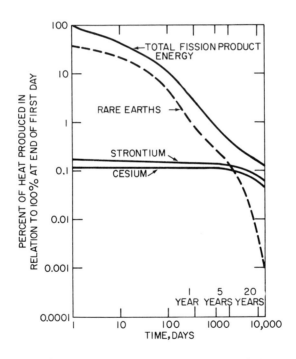

Fig. 12-6. Heat production in high-level wastes from spent fuel processing. After 5 years, strontium and cesium account for most of the heat production. Note that removal of these isotopes before several years of aging would have little effect on heat production of the remaining mixture (U.S. Atomic Energy Commission, 1968a).

WASTE DISPOSAL BY HYDRAULIC FRACTURING

The Oak Ridge National Laboratory has for some years disposed of intermediate-level wastes by injection into hydraulically produced fractures in shale at a depth of 1000 to 1500 ft. About 600,000 gal of liquid waste containing nearly 400,000 Ci of fission products, largely [137]Ce, had been disposed of in this manner by the end of 1969 (de Laguna, 1970; de Laguna et al., 1971). The technique of hydraulic fracturing for radioactive waste disposal is an outgrowth of hydraulic fracturing for oil well stimulation and is undoubtedly feasible in certain specialized situations. However, it has not developed the same degree of support as the solidification program.

The Special Problems of Tritium and Krypton-85 Disposal

Disposal of tritium and [85]Kr presents potential long-range problems that require attention. Before the advent of nuclear energy, the environmental tritium levels were in equilibrium with the rates of cosmic-ray production and decay, but we will see in Chapter 14 that the world's hydrogen pool has become contaminated with increased levels of tritium as a result of weapons testing. It is also to be expected that the tritium levels will slowly increase due to the release of this nuclide by power reactors and the reprocessing of spent fuel. Since the tritium exists in the form of water, there is little that can be done to separate it from the waste streams.

Based on estimates of the future nuclear electrical generating capacity in the western world, Jacobs (1968) has estimated that by the year 2000 the worldwide inventory of waste tritium will be 96 MCi, and that if the accumulation increases thereafter at the year 2000 rate of production (15 MCi/year), a steady-state inventory of 260 MCi will be reached, based on a nuclear power generating capacity of a little more than 1 million MW. The tritium would mix throughout the hydrosphere, with the oceans and seas representing the largest reservoirs. Jacobs has calculated that the average dose from tritium to the world's population will be less than 1 μrem by the year 2000. A study conducted by the EPA (1972) estimated the dose will reach 40 μrem. The steady-state tritium inventory, assuming its production continues at the year 2000 rate, would be reached in 2060, at which time the dose would be between 2 and 3 μrem/year, according to Jacobs. Although it is undoubtedly true that the tritium will not be uniformly mixed and that some people will receive more than the average dose, there appears to be little reason to anticipate significant levels of exposure, except possibly in the immediate vicinity of fuel-reprocessing plants if proper precautions are not taken. In any case, the buildup of tritium should be carefully monitored.

Coleman and Liberace (1966) and Hendrickson (1971) have estimated the dose from the anticipated future accumulations of [85]Kr in the atmosphere, which has been monitored (Shuping *et al.*, 1970) by analyses of dated krypton samples from commercial suppliers. By late 1970, the [85]Kr concentration had increased to 16×10^{-6} pCi/cm³. If no provisions are made for noble gas removal in fuel reprocessing and based on an estimated worldwide nuclear generating capacity of 50 billion kW in the middle of the 21st century, it is estimated that the [85]Kr concentration could increase by several orders of magnitude, possibly reaching a concentration of 0.3 pCi/cm³. The skin is the critical organ for an individual immersed in an atmosphere containing this pure β-emitting nuclide, and at a concentration of 0.3 pCi/cm³ the maximum dose received by skin cells would be 300 mrem/year (Hendrickson, 1971). The range of [85]Kr β particles in skin is 2.3 mm, and the dose varies over several orders of magnitude between the base of the dead layer of skin tissue (at a depth of 0.07 mm) and the cells (at a depth of 2.3 mm). The average dose to the skin would be very much less than the dose to the layer that receives the maximum dose. The EPA (1972) estimated that the skin dose would average 1.6 mrem/year by the year 2000, by which time the whole-body dose from [85]Kr would be 10 μrem/year.

Thus, unless future research on the biological effects of low-level radiation justifies a relaxation of present concepts of the need to achieve the "lowest practicable dose," some method of removing krypton from waste gases may be required. It is of course possible that the role of the ocean as a sink for atmospheric krypton is more important than we now think, in which case the atmospheric krypton might not accumulate to the extent now forecast. However, this seems unlikely on the basis of what is known about the solubility of noble gases in sea water (Bieri *et al.*, 1966).

Contemporary technology permits removal of up to 99.9% of radioactive noble gases from the effluents of reactors and fuel-processing plants (Nichols and Binford, 1971). The basic techniques are based on cryogenic separation, or fluorcarbon absorption that takes advantage of the relative solubilities of the noble gases in the commonly used refrigerant dichlo-difluoro methane, the trade name of which is Freon-12. Adsorption on charcoal at low temperatures is also feasible (Slansky, 1971; Keilholtz, 1971). However, since contemporary worldwide [85]Kr exposure is rising very slowly, a prudent policy would be to periodically examine the need for removal of this nuclide from waste gases, perhaps at intervals of 5 to 10 years, so that the need can be decided in the light of the accumulated body of evidence.

Chapter 13

Fallout from Nuclear Explosions
I. Short-Term Effects

In previous chapters, we have discussed the manner in which the *civilian* uses of atomic energy can cause environmental contamination, and we have seen that one need be concerned for the most part with contamination only in the immediate environs of the site at which radioactive materials are being utilized. Except in the improbable event of the release of a major fraction of the radioactivity from the core of a large reactor or chemical-reprocessing plant, the quantities of radioactive materials available for off-site dispersion are relatively small and can be controlled readily. The problem reduces to one of economics, and as long as there is a willingness to operate the civilian nuclear industry safely the techniques are there to be used.

In contrast, the use of nuclear weapons produces radioactive contamination on a scale that, when compounded with the effects of blast and fire, would create wartime problems for which at present there is no foreseeable solution. Moreover, as is well known, the peacetime testing of nuclear weapons is capable of producing worldwide contamination and was the subject of intense worldwide concern between 1954 and the signing of a test-ban agreement in 1963. In this chapter, we will first discuss the subject of nuclear and thermonuclear explosions in general and then review the principal radiological consequences of the use of such weapons in war. The subject of global fallout from weapons tests will be reviewed in Chapter 14.

Physical Aspects of Nuclear Explosions

Nuclear energy can be released from a bomb by means of either the fission or fusion process. It has become common practice to equate the explosive yields of nuclear and thermonuclear detonations to the equivalent amount of TNT. Thus, a 20 kiloton nuclear bomb is said to have an explosive yield equivalent to 20,000 tons of TNT, and a bomb having an explosive yield equivalent to 1 million tons of TNT would be said to have a yield of 1 megaton.

The pure fission (nuclear) bomb obtains its energy from either ^{235}U or ^{239}Pu. The assembly of fissionable material may be made critical by quickly joining two subcritical masses which then become supercritical or by the implosive compression of a hollow subcritical mass. Both processes may be accomplished with the aid of chemical explosives. The complete fission of about 56 g of material will produce an explosion equivalent to 1 kiloton of TNT with a thermal yield of 10^{12} cal.

The nuclear bomb can be thought of as a fast reactor in the prompt critical condition in which neutron multiplication proceeds with a generation time of about 10^{-8} sec; 99.9% of the explosive yield is developed in the last 0.07 μsec, and in order to achieve maximum yield, it is important that the shape of the supercritical mass be preserved for as long a time as possible by means of a surrounding tamper. In this way, the maximum number of fission generations can be achieved before the explosion effects change the shape of the device to the extent that criticality can no longer be maintained.

Approximately 50% of the energy from a nuclear explosion is released in the form of blast effects, 35% as thermal radiation, and the remaining 15% as ionizing radiation. In considering the total consequences of a nuclear explosion, the effects of blast and fire may be of even greater importance than the effects due to ionizing radiations, but only the latter aspect will be discussed in detail in this text. The reader is referred elsewhere (Glasstone, 1962) for a review of the effects due to blast and fire.

Of the ionizing radiations, one-third is prompt radiation produced within a few seconds after the detonation, and two-thirds, or 10% of the total energy of the explosion, is delayed ionizing radiation produced by decaying fission products and induced radionuclides. The prompt ionizing radiations, consisting of γ rays and neutrons which are released at the time of detonation, produce their effects in the same zone in which widespread blast and thermal effects also occur.

The fusion (thermonuclear) bomb utilizes the fusion reactions of light elements such as deuterium and tritium. Several different reactions may occur, but all require that the nuclei have energies which can only be obtained with the aid of temperatures of several million degrees. Since this heat can be achieved in a fission bomb, such a device may be used as a trigger for a thermonuclear explosion. The thermonuclear reactions which can take place include the following:

$$^3H + {}^2H \rightarrow {}^4He + n + 17.6 \text{ MeV}$$

$$^3H + {}^3H \rightarrow {}^4He + 2n + 11.3 \text{ MeV}$$

These reactions are sources of fast neutrons which have sufficient energy to cause fission of ^{238}U and which can, therefore, be utilized to increase the explosive yield by surrounding the fusion weapon with natural uranium. The fission of ^{238}U contributes in a major way to the overall energy released in thermonuclear weapons. As a rule of thumb, it had been general practice prior to about 1958 to assume that on the average the energy from thermonuclear bombs was derived both from fission and fusion in equal proportions. However, based on United States studies of the 1961 series of the Soviet Union, it was announced that the fission component of that series amounted to about 21% (Dunning, 1962).

The terms "clean" and "dirty" have been sometimes used to describe the relative amounts of radioactivity produced by bombs. Those in which the energy is obtained primarily from fusion yield comparatively less radioactivity than weapons whose energy is derived primarily from the fission reactions. However, even a pure fusion device will produce some radioactivity by means of neutron activation.

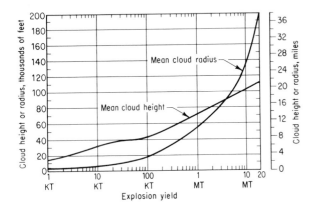

Fig. 13-1. Cloud radius and height of stabilization for various explosive yields. Average values are given. Note that bombs below 100 kilotons tend to stabilize within the troposphere (Glasstone, 1962).

Fig. 13-2. Explosion of a nuclear weapon showing the toroidal structure of the fireball shortly after its formation. The fireball did not touch the ground, but it is sufficiently low so that dust dislodged from the surface by the blast wave is being sucked into the fireball (U.S. Atomic Energy Commission).

Explosion of a nuclear or thermonuclear device produces a cloud of incandescent gas and vapor that has come to be known as the fireball, which increases in brightness for less than 1 msec after detonation, at which time it is manyfold brighter than the noonday sun. Although the brightness then begins to diminish, the fireball grows in size for several seconds, reaching a final diameter equal to $D = 460W^{2/5}$ (Glasstone, 1962), where D is in feet and W is the energy in kilotons.

In about 10 sec, when the fireball from a 1 megaton explosion has reached its maximum size, it will be about 7200 ft in diameter and will be rising at

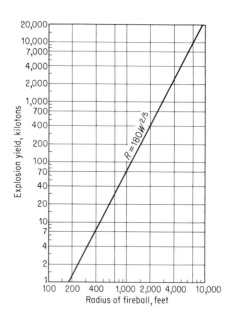

Fig. 13-3. Approximate height of burst below which a bomb of given yield will produce a fireball that touches the ground (U.S. Atomic Energy Commission).

a rate of 250 to 350 fps. In 1 min, the fireball will have cooled sufficiently so that it no longer glows, and by this time it will have risen about 4.5 miles. The rise of the cloud will cease at about 15 miles, about 7 min after detonation. The size and height of stabilization are given as a function of yield in Fig. 13-1.

Convective forces initiated by the fireball result in enormous amounts of air and debris being sucked upward. Figure 13-2, which is a photograph of a relatively small detonation over the Nevada desert, shows the way desert sands are being sucked upward into the fireball which has assumed a toroidal shape. With a detonation elsewhere, water might be included in the convected material.

Debris which enters the fireball sufficiently soon after its formation is vaporized and admixed within the fireball. Later, as the fireball cools appreciably, particles of debris will no longer be volatilized but may serve as nuclei on which condensation of the radioactive constituents of the fireball can occur. Some of these particles may be as large as grains of sand which possess sufficient mass so that fallout will occur promptly. However, if the fireball is sufficiently high off the ground so that large particles are not sucked into it, the vapors will condense into a fume in which the particles are very small and will fall more slowly.

The physical and chemical characteristics of the particles have been observed to be highly variable in other respects as well (Crocker *et al.*, 1966). Depending on the temperature–time history of the particle, the radioactivity can be coated on the surface or distributed throughout. Thus the extent to which a given explosion will produce radioactive fallout close to the place of detonation depends on the size of the explosion and its height above ground. From the point of view of the amounts of fallout produced, the worst explosions will be those in which the fireball actually touches the ground. If only 5% of a 1 megaton bomb is spent in volatilizing sand with which the fireball is in contact, about 20,000 tons of debris will be added to the fireball (Glasstone, 1962). Figure 13-3 shows the relationship between explosive yield in kilotons and the height about ground below which a detonation may be expected to touch the ground and produce heavy local fallout.

The fireball will rise to a height which is determined by the explosive yield, the height at which the detonation occurred, and the meteorological conditions existing at the time of detonation. This is illustrated in Fig. 13-1, which shows that in temperate latitudes detonations under 100 kilotons stabilize below the tropopause, and detonations in the megaton range penetrate well into the stratosphere.

Whether or not underground bursts produce fallout depends on whether the explosions penetrate the surface of the earth. Many explosions have occurred in Nevada in which all the radioactive debris has been contained below the surface. On the other hand, underground bursts that penetrate above ground have a very great potential for surface contamination.

TABLE 13–1

APPROXIMATE YIELDS OF THE PRINCIPAL NUCLIDES PER MEGATON OF FISSION

Nuclide	Half-life	MCi
^{89}Sr	53 days	20.0[a]
^{90}Sr	28 years	0.1[a]
^{95}Zr	65 days	25.0[a]
^{103}Ru	40 days	18.5[a]
^{106}Ru	1 year	0.29[a]
^{131}I	8 days	125.0[b]
^{137}Cs	30 years	0.16[a]
^{131}Ce	1 year	39.0[a]
^{144}Ce	290 days	3.7[a]

[a] Klement (1965).
[b] Knapp (1963).

The radioactive debris from a nuclear detonation originates in a number of ways. The principal source is the production of fission products in the relative amounts given by the fission-product yield curves in Fig. 9-2. The initial fission–product mixture contains more than 200 isotopes of 35 elements. Most of the isotopes are radioactive, and most of them have very short half-lives, so that the diminution in radioactivity is very rapid immediately after their formation. The yields of the principal fission products of concern to the environmentalist are listed in Table 13-1.

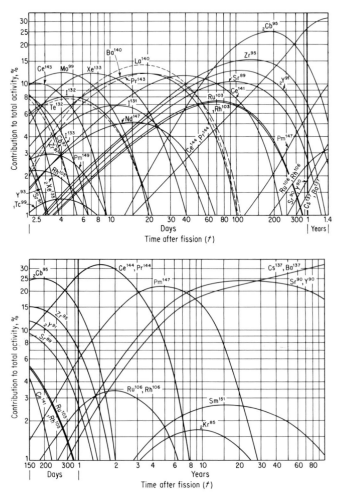

Fig. 13-4. Yields of the principal radionuclides from the slow-neutron fission of ^{235}U (Hunter and Ballou, 1951).

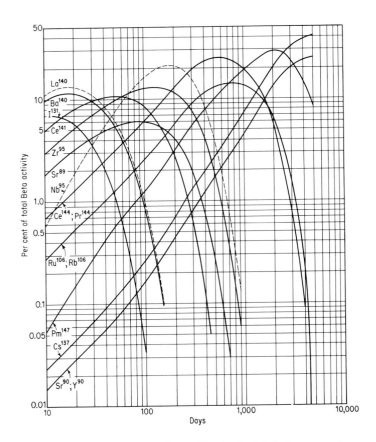

Fig. 13-5. Yields of the principal radionuclides in the debris from megaton weapons. The differences from Fig. 13-4 occur because in such weapons fission occurs from both fast and thermal neutrons and in ^{238}U and ^{239}Pu as well as ^{235}U (Hallden *et al.*, 1961).

The radioactivity A at any given time t after a nuclear explosion may be approximated if the radioactivity at unit time A_0 is known.

$$A = A_0 t^{-1.2} \qquad (13.1)$$

This familiar equation provides a satisfactory estimate of the radioactivity for periods of time less than 6 months. When the radioactivity is decaying according to this law, the levels will diminish approximately tenfold for every sevenfold increase in time since the explosion. Thus, if the radiation level is 3.0 R/hr at 4 hr after the burst and if all fallout has ceased, the radiation levels can be expected to diminish to about 0.3 R/hr at the end of 28 hr.

The composition of fission-product mixtures produced by the slow neutron fission of ^{235}U at various times after production is given in Fig. 13-4. Hallden *et al.* (1961) have reported on the isotopic composition of the debris of megaton weapons in which fast fission of ^{238}U is presumed to have occurred along with fission of ^{235}U and ^{239}Pu. Their data, which cover the period from 10 to 5000 days and which are based to some extent on radiochemical analysis of the debris from megaton tests, are given in Fig. 13-5. Although some differences are found to exist between the distributions in Figs. 13-4 and 13-5 when the various radionuclides in each case are summed, the total β activities at any given time and the rates of decay are almost identical.

The manner in which the radioactivity of fission-product mixtures may depart from the rate described by Eq. (13-1) is shown in Fig. 13-6. It is seen that the actual decay is more rapid than that described in Eq. (13-1).

In addition to fission products, a number of induced radionuclides are produced by nuclear bombs, including ^{239}Pu, which may be formed from ^{238}U in a ratio of about 4 atoms of plutonium per fission. Other radionuclides

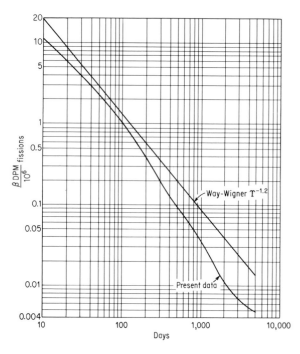

Fig. 13-6. Comparison of observed and theoretical decay of total β activity of the debris from megaton weapons (Hallden *et al.*, 1961).

TABLE 13–2

PRINCIPAL RADIONUCLIDES INDUCED IN AIR[a]

Isotope	Half-life (years)	Ci/megaton
^{3}H	12.3	<1
^{14}C	5600	3.4×10^4
^{39}A	~260	59

[a] Adopted from Klement (1959).

are produced by neutron interactions with nonradioactive elements of the bomb, the atmosphere, and in some instances with sea water or soil. Libby (1958) estimated that the interactions of neutrons with atmospheric nitrogen produces 3.2×10^{23} ^{14}C atoms per kiloton of yield. We will see in Chapter 14 that these reactions have been sufficient to produce a marked increase in the natural background of radiocarbon and tritium.

The complex neutron spectrum is capable of producing a variety of other induced activities in the atmosphere, and some of these are shown in Table 13-2.

When nuclear weapons are detonated close to the ground, an appreciable number of neutrons are available for reactions with soil constituents, and a number of radionuclides may be produced in this way. These are summarized in Table 13-3, which shows that insofar as long-lived radionuclides are concerned only ^{45}Ca and ^{55}Fe are present in significant amounts.

A number of attempts have been made to estimate the extent of induced radioactivity at Hiroshima and Nagasaki. From the known neutron spectrum of the two bombs and radiochemical analyses of soils and building

TABLE 13–3

PRINCIPAL RADIONUCLIDES INDUCED IN SOIL[a]

Isotope	Half-life	Ci/megaton
^{24}Na	15 hr	2.8×10^{11}
^{32}P	14 days	1.92×10^8
^{42}K	12 hr	3×10^{10}
^{45}Ca	152 days	4.7×10^7
^{56}Mo	2.6 hr	3.4×10^{11}
^{55}Fe	2.9 yr	1.7×10^7
^{59}Fe	46 days	2.2×10^6

[a] Adopted from Klement (1959).

materials from the two cities, it has been estimated that the infinity dose to individuals who were at the hypocenters starting 1 day after the bombing could have been 80 rad in Hiroshima and 30 rad in Nagasaki (Hashizume *et al.*, 1969). The two bombs were detonated about 1500 m above ground level. The dose at 500 and 1000 m from the hypocenters would have been 18 and 0.07% of the hypocenter doses in Hiroshima and Nagasaki, respectively. The dose to an individual who entered the Hiroshima hypocenter 1 day after the bombing and remained for 8 hr would have been 3 rad.

It is possible that under certain conditions fractionation of fission products will occur so that debris falling out in different places will be enriched or depleted in certain of the radionuclides. One reason why this may occur is that the fission products include noble gases such as xenon and krypton which, though short-lived, exist sufficiently long so that the radionuclides to which they give rise may be formed relatively late in the history of the fireball and, in some cases, after condensation of debris has already begun to take place. For example, the several isotopes of krypton and xenon decay to rubidium and cesium, respectively, and these, in turn, decay to strontium and barium. Thus, ^{90}Sr, which is the granddaughter of ^{90}Kr, may exist in less than theoretical amounts in fallout close to the site of detonation, and there may be a corresponding enrichment in ^{90}Sr in the debris that falls out at a later time. The decay scheme of ^{90}Kr is as follows:

$$\underset{(33\ sec)}{^{90}Kr} \rightarrow \underset{(2.7\ min)}{^{90}Rb} \rightarrow \underset{(28\ years)}{^{90}Sr} \rightarrow \underset{(65\ hr)}{^{90}Yr} \rightarrow {^{90}Zr}\ (stable)$$

It is seen that the ^{90}Sr will not be formed completely for several minutes after detonation.

From the data obtained during weapons tests, it is generally accepted that about 90% of the radioactivity produced in a detonation is to be found in the head of the mushroom-shaped cloud formed by the fireball as it cools. The remaining 10% is contained in the stem of the mushroom.

The particle size of the debris is highly dependent on the type of explosion. Near-surface detonations will produce large glassy masses that are highly radioactive, as well as a log normally distributed spectrum of smaller particles. Bursts that take place high in the atmosphere produce smokelike particles that remain suspended for considerable periods of time. Analyses of many samples from air-burst clouds indicated a modal particle diameter less than 1 μm with few particles greater than 20 μm in diameter (Freiling and Kay, 1965). It has been estimated on theoretical grounds that air-burst particles cannot grow to more than 0.3 μm by condensation, and that larger particles are probably the results of coalescence of smaller ones.

The manner in which the radioactivity varies with particle size is likewise dependent on the conditions at time of burst. If the particle is formed by condensation within the fireball, the radionuclides will be distributed throughout the mass of the particle, and the radioactivity would be proportional to d^3. If the radionuclides condense on inert particles, relatively late in the fireball history, the activity will be proportional to d^2. Intermediate distributions are possible owing to mixtures of the two kinds of particles or to gas inclusions in the larger particles.

The radioactivity of the debris produced in a nuclear explosion diminishes by a factor of about 20 from the first hour to the end of the first day. In addition, the radioactivity of the cloud becomes more diffuse during this period as a result of meteorological dispersion. The potential hazards of radioactive fallout, therefore, vary greatly, depending on whether the debris falls out within a few hours after the detonation or over a longer period of time. Localities receiving fallout within the first day may be subjected to lethal radiation hazards.

The extent to which "local" fallout occurs after a detonation depends on the manner in which the debris distributes itself among three major fractions. The larger particles fall out close to the site of detonation within a matter of a few hours and produce intense radiation fields.

A second fraction penetrates the stratosphere and produces worldwide fallout over a period of many months. An intermediate type of fallout is produced by the debris dispersed into the troposphere, but it does not fall out during the first day because these particles are sufficiently small to behave somewhat like aerosols and be subject to the laws of diffusion and rainout that govern small particles. The tropospheric fallout tends to be distributed in bands at the latitude of detonation, whereas the stratospheric debris distributes itself globally, as will be described in the following chapters.

From Fig. 13-1, it is seen that the debris from bombs smaller than 100 kilotons detonated in temperate latitudes tends to remain in the troposphere, whereas stratospheric injection is almost complete for detonation greater than 500 kilotons.

Most of what we know of nuclear weapons has been derived by studying the effects of nuclear explosions that have taken place in various parts of the world since the first atomic bomb was detonated on a New Mexico desert in July, 1945. In the intervening 25 years, there have been hundreds of explosions conducted by five nations, the United States, United Kingdom, Russia, France, and Communist China. The vast majority of these tests have been conducted by the United States and Russia. Table 13-4 summarizes the yields of nuclear explosions that have been announced by

TABLE 13-4

APPROXIMATE FISSION AND TOTAL YIELDS OF NUCLEAR WEAPONS TESTS
CONDUCTED IN THE ATMOSPHERE BY ALL NATIONS (YIELD IN MEGATONS)[a]

	Fission yield		Total yield	
Inclusive years	Air	Surface	Air	Surface
1945–1951	0.19	0.52	0.19	0.57
1952–1954	1	37	1	59
1955–1956	5.6	7.5	11	17
1957–1958	31	9	57	28
Subtotal	37.8	54	69.2	104.6
1961	25[b]	—	120	—
Subtotal	63	54	189	105
1962	76[b]	—	217	—
Total	139	54	406	105

[a] Federal Radiation Council (1963).
[b] The small yield tests conducted in Nevada do not contribute significantly to the worldwide distribution of ^{90}Sr to which this summary is related.

the various nations (Federal Radiation Council, 1963; Edwards, 1966). The tests have been conducted hundreds of miles above the earth's surface and thousands of feet underground.

Testing nuclear weapons in the open atmosphere began to arouse worldwide concern in the mid-1950's, and fallout became a highly emotional and controversial subject. It was unquestionably the widespread interest in the subject of nuclear fallout that alerted the world to problems of global pollution generally and led in the 1960's to such widespread public and official concern with pollutants such as DDT, lead, petroleum oil, detergents, and other products of our chemical technology.

Responding to worldwide popular pressure, the United States, United Kingdom, and Russia declared a moratorium in the fall of 1958 on further weapons testing. By that time the three nuclear powers had conducted 38 separate series of tests having a total of at least 227 detonations. The French, who were not a party to the moratorium decalaration, conducted a small series of nuclear tests on the Sahara Desert in 1960 and became the fourth of the nuclear powers. In 1961, without advance warning, the Soviet Union unilaterally broke the moratorium agreement and exploded about 50 devices. The United States responded, and the two major powers began a frenetic competition which led to additional worldwide concern, particularly in view of the greatly accelerated testing schedules. The pace of weapons testing can be seen from Fig. 13-7 in which the rapidly increasing quantities of ^{90}Sr injected into the earth's atmosphere is shown.

A nuclear weapons test-ban agreement was signed by the United States, United Kingdom, and Soviet Union early in 1963 and to this writing has succeeded in eliminating further testing in the open atmosphere by the three signatory powers. However, France did not sign this agreement, nor did China, which exploded its first nuclear bomb in October, 1964, thus becoming the fifth nuclear power.

The test-ban agreement did not rule out underground explosions which do not vent to the atmosphere. Accordingly, there began with the signing of the agreement a new era of nuclear weapons technology in which a variety of devices have been tested in underground cavities by methods which with rare exceptions have prevented atmospheric pollution by radioactive debris. At least 220 underground tests had been conducted by the United States up to 1971, of which 17 resulted in some degree of venting (Allen, 1971). Whether the underground accumulations of radioactive debris will in time prove significant as a form of environmental pollution remains to be seen. The quantities of debris so involved are huge, but objective evaluation of potential long-range risks has not been possible because little of the basic data have been made available.

Fallout of radioactive debris occurred after the 1945 New Mexico test in which a 19 kiloton device was fired from a 30 m steel tower, and some of the larger particles of radioactive debris fell out on cows grazing about 20 miles downwind and produced skin burns (Lamont, 1965). The finer particles drifted across the Middle West, and enough fallout occurred in Indiana (Webb, 1949) to contaminate cornstalks that ultimately found their way into a paper-making process used by the photographic industry.

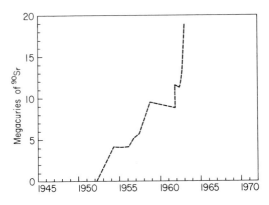

Fig. 13-7. Production of ^{90}Sr by nuclear weapons tests, 1952–1963.

Radioactive particles in the finished product eventually caused damage to X-ray film that had been packaged with contaminated interleaving paper (Fig. 13-8).

Our knowledge of fallout has been gathered from weapons tests and is, therefore, limited by the fact that conditions were necessarily very different from those that would be faced in time of war. Tests within the United States were conducted near Las Vegas where, for reasons of public safety, it was necessary to restrict the size of atmospheric tests to less than 100

Fig. 13-8. X-ray film marred by exposure to contaminated interleaving paper processed August 6–10, 1945 (Eastman Kodak Company).

kilotons. Moreover, it was the practice to detonate many of the devices high above the ground so as to minimize fallout. Bombs larger than 100 kilotons were detonated in the Marshall Islands, where much information about fallout was obtained, but the circumstances were somewhat less than ideal because of the enormous expanses of water over which it was necessary to make observations. The Marshall Islands are thin coral reefs, and large land surface bursts could not be simulated. In fact, many of the devices were mounted on barges floating in lagoons. Other detonations were on reefs just barely above the surface of the water, with cratering extending well below the water line.

As a result of these limitations, many questions remain unanswered. Very little is known about the physical and chemical properties of the fallout that would be encountered under the somewhat wide range of conditions that might prevail during a nuclear attack. Bombs of various sizes would be detonated at various heights above the ground, the terrain would vary from place to place, and it might be raining or snowing. The only thing about which one could be quite certain is that the conditions would be very unlike those in either Nevada or the Marshall Islands.

Nevertheless, much has been learned. The sequence of events from the moment of detonation until the time of fallout is known, and a first approximation can be made of the probable extent of the area to be affected by fallout if the conditions of detonation are known. The remainder of this chapter will be concerned with fallout effects in the immediate vicinity of the explosion. More remote effects or "worldwide fallout" will be discussed in Chapter 14.

The levels of radioactivity that will exist at any given location will depend on the amount of radioactivity produced in the burst, the cloud height, the distribution of radioactivity within the cloud, and the manner in which this radioactivity is distributed as a function of particle size. Needless to say, the fallout pattern will also be influenced markedly by the wind velocity and direction along the full height of the cloud, as well as by the presence of precipitation.

The United States Weather Bureau operates a network of stations which provides forecasts of wind directions and velocities at various altitudes up to 80,000 ft. These stations normally broadcast their observations over the regular Weather Bureau teletype network at regular intervals, but forecasts could be made more frequently in the event of an alert. The information transmitted includes the upper-wind observations as well as forecasts of the directions at ground level along which radioactive particles from various altitudes would fall. The teletyped information can be converted to plotted fallout forecasts using relatively simple graphical procedures that have been developed. Figures 13-9 and 13-10 illustrate the types of fallout

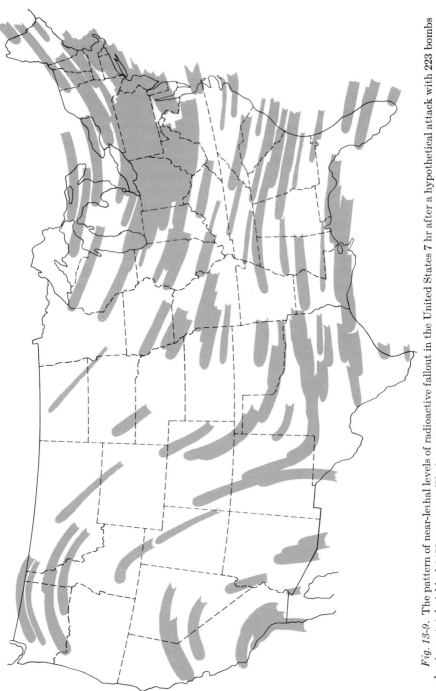

Fig. 13-9. The pattern of near-lethal levels of radioactive fallout in the United States 7 hr after a hypothetical attack with 223 bombs having a total yield of 1453 megatons (Shafer, 1959).

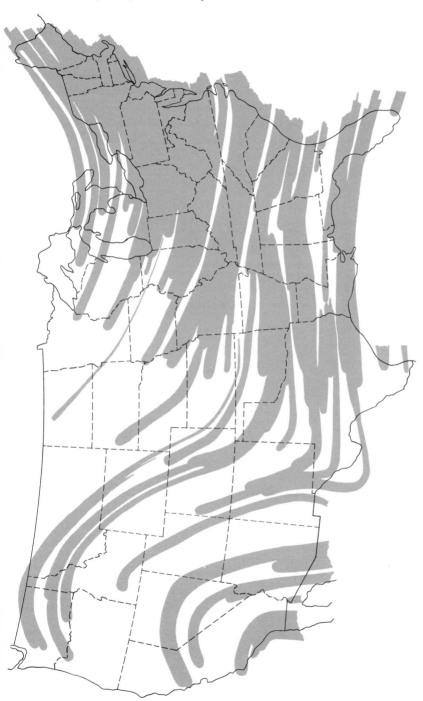

Fig. 13-10. The fallout patterns 48 hr after hypothetical attack of Fig. 13-9 (Shafer, 1959).

patterns that were developed on a country-wide basis during a civil defense drill in 1959.

The predicted fallout patterns may be regarded as potentially useful for forecasting the general areas in which fallout is most likely to occur. However, any reliance on fallout predictions beyond the need to decide that fallout may occur in a given sector and that the public should seek the protection of shelters would be extremely dangerous in view of the lack of precision of the fallout prediction methods even under the best conditions. The empirical information now available was obtained under the somewhat special conditions that prevailed during tests in Nevada and the Marshall Islands and, as noted, the conditions existing at those locations would probably not be applicable elsewhere. The fallout forecasts made during the weapons tests were based on far more exact knowledge of the explosive yield, the conditions of detonation, and the prevailing meteorology than would be available under attack conditions, when it might be necessary to utilize meteorological data several hours old that might have been collected at some location other than the site of attack.

Even under the comparatively ideal conditions of the weapons tests, the fallout predictions gave only an approximation of the actual fallout patterns. It seems doubtful that under wartime conditions a civil defense commander could make any decision that would require more than the simple knowledge that fallout is most likely to occur in a given general area. Information as to exactly when, where, and how much fallout will occur would have to await actual observations.

The problem becomes even more complicated when one considers the effects produced by several overlapping fallout patterns. Figure 13-11 presents an idealized fallout-pattern forecast by the Office of Civil and Defense Mobilization following a hypothetical burst in the state of Washington. Also given is the more "realistic" pattern which might prevail (Machta, 1959). An example of the complexity of the fallout patterns actually observed is given in Fig. 13-12 in which the fallout patterns observed after a 5 megaton land burst is shown. Note the irregularity of the isodose contours and the two "hot spots" 40 and 60 nautical miles north of the hypocenter (Triffet, 1959).

In general, one may expect that relatively small detonations in the kiloton range will result in fallout patterns that more closely resemble the idealized cucumber-shaped patterns. However, fallout from megaton bursts may be expected to deviate from the ideal to a great extent because the greater cloud height subjects the debris to more wind shear.

There is as yet no way of forecasting the effects of natural terrain features or man-made structures on fallout. For example, it is possible that great

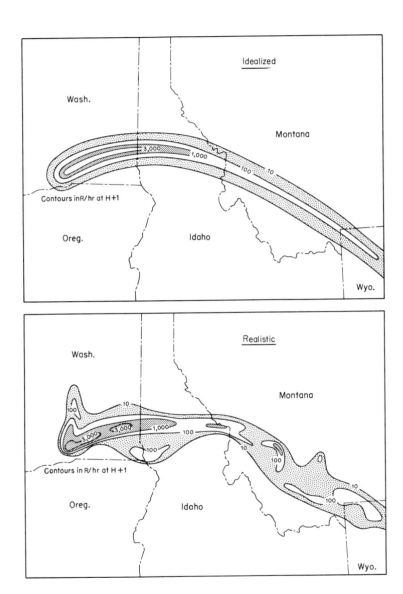

Fig. 13-11. Idealized forecast of fallout pattern following a burst in the state of Washington (Machta, 1959).

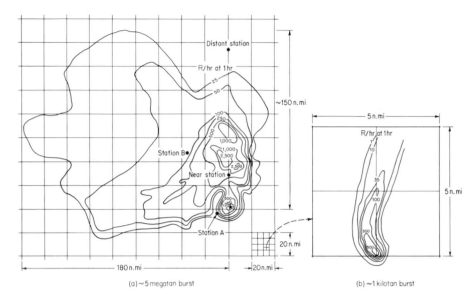

(a) ~5 megaton burst   (b) ~1 kiloton burst

Fig. 13-12. Observed fallout following a (a) 5 megaton and a (b) 1 kiloton burst on a Pacific island (n.mi = nautical miles) (Triffet, 1959).

differences may exist from one side of a building to another, particularly if the buildings are multistoried, reinforced-concrete structures.

METHODS OF DOSAGE CALCULATION

For the purpose of plotting data and standardizing the types of fallout calculations that must be made in order to estimate dose rates and dosages over various intervals of time, it is customary to forecast the fallout levels at a given point as though all the fallout occurred 1 hr after the detonation. Thus, the fallout patterns in Fig. 13-12 are plotted in this way despite the fact that the contours cover an area in which some of the points are over 150 miles from the point of detonation, and fallout at these distant points did not begin until 6 or 7 hr after the detonation.

The plots of radiation levels based on the activity at 1 hr grossly overestimate the magnitudes of levels actually encountered, but they do serve the purpose of facilitating computation of dose rates and integral doses at various times.

The dose calculations are based on the approximation that 1 MCi/square mile of fission products will result in a dose of about 4 R/hr 3 ft above the ground.

Given the dose rate at any time after a detonation and assuming radioactive decay occurs at a rate proportional to $t^{-1.2}$, the various required calculations can be performed quickly with the aid of nomograms, curves, and slide rules. Some useful examples of these calculations are the following which have been adopted from The Effects of Nuclear Weapons (Glasstone, 1962).

Example 1

Given: The radiation dose rate due to fallout at a certain location is 8 R/hr at 6 hr after a nuclear explosion.

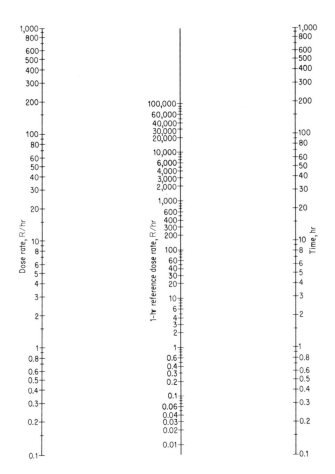

Fig. 13-13. Nomogram for the calculation of dose rate from fission products in fallout (Glasstone, 1962).

Find: (a) The dose rate at 24 hr after the burst. (b) The time after the explosion at which the dose rate is 1 R/hr.

Solution: By means of a straight edge, join the point representing 8 R/hr on the left scale of Fig. 13-13 to the time 6 hr on the right scale. The straight line intersects the middle scale at 70 R/hr; this is the 1-hr reference value of the dose rate.

a. Using the straight edge, connect this reference point (70 R/hr) with that representing 24 hr after the explosion on the right scale and extend the line to read the corresponding dose rate on the left scale, i.e., 1.5 R/hr.

b. Extend the straight line joining the dose rate of 1 R/hr on the left scale to the reference value of 70 R/hr on the middle scale out to the right scale. This is intersected at 34 hr after the explosion.

Example 2

Given: The dose rate at 4 hr after a nuclear explosion is 6 R/hr.

Find: (a) The total dose received during a period of 2 hr commencing at 6 hr after the explosion. (b) The time after the explosion when an operation requiring a stay of 5 hr can be started if the total dose is to be 4 R.

Solution: The first step is to determine the 1-hr reference dose rate. From Fig. 13-13, a straight line connecting 6 R/hr on the left scale with 4 hr on the right scale intersects the middle scale at 32 R/hr per hour; this is the reference dose rate at 1 hr.

a. Enter Fig. 13-14 at 6 hr after the explosion (vertical scale) and move across to the curve representing a time of stay of 2 hr. The corresponding reading on the horizontal scale, which gives the multiplying factor to convert the 1-hr dose rate to the required total dose, is seen to be 0.19. Hence, the total dose received is

$$0.19 \times 32 = 6.1 \text{ R}$$

b. Since the total dose is given as 4 R and the 1-hr dose rate is 32 R/hr, the multiplying factor is $4/32 = 0.125$. Entering Fig. 13-14 at this point on the horizontal scale and moving upward until the (interpolated) curve for 5-hr stay is reached, the corresponding reading on the vertical scale, giving the time after the explosion, is seen to be 19 hr.

Example 3

Given: On entering a contaminated area 12 hr after a nuclear explosion, the dose rate is 5 R/hr.

Find: (a) The total radiation dose received for a stay of 2 hr. (b) The time of stay for a total dose of 10 R.

Solution: Start at the point on Fig. 13-15 representing 12 hr after the

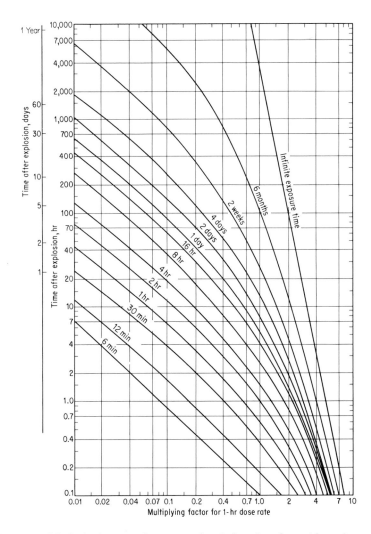

Fig. 13-14. Method of computing accumulated dose based on 1-hr reference dose rate. See Example 3 in text (Glasstone, 1962).

explosion on the vertical scale and move across to the curve representing a time of stay of 2 hr.

a. The multiplying factor for the dose rate at the time of entry, as read from the horizontal scale, is seen to be 1.9. Hence, the total dose received is

$$1.9 \times 5 = 9.5 \text{ R}$$

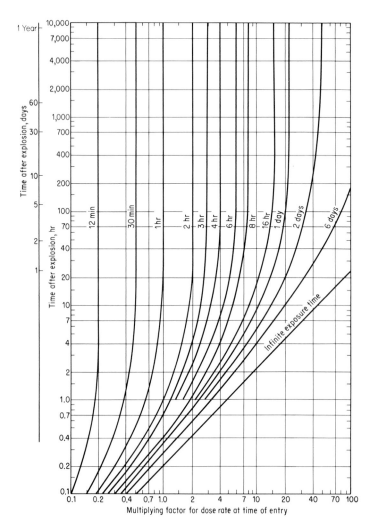

Fig. 13-15. Method of computing accumulated dose based on dose rate at time of entry into contaminated area. See Example 2 in text (Glasstone, 1962).

b. The total dose of 10 R and the dose rate at the time of entry is 5 R/hr; hence, the multiplying factor is $10/5 = 2.0$. Enter Fig. 13-15 at the point corresponding to 2.0 on the horizontal scale and move upward to meet a horizontal line which starts from the point representing 12 hr after the explosion on the vertical scale. The two lines are seen to intersect at a point indicating a time of stay of about $2\frac{1}{3}$ hr.

As a general statement, it can be said that the above computation techniques are potentially valuable, but like the fallout pattern computing techniques, they tend to oversimplify the matter. The $t^{-1.2}$ law on which these computations are based implies that the fission products are of a known age, having originated from a single burst. In a nuclear war, a given location might receive fallout from many bursts detonated at different locations and at different times. Thus, even before the fallout from one burst has completely deposited, the fallout from another burst would begin. Depending on the assumptions that are made, one might visualize radioactivity levels building up and falling off in an irregular fashion determined by the number and location of bombs contributing to the fallout in a given area.

The Short-Term Effect of Fallout

The Joint Committee on Atomic Energy (JCAE) of the United States Congress held several days of hearings in 1959 in which many specialists presented their estimates of the consequences of an attack on the United States (JCAE, 1959c). A hypothetical attack was considered in which a total of 223 targets in this country were struck by nuclear and thermonuclear bombs, each of which had the explosive force of from 1 to 10 million tons of TNT. The total explosive yield of all bombs dropped on the United States was equivalent to 1453 megatons of TNT, and it was assumed that additional bombs delivered outside continental United States had a total explosive yield equivalent to 2500 megatons of TNT. The fallout patterns in the continental United States are shown in Figs. 13-9 and 13-10.

This hypothetical attack, which may be presumed to be a realistic appraisal of the ability of a potential enemy to deliver nuclear weapons within our borders in 1959, would have resulted in death to an estimated 42 million Americans and injury to an additional 17 million. Almost 12 million dwellings would have been so badly damaged that they could not have been salvaged, and an additional 8 million dwellings would have had to be evacuated for major repairs. Thus, more than 30% of our population would have been killed or injured, and more than 40% of our dwellings would have been destroyed or badly damaged. A more recent analysis of the status of civil defense in the United States (National Academy of Sciences—National Research Council, 1969) refers to 12,000 megaton attacks, an order of magnitude greater than the basis on which Figs. 13-9 and 13-10 were drawn. The fallout patterns shown must, therefore, be taken as a lower estimate of the radiological problems that would be faced in the event of a massive nuclear attack.

The above casualty estimates are based on an analysis of the radiation levels, the blast effects, and the effects of fires produced by the bombings. The extent to which additional casualties would be produced among survivors by infectious disease, starvation, deterioration of law and order, and excessive exposure to the elements was not assessed. The vast numbers of injured and diseased people, the enormous destruction of production facilities and housing, interruptions of power communications, and disruption of the methods of distributing essential goods and services would be only a few of the factors that would interact in so complicated a manner as to place the true consequences of a nuclear attack beyond comprehension. In the above hypothetical attack on the United States, about 96 million people lived outside the areas of likely blast damage. These people would be exposed to radioactive fallout at various times, from a few minutes to a few hours after the detonation, depending on their location with respect to the bursts. The principal immediate problem which would face this portion of the population would be the potentially lethal levels of radioactive fallout. However, although they would be beyond the range of the effects of blast and thermal radiation, the problems of dealing with the radiation effects would nevertheless be complicated greatly by the enormous social and ecological stresses that would be produced throughout the nation as a whole.

The primary short-term effect of fallout results from the fact that large areas downwind of the detonation are blanketed with potentially lethal amounts of radioactive dust. The extent of fallout will depend on the meteorological conditions, size of detonation, type of detonation, and type of terrain. Much of our civil defense planning is based on the types of weapons with which our government has had experience. Thus the computational aids provided for civil defense planning, and almost all discussions of nuclear weapons effect, are limited to thermonuclear weapons under 20 megatons in which the energy is derived about half from fusion and half from fission, although in 1961 the Soviet Union conducted a series of nuclear explosions in which the fission components averaged about 25% of the total yield (Dunning, 1962).

It is most likely that in areas where the fallout is acutely hazardous the particles would be so large as to be visible. Japanese fishermen who were trapped by fallout on their boat, the Lucky Dragon, about 80 miles downwind of the large thermonuclear detonation on March 1, 1954 reported that the fallout of white dust resembled snow, and the deck was said to have been covered to such an extent that footprints were clearly visible (see Chapter 16). Japanese physicists who investigated this accident (Japan Society for the Promotion of Science, 1956) reported that the fallout at the time of deposition amounted to 38 to 85 g of dust/m² of deck surface. The fallout was observed to be in the form of dust having a particle size of 0.1 to 3 μm, but agglomerated into granules of about 300 μm. At closer distances the particles might range up to several millimeters in size (Triffett, 1959).

These particles become sources of external γ radiation, and, in addition, they can produce β burns on the skin or can be inhaled or ingested.

γ *Radiation.* The rate of rise of the γ-radiation levels is most likely to be variable, depending on distance from the detonation. Experiments during United States nuclear weapons testing programs in the Pacific provided data such as are shown in Fig. 13-16 (Triffett, 1959) in which the γ-radiation levels as a function of time after burst are plotted. The data were collected at two typical stations; one just beyond the zone of blast and thermal damage and the other sufficiently far downwind so that it took several hours for the fallout cloud to arrive.

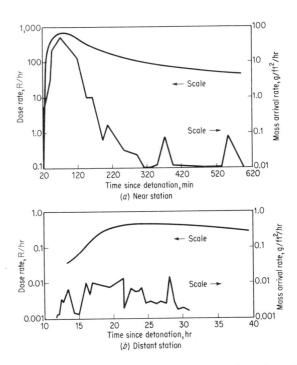

Fig. 13-16. Arrival times of fallout at two stations downwind of a large thermonuclear explosion. This is only one example, and it cannot be assumed that the curves would be similar for another explosion. The shapes of the curves will depend on many factors, including size and type of burst, azimuth, and meteorology (Triffet, 1959).

The short-term biological effects to be expected from fallout would depend in a highly complex way on the rate at which the total dose was delivered and on whether there were any concomitant effects such as those from blast and thermal radiation. The effects of acute exposure were discussed in Chapter 2 and summarized in Table 2-1.

β Burns of the Exposed Skin. In a subsequent chapter will be described the detailed findings of investigators who studied the results of the 1954 accident in which a group of Marshallese inhabitants of the Rongelap Atoll and 22 Japanese fishermen were inadvertently exposed to heavy fallout. In both series of cases, skin burns were produced by the β activity of dust which had been deposited on the skin. This effect was limited to exposed portions of the body and particularly to those portions on which the dust particles tended to collect. The wearing of clothing, the practice of remaining indoors during periods of fallout, and reasonable standards of personal hygiene should make it possible to avoid injuries of this type.

Inhalation and Ingestion of Radioactivity. The 1954 accident also provides useful information about the potential danger of contaminated food and water, because the Japanese and Marshallese victims lived under comparatively primitive conditions at the time of the fallout and were in intimate contact with the debris for lengthy periods of time. We shall see later that except for radioiodine, inhalation and ingestion did not contribute in a major way to the internal dose received by the victims.

The question of what level of radioactive contamination of foods or water would become significant in the postattack period is one which cannot be answered in advance. It is only proper that in peacetime we be conservative in our attitude toward radiation injury, and the levels of contamination that are permissible under normal circumstances are presumably safe for an indefinite exposure or, if not absolutely safe, would result in injury on a scale that would be undetectable because of the infrequency with which the effects would occur in the population. In contrast, the levels of permissible contamination in a postattack period would vary, depending on various factors that would enter into the decision-making process. Under many conceivable circumstances, the maximum allowable level of radioactivity would be limited only by the requirements that severe *acute* effects be avoided. The harsh realities of the postattack period might make it relatively unimportant that delayed effects may occur in a substantial fraction of the population. The risk of not surviving the postattack period owing to factors totally unrelated to radioactive contamination might be so great that an additional risk of developing leukemia or thyroid cancer in 5 to 15 years would hardly add to the total social cost of such an attack.

It would be necessary for local authorities to determine, on a day-to-day basis, what levels of contamination should be permitted, depending on the number of survivors and the amount of food that is known to be available.

In the immediate postattack period, there would be two principal dangers from contaminated food (Eisenbud, 1964). Contamination of fresh foods by radioiodine could produce thyroid injury, and contamination of leafy vegetables or water by intensely radioactive fallout particles could constitute a hazard to the lining of the intestinal tract. It is most likely that these would be the controlling factors that would limit the amount of radioactivity that could be ingested in food or water.

Of the several radioactive isotopes of iodine that occur in the debris of nuclear explosions, the most significant is ^{131}I, with a half-life of 8.1 days. In the United States the most important way in which radioiodine passes to man is by fresh dairy products, because of the relative speed with which the short-lived isotope can move from contaminated forage to man via this route.

In wartime, the thyroid doses among the general population could exceed thousands of rads, assuming that no countermeasures are taken. Weathering and radioactive decay combine to reduce radioiodine deposited on foliage with a half-time of about 5 days (Thompson, 1965). The requirement for restrictions and countermeasures might thus be limited to the first few weeks of the postattack period, so far as ^{131}I is concerned.

Little is known about the effects of ingesting relatively insoluble radioactive particles. Tajima (1956) has reported that fallout on the deck of the Lucky Dragon contained granules as large as 300 μm in diameter. From his estimate that the specific activity of this fallout was initially 1 Ci/g, one can calculate that a 300 μm particle contained about 70 μCi of activity. A particle such as this would be capable of delivering a heavy dose to the lining of the intestine.

Individuals who survived the period of acute radiation danger would emerge into an environment that would be contaminated with radioactivity for the rest of their lives, but most of the fission products that have intermediate or long half-lives are relatively inactive biologically and would not appear as significant contaminants of foods. These isotopes, which include ^{95}Zr, ^{106}Ru, and ^{144}Ce, would result in elevation of the external γ-radiation background, but would contaminate food only by being deposited physically on the foods and would not be absorbed significantly by people ingesting these nuclides. However, as noted earlier, if present in particles of sufficient β activity, serious injury to the intestinal lining could occur.

Considerably more significant would be the ^{89}Sr and ^{90}Sr isotopes. ^{89}Sr has a half-life of only 59 days, but in fresh fallout the ratio of ^{89}Sr to ^{90}Sr may be as much as 200 to 1, diminishing to about 10 to 1 at the end of 1 year and to an insignificant fraction after 2 years. These strontium isotopes can enter man's food supplies either by direct foliar deposition or by absorption from the soil. The former mechanism is dominant when the rate of fallout is relatively high but becomes less important when the rate of fallout diminishes, and the radiostrontium in food is primarily due to absorption from the soil (see Chapter 5).

The dose from ^{89}Sr is self-limiting because of its short half-life, but it has been estimated (Dunning, 1962) that during the first year of exposure this nuclide may deliver as much as 3.4 times the dose delivered by ^{90}Sr. At the end of the second year this isotope will have decayed to such an extent that the ^{89}Sr component of the skeletal dose will become relatively unimportant for the whole lifetime of the individual.

The extent to which fallout would result in significant contamination of water supplies would depend on such factors as the amount of fallout on the reservoir, the rate of runoff from the watershed, the reservoir volume, the time it takes the water to pass from the reservoir to the consumer, and the extent to which decontamination takes place. Factors that assist decontamination would include the settling of suspended solids to the bottom of the reservoir, adsorption of radionuclides onto suspended solids, and water-treatment processes.

If one makes the assumption that uniform mixing of the fallout occurs and that no decontamination processes take place, and if it is further assumed that there is no delay between the reservoir and the consumer, it is possible to calculate the dose a consumer would receive from any given fallout level. Thus, if the fallout level on an infinite smooth plane is 3000 R/hr at 1 hr and this amount of fallout is received by a reservoir 10 ft deep, a dose of 305 rad would be received by the thyroid of an average person drinking the water for 30 days (Hawkins, 1961). A dose of 142 rad would be received as a result of ingestion during the first day. If the consumer should abstain from drinking the reservoir water for 9 days, the dose would be less than 15 rad.

The extent to which decontamination would take place in water-treatment plants of various conventional types is shown in Fig. 13-17 (Lacy and Stangler, 1962). The conventional process of coagulation, chlorination, and filtration removes from 70 to 80% of the radioactivity. The effectiveness of removal can be increased to as much as 99.9% by the addition of ion-exchange columns. A distillation process would have an efficiency of 99.999% in removing the contamination, but this process would presumably be impractical on a large scale.

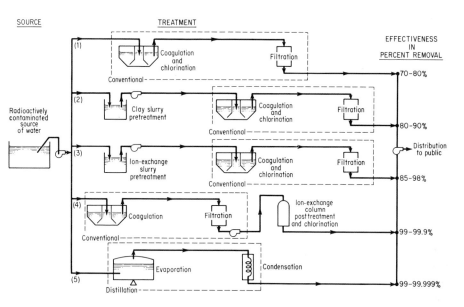

Fig. 13-17. The decontaminating effectiveness of various water-treatment processes (Lacy and Stangler, 1962).

On balance, the problem of water contamination is one which must certainly be considered, but the postattack problems created by this factor would be secondary, to the extent that a population could survive on contaminated water, assuming other sources of exposure were under control.

It is generally accepted that the long-term risk of somatic injury from residual radiation would result mainly from ^{90}Sr. The 1959 hearings before the Joint Committee on Atomic Energy considered the effects of a hypothetical attack on the United States, equivalent to a total explosive yield of 1453 megatons of TNT. It was further assumed that the total explosive yield of bombs delivered outside of the United States was equivalent to 2500 megatons of TNT. Thus, the total explosive yield was equivalent to a little less than 4000 megatons of TNT, assumed to be equally divided between fission and fusion. The total yield of ^{90}Sr can be estimated on the assumption that the nuclide is produced at a rate of 0.1 MCi/megaton of fission. Since the fission yield was about 2000 megatons, the ^{90}Sr production would be 200 MCi.

Four thousand megatons of explosives is an enormous yield which could result in hundreds of millions of casualties and which could destroy civilization as we know it. The dangers owing to ^{90}Sr contamination of food following such an attack would be one of the minor problems when considered against the total consequences of a war of such magnitude. One can gain perspective from experience with the ^{90}Sr produced in weapons tests, the fission yield of which we have seen is about 200 megatons, if we limit the tests to those exploded in the atmosphere. The tests have thus injected about 20 MCi of ^{90}Sr to the atmosphere. We will see in Chapter 14 that the global fallout of ^{90}Sr has totalled about 12 MCi, or about 55% of the total. The remainder can be assumed to have been deposited in the immediate vicinity of the explosion.

We will also see in Chapter 14 that the "average" skeletal dose commitment from this deposit is estimated to be 130 mrem in the North Temperate Zone. Thus, the skeletal dose from ^{90}Sr might reach 1300 mrem following a 2000-megaton attack. This figure applies to areas beyond a few hundred miles of the targets and is an average figure. Since this dose commitment is for a 50-year period, the 1300 mrem would be superimposed on a dose of more than 5000 mrem owing to the natural background.

The ^{90}Sr levels in the most contaminated areas immediately downwind of the targets would reach about 300,000 mCi/square mile, about 100 times the deposition in the regions just discussed (Machta, 1959). Assuming that these areas were farmed following the attack, it can be estimated that the skeletal burden of children deriving their calcium from such food would be about 12,000 pCi/g Ca, equivalent to about 36 rem during the first year and about 520 rem over a 70-year lifetime. An additional bone dose would be received during the first year because of the presence of ^{89}Sr, and this we estimate to be about 3.4 times the dose from ^{90}Sr, or about 120 rem. The total dose to the bone during the first year would thus be about 150 rem, and over a 70-year lifetime it would be about 640 rem. To these bone doses, derived from contaminated food, it is necessary to add the dose from external sources, which might be larger than the dose from ^{90}Sr. In these heavily contaminated areas, severe restrictions on agricultural practices would be required. These would consist primarily of prohibiting the use of the land for dairy pastures or crops consumed directly by man. It would, however, probably be permissible to grow corn or other foods for hogs, steers, or other animals used for meat.

LIVESTOCK AND WILDLIFE DAMAGE (BENSEN AND SPARROW, 1971)

Since estimates of the lethal dose for mammals and poultry show them to be about as radiosensitive as man, it is clear that radiation alone would decimate the food animals of the nation. The estimated lethal response of farm animals is given in Table 13-5. The lethality would in many cases be greater than indicated owing to additive effects of blast, fire, food shortages, disease, the effects of internal exposure to radionuclides ingested by foraging

TABLE 13-5

LETHAL RESPONSE OF MAMMALS AND POULTRY TO BRIEF EXPOSURES TO NUCLEAR RADIATIONS[a]

Species	$LD_{50/30}$[b]	Rate (R/hr)
Burro	784	50
Burro	651	18–23
Burro	585	19–20
Swine	618	50
Sheep	524	20
Cattle	540	25
Swine	555	180
Swine	388	90
Burro	402	—
Poultry		
Males	600	50
Females	1000	50
Chicks	900	Very short

[a] National Academy of Sciences—National Research Council (1963).
[b] The dose that produces 50% mortality in 30 days.

animals, and skin burns caused by fallout deposited directly on the animals. For man and animal alike, the tables of lethality may underestimate the effects of radiation because they do not take into consideration the effects of other concomitant types of stress.

Although decontamination of heavily contaminated farm lands would be desirable, it is doubtful that under the conditions that exist in the first few years after attack manpower would be profitably expended if used in this way. It would seem far wiser to manage these contaminated farm lands in some other way, for example, by denying the land to agriculture, or by growing foods low in calcium, or growing fodder for stock animals that are raised for meat.

Some Problems of Recovery from Nuclear Attack

Although nuclear attacks may produce complete destruction by blast and fire over an area many miles in diameter, even a saturation attack would leave much of the country relatively intact except for the effects of fallout. For example, woodframe buildings would be severely damaged as far as 40,000 ft from the detonation of a 1 megaton bomb, and fires might be produced as far away as 20 miles. Although the combination of blast, thermal, and radiation effects would make any discussion of counter-measures in the immediate target area of somewhat doubtful value, there is much that could be done to ameliorate the problems that would be faced by the large numbers of people outside the zone of blast and thermal damages but well within the region of potentially lethal fallout. During Operation Alert, 1959, a nationwide civil defense exercise in which a saturated attack on the United States was simulated, there were 5 million estimated fatalities in the state of New York. Of this number, approximately 1 million would have been killed by blast and heat, whereas 4 million fatalities would have been caused only by fallout and not the primary blast or thermal effects. Although 1 million fatalities together with the enormous economic loss resulting from the destruction of our major cities would be a calamity so enormous as to be unimaginable, the loss of 4 million additional people is even more calamitous.

Home shelters were widely recommended up to 1961 (National Academy of Sciences—National Research Council, 1964; New York State, 1959) when there developed a greater recognition of the advantages of community shelters (United States Department of Defense, 1961). The advocates of shelters note that millions of people would find themselves in localities in which the radiation levels are lethal but which are located beyond the zone of destruction. These people would require radiation shielding for their survival. Thus, the 4 million radiation fatalities in New York State during the civil defense exercise discussed earlier would have been saved from radiation injury had they had access to shelters which could attenuate the radiation levels by a factor of 100. Assuming that these inhabitants took no precautions other than to remain indoors in frame houses, thereby attenuating the radiation by a factor of 2, the vast majority of inhabitants would have received more than disabling doses during the first 24 hr after the attack. By taking advantage of the basements of ordinary houses, an attenuation factor of 10 would have been gained, and the number of casualties would have been reduced by 50%. By using improved basement shelters offering an attenuation factor of 100, almost all the 4 million radiation casualties would have been prevented.

After a full-scale nuclear onslaught, survival for millions of families would require that each be protected for as long as several weeks after the fallout begins. It would be necessary that adequate shelter be made available and that they be provided with food, water, essential drugs, sanitary facilities, and radio communication. Some people would be required to remain in shelters for longer periods of time than others. At first, it might be safe to leave the shelter for a few minutes a day. In time, longer periods out of doors would be permitted, and as these excursions were extended in time, the survivors could begin to assume duties that would assist the community as a whole to begin functioning once again.

In some areas, the radiological situation would be such that people would be instructed to leave their shelters to be transported to less contaminated areas. It is possible that large areas may be uninhabitable for many months. However, the most crucial survival problems would be those encountered during the first hours or days following attack. It is in this, the acute phase, that small groups of people must be self-sufficient and equipped with the basic tools for survival.

The acute phase following attack would present government with problems of unprecedented complexity. It would be necessary for officials to assume the most complicated tasks in all history, but these tasks would be faced at a time of unprecedented disruption of manpower, communications and transportation, and overwhelming disruption of the material resources of government.

Although in such a situation it would be necessary that the shelter inhabitants be self-sufficient so far as food, clothing, water, and drugs are concerned, there are a number of things that government must do to ensure their survival. The civil defense organization must have the capacity to assemble the information needed to reach decisions that will affect everyone. If able-bodied people are kept in their shelters too long, they cannot be used to assist the stricken areas. If the people are told to leave their shelters too soon, they will become casualties. Thus, the civil defense officials must have accurate knowledge of radiological conditions throughout the area for which they are responsible. Use of radiation detectors by the family should be discouraged except in unusual cases. Particularly in urban areas, the radiation status of a neighborhood should be evaluated by skilled monitors rather than many individuals using many instruments of doubtful performance under conditions in which it would be very difficult to integrate the changing dose rates.

Rapid radiation monitoring in the postattack period would be essential to competent decision making. These would be the decisions such as whether a population in a given sector can safely leave their shelters or whether that section of the community can serve as a receiving area for people to be evacuated from regions of blast or fire. Accurate information would be required on which would be based difficult decisions such as whether the children in a given school may be released to rejoin their families or whether or not power lines can be serviced, streets can be patrolled, or warehouses filled with food can be made available as sources of relief.

Unfortunately, these decisions could not be made competently on the basis of contemporary methods of collecting radiological data. Present techniques for gathering radiation data would be ineffective, because they depend largely on manual observations which would result in radiation monitors receiving lethal doses of radiation in only a few minutes. Men on foot or in conventional vehicles could not make measurements in heavily contaminated communities, as is now intended by most civil defense organizations.

In this era of mechanization and automation, a network of automatic, continuously recording radiation monitors would be required that would transmit their data to a central headquarters where the data could be fed into computers which would provide continuous and accurate radiological information. Only in this way could a civil defense commander obtain the information he would need in sufficient time to make the decisions that would be necessary for the management of large populations beyond the blast area. Until this is accomplished, the shelters cannot be fully effective.

Following an explosion, so long as the radioactive cloud has not reached a given community, the people could prepare themselves for the days of shelter life that may be ahead of them. These would be difficult minutes in which one could only plan if the community authorities have the means with which to maintain discipline among the populations and the knowledge and ability to keep the people continually informed. If the radiological situation permits during the first hour or two after attack, people could be permitted to remain above ground to assemble their families and to gather the various paraphernalia that may assist their survival.

The arguments for and against shelters as part of a civil defense program are complicated by political, moral, and sociological questions as well as by the incomprehensible consequences of nuclear war. These factors are outside the scope of this text which has dealt only with the rationale based on the need to protect people who would be outside the zones of destruction and for whom protection against radiations from fallout would be an essential part of a survival program.

The foregoing has constituted a straightforward summary of what is known about the immediate radiological effects of nuclear weapons, particularly when the explosions occur at or near ground level. It is in all respects a superficial treatment that deals only with certain very obvious aspects of the radiological effects of nuclear weapons. Little or nothing has been said about the social, logistic, and medical implications of massive fallout. The problems that would face people at every level of the social structure would be so complicated as to defy meaningful analysis. The problems become more complex by additional orders of magnitude when we consider the comcomitant effects of blast damage and fire when superimposed on

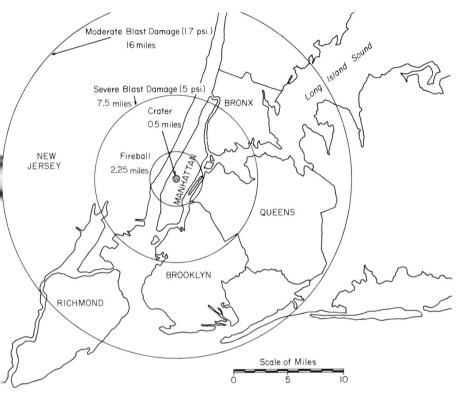

Fig. 13-18. Consequences of a 20 megaton ground burst in mid-Manhattan (Stonier, 1964).

the radiological problems. The physical dimensions of the matter are illustrated by Fig. 13-18, which depicts the extent of the damage that would be produced if a 20 megaton bomb were exploded in mid-Manhattan. Stonier (1964) has considered the consequences of such a catastrophe in vivid detail.

The radiation problems cannot be considered without understanding the social effects of blast damage and extensive conflagrations. A wide spectrum of individual factors such as low water pressure, radiation damage to livestock, burned fields and forest subject to rapid erosion, acute food shortages, social disorganization, mass casualties, severe ecological disruption, disease, and pollution would create a hundred new problems for everyone foreseen. The disastrous immediate consequences of such a war would be followed by a prolonged period of social retrogression from which recovery would be slow and uncertain.

Chapter 14

Fallout from Nuclear Explosions
II. Worldwide Effects

The behavior of the radioactive dust that remains airborne for more than a few hours after a nuclear explosion depends on whether the material is contained in the stratosphere or the troposphere, the latter component being deposited within a matter of weeks after the explosion, whereas the stratospheric component remains in the upper atmosphere for many months.

The importance of the fact that the radioactive debris is partitioned between the troposphere and the stratosphere was not appreciated until the advent of thermonuclear explosions equivalent to megatons of TNT in the mid-1950's. Prior to 1952, all the nuclear explosions were in the kiloton range, and it was observed that after each series of weapons tests the atmospheric radioactivity diminished at a rate corresponding to the half-life of dust in the lower atmosphere, which was shown to be about 20 days (Stewart *et al.*, 1957). Essentially all the debris from kiloton bombs is deposited within 2 months or so following injection into the atmosphere. The dust from such explosions is confined to the troposphere and is carried by the winds characteristic of the latitude in which injection takes place, and deposition ultimately takes the form of bands, as shown in Fig. 14-1. The fallout from any one explosion is often spotty, because rainfall some-

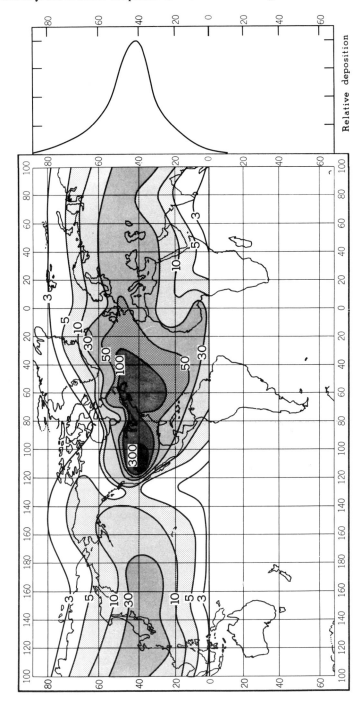

Fig. 14-1. Worldwide fallout of radioactive fallout from nuclear weapons tests in Nevada in 1953. The explosions were in the kiloton range of yields, and debris was confined to the troposphere. The intensity of fallout is shown in relative units (L. Machta).

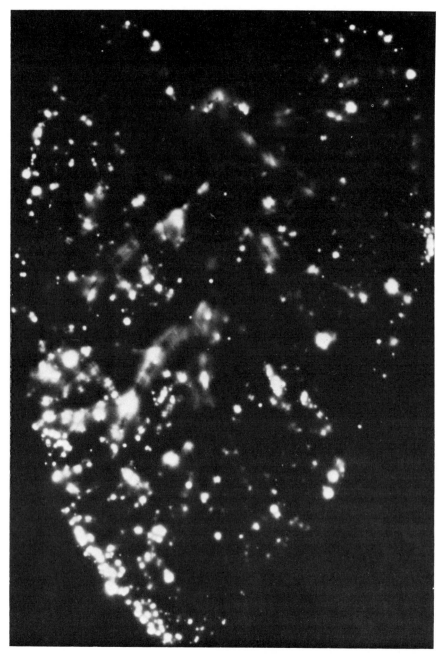

Fig. 14-2. Autoradiograph of a leaf following a fallout of radioactive dust in Troy, New York. The dust originated from an explosion in Nevada about 36 hr previously (Herbert M. Clark).

times coincides in time and place with passage of the clouds of radioactive dust. After one Nevada explosion in April, 1953, the highest fallout recorded anywhere in the United States was at Troy, New York more than 2000 miles from the Nevada test site, where a storm caused precipitation of the radioactive dust to the extent that it was estimated (Clark, 1954) that the cumulative γ dose received by the inhabitants was about 100

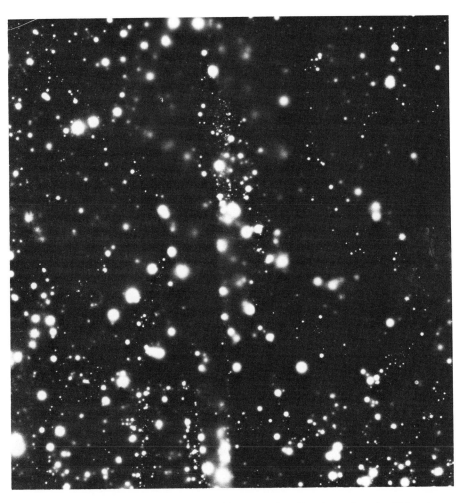

Fig. 14-3. Autoradiograph of adhesive film exposed to the atmosphere for 24 hr at a fallout monitoring station operated by the Health and Safety Laboratory of the United States Atomic Energy Commission in 1953. The sample was collected several hundred miles from the test site.

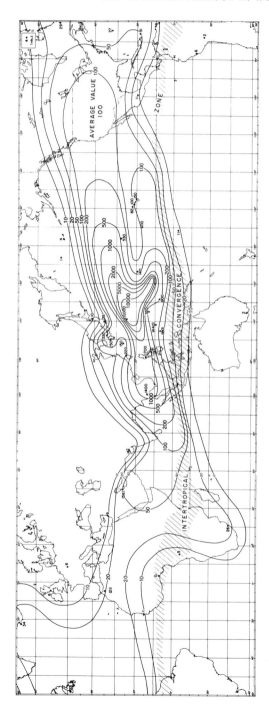

Fig. 14-4. Radioactive fallout from a multimegaton thermonuclear explosion in the Marshall Islands in November, 1952. The values shown are millicuries per 100 square miles between 2 and 35 days following the explosion (Machta *et al.*, 1956).

mrad. This proved to be a higher dose than was received anywhere in the United States, except in the immediate vicinity of the test site for the entire 1953 series of tests. Figures 14-2 and 14-3 illustrate the extent to which radioactive particles deposited on surfaces at great distances from the nuclear explosions during the period when tests of nuclear weapons were conducted above ground in the United States.

As is shown in Fig. 14-4, even the intermediate fallout from a thermo- nuclear explosion in the Pacific Ocean is, within 35 days, distributed widely in both hemispheres. The fraction of the debris that is injected into the stratosphere is eventually deposited from pole to pole.

Amounts and Distribution of Radioactive Debris Produced in Weapons Tests

The first nuclear test explosion took place in the summer of 1945, and from that time until 1963, when a ban on atmospheric testing of nuclear weapons was agreed to by the major nuclear powers (Chapter 13), radio- active contamination from weapons debris took place on a global scale. This chapter will review the state of our knowledge concerning the physical and biological behavior of the more important radionuclides dispersed in this way.

As was noted in Chapter 13, France and China were not signatories to the agreement that banned nuclear weapons testing in the atmosphere, and those countries have exploded a number of devices since that time, but not on a large scale compared to the tests conducted by the United States, Soviet Union, and United Kingdom prior to 1962. Moreover, the test-ban agreement does permit a nation to test underground, provided the tests are controlled so that vented debris is not detected beyond the national borders. Both the United States and the Soviet Union have tested nuclear weapons extensively underground, and although some venting has been reported, it has not produced significant fallout compared to that which occurred when open-air tests were in progress.

Of the many radionuclides produced in nuclear and thermonuclear explosions, the nine listed in Table 13-1 account for most of the exposure received by man. Most of these radionuclides are short-lived and are, thus, of significance mainly in the intermediate (tropospheric) component of the fallout.

The average life of dust in the troposphere is about 1 month, which is a sufficiently brief period to permit substantial residues of these nuclides to deposit on the surface of growing crops and in this way enter man's food supply. However, these nuclides are relatively unimportant when deposited in soil because they are short-lived in relation to the growing season of most crops.

^{90}Sr and ^{137}Cs (28 and 30 years, respectively) are two of the most important long-lived fission products that occur in fallout. ^{239}Pu, with a half-life of about 24,000 years, is produced by neutron capture in ^{238}U. Although this nuclide is formed in relatively large amounts and will persist in the environment for a long time because of its very long half-life, we will see later in this chapter that its relative insolubility prevents it from being assimilated by plants or animals (Langham and Anderson, 1958). ^{14}C, with a half-life of 5600 years, is produced by the interaction of bomb-produced neutrons with the nitrogen nuclei of the atmosphere. This isotope has been produced in nuclear explosions in such quantities as to result in an appreciable increase in the ^{14}C reservoir of the atmosphere, oceans, and biosphere.

Our knowledge of global fallout phenomena has benefited enormously from the collaborative efforts of many nations. In particular, the United Nations Scientific Committee on the Effects of Atomic Radiation (UNSCEAR), a 15 nation committee that reports to the General Assembly of the UN, has met regularly to review new data as it was accumulated in many countries. The several reports of UNSCEAR (1958, 1962, 1964, 1966, 1969), and the 1969 report in particular, present an excellent summary of the subject of worldwide fallout from tests of nuclear weapons.

Methods of Estimating How the Radioactive Debris Is Partitioned among the Three Components of Fallout

In this chapter we are interested in the behavior of the radioactive debris from nuclear explosions so that we may understand the mechanisms that affect human exposure at large distances from the explosions. We wish to learn how to predict the manner in which the radioactive debris from a given explosion is transported by the atmosphere, the mechanisms by which it is deposited on the surface of the earth, and the manner in which such deposits result in exposure of ionizing radiations. One of the first steps requires that we estimate how the total amount of radioactive debris produced in an explosion is apportioned among three fractions: (1) the fallout in the immediate vicinity of the explosion, (2) the debris injected into the troposphere, and (3) the debris injected into the stratosphere. Surprisingly, there is comparatively little reliable information on this subject, except for extreme cases such as very high air bursts and small surface bursts. Thus, it is well established that all the debris will remain in the troposphere when a kiloton weapon is exploded well below the tropopause but sufficiently high off the ground that the fireball does not touch the surface; essentially all the fallout from such explosions will be intermediate fallout. It is also known that bombs exploded high in the stratosphere produce long-delayed fallout. However, there is little information about the behavior of the debris from surface or near-surface explosions, particularly of megaton weapons. Surface explosions in the Marshall Islands are thought to have deposited as much as 80% of the fallout in the immediate vicinity, but this fraction is subject to great uncertainty, depending on the size of explosion, height above ground, type of terrain, and meteorological factors. The lack of adequate information can be explained by the great operational difficulties that handicapped fallout studies in the Pacific, where the large United States test explosions have taken place. For the period up to 1958, it was estimated (Joint Committee on Atomic Energy, 1959a) that about one-third of the total quantity of radioactive materials produced by bombs exploded in the Marshall Islands was deposited in the immediate vicinity of the explosions.

The tropospheric component is thought to be about 5% of the radioactive yield of megaton surface explosions (Machta and List, 1959) and less in the case of air bursts. The principal source of worldwide contamination by long-lived radionuclides is thus the component of the debris that is injected into the stratosphere. Tropospheric fallout contributes to the total in a minor way and can be neglected. Thus, a quantitative treatment of the subject of worldwide exposures from nuclear explosions having yields measured in megatons of TNT must start with knowledge of the total quantity of debris produced and the fraction that deposits close to the site of explosion. The difference represents the potential for worldwide exposure to the long-lived radionuclides. Although the 5% or so that remains in the troposphere contributes in only a minor way to the worldwide deposition of long-lived debris, significant exposure results from the fallout of short-lived tropospheric debris, notably [131]I.

A quantitative assessment of global fallout by material balancing techniques is handicapped by certain difficulties. Estimates of the yields from United States explosions are available in this country, but the data on individual bursts are in many cases classified and, therefore, accessible to only a limited number of individuals. The estimates of detonations by the Soviet Union and the United Kingdom are presumably available, but, particularly in regard to the detonations by the Soviet Union, the data are even less accessible than information from the United States. Only a few individuals have access to such data or to data on the amounts of close-in fallout obtained during the Pacific weapons tests.

The first estimates of the amounts of radioactive dust originally injected into the stratosphere were made by subtracting the estimated amount of close-in fallout from the total amount of debris thought to have been produced. Based on field studies, detonations in the megaton range were assumed to deposit from 10 to 80% of the radioactive debris in the immediate vicinity of the detonation, depending on the type of burst. Local fallout was assumed to be 80% for land surface explosions, 20% for explosions on the surface of water, and 10% for explosions in the air. The stratospheric inventory was not believed to be influenced significantly by the many detonations in the kiloton range since, for the most part, these did not penetrate the stratosphere.

On examination, the material balance studies of debris produced in the Pacific detonations from 1952 to 1958 prove to be subject to many uncertainties. In order to ascertain the amounts deposited "close" to the site of detonation, it was necessary to collect fallout samples over an area of many thousands of square miles of the Pacific Ocean under operating conditions that defied the ingenuity of a number of well-equipped teams of investigators. Even the best data left considerable uncertainty.

Given an estimate of the amount of a given isotope produced in a megaton-size detonation and an estimate of the amount that is deposited in the immediate vicinity, the balance, within the limits of uncertainty of the estimates, can be said to have been injected into the stratosphere. This follows from the observation that the tropospheric component from very large near-surface explosions is only about 5% of the total. If an estimate is available of the total amount of debris that has been deposited globally at any given time following an explosion, the stratospheric inventory can thus be approximated. This is the method that was used prior to 1959 by American investigators.

Recognizing that the method of material balancing resulted in very uncertain estimates of the stratospheric inventory, a number of attempts were made to collect samples of stratospheric dust. The AEC in 1953 (Holland, 1959b) succeeded in collecting about 12 dust samples at an altitude of 80,000 to 100,000 ft using elastrostatic precipitators carried aloft by balloons. Although fission products were found in these samples and served for the first time to demonstrate the presence of radioactive dust in the stratosphere, the data were at best only semiquantitative because little was known about the performance of the elastrostatic precipitator at the altitudes involved (Loysen et al., 1956).

Beginning in 1956, the AEC undertook Project Ashcan in which balloonborne filters sampled the stratosphere at Minneapolis, Minnesota; San Angelo, Texas; Panama; and Sao Paulo, Brazil (Holland, 1959a). Samples at each of these stations were collected monthly at four altitudes, from 50,000 to 90,000 ft, and the filters analyzed for gross β activity and six radionuclides; ^{140}Ba, ^{95}Zr, ^{144}Ce, ^{137}Cs, ^{89}Sr, and ^{90}Sr. The shorter-lived isotopes made it possible to ascertain the age and origin of the debris.

In August, 1957, the United States Air Force began to undertake a series of long-range high-altitude flights with Lockheed U-2 aircraft equipped with filtering equipment (Feely, 1960). The planes monitored the atmosphere along the 70th meridian West, from 66°N latitude to 6°S latitude, and by mid-1959 more than 1400 samples had been collected and analyzed (Stebbins, 1961).

Utilizing the data from Ashcan and the high-altitude sampling program (Project HASP), it became possible for the first time to make reliable estimates of the stratospheric inventory of bomb debris. The most striking finding was that the inventory of ^{90}Sr was very much less than had been previously estimated. At the time of the October, 1958 moratorium on weapons testing, immediately following heavy tests series by both the United States and the Soviet Union, the stratospheric inventory was estimated by Eisenbud (1959) to be about 4.3 MCi, and Libby (1959) estimated that it was between 4.5 and 5.5 MCi. The HASP investigators (Shelton, 1959; Feely, 1960; Stebbins and Minx, 1962) concluded from an analysis of data collected from January to August, 1959 that the stratospheric inventory during that period was only 0.8 MCi, less than 20% of the previous estimates. By examining the rate of fallout from the tests by the Soviet Union in the fall of 1958 and from the measured ^{90}Sr stratospheric inventory in the early fall of 1958, prior to the tests by the Soviet Union, Feely concluded that more than half of the debris from the tests by the Soviet Union fell out within 6 months and that at least half of the debris injected by the United States and Great Britain during 1958 fell out within 12 months. These values would be equivalent to a mean residence time of about 8 months in the case of the Soviet debris and about 18 months in the case of the debris injected into the stratosphere near the equator. These residence times were very much less than the estimates of 5 to 7 years that had been made by Libby (1956) and others (UNSCEAR, 1958).

Martell (1959) came to a similar conclusion by studying the relative contribution of debris of tests made by the Soviet Union and the United States during late 1958 and 1959. He found it possible to apportion the radioactivity in rain water between the United States tests in the spring of 1958 and the Soviet tests in the fall of 1958. This was done by taking advantage of the fact that nuclear explosions in the series of United States tests in the Marshall Islands between May and July, 1958 were uniquely labeled by the presence of ^{185}W in the debris. About 250 MCi of this 75-day half-life isotope were formed as of August 1, 1958 (Libby, 1959). Martell estimated that 40% of this total was retained in the stratosphere, where it coexisted initially with other Hardtack debris, and by studying the ratio of ^{185}W to other isotopes such as ^{89}Sr, ^{90}Sr, and ^{145}Ba, Martell was able to trace the Hardtack debris and differentiate it from the debris of the Soviet Union explosions which was characterized by similar fission product spectra but which did not contain ^{185}W. Kuroda et al. (1962), by analyzing the diminution of ^{90}Sr in rainfall during the test moratorium, concluded that the overall stratospheric residence time during the period 1958 to 1960 was 0.7 ± 0.1 year.

The reason why the stratospheric inventory was overestimated initially is not known but may have been due to underestimates of the amount of fallout that occurred in the vicinity of the detonations. As noted earlier, the fraction of close-in fallout may vary considerably, depending on the type of burst, and the estimates of the total amount of close-in fallout from the various shots were based on very scanty data.

Behavior of Individual Radionuclides Produced in Weapons Tests

STRONTIUM-90

Because of its long life and the fact that it is an isotope of a bone-seeking element, it has been suggested that worldwide exposure to ^{90}Sr is the limiting factor of the extent to which human beings can tolerate global contamination by fission products (Rand Corporation, 1953). Prior to about 1953, studies of the behavior of ^{90}Sr from bomb debris were undertaken by the AEC as part of a classified program for determining the long-range effects of nuclear weapons. This program was known as Project Sunshine and for this reason the unit "picocuries of strontium 90 per gram of calcium" was originally designated the Sunshine Unit, but this was gradually replaced by strontium unit, retaining the same abbreviation SU.

Extensive investigations of the fate of the ^{90}Sr produced in nuclear explosions have made it possible to predict the manner in which this nuclide is transported through the atmosphere and enable one to understand how global contamination results from the release of a given amount of ^{90}Sr into the atmosphere. It is also possible to relate this global contamination to the contamination of man's foods, and, finally, given a knowledge of the amount of contamination in food, it is possible to forecast the skeletal burden of ^{90}Sr in individuals consuming the food.

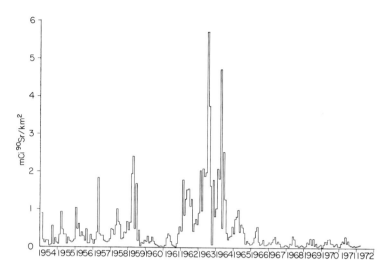

Fig. 14-5. Monthly deposition of ^{90}Sr in New York City, 1954–1971 (U.S. Atomic Energy Commission, Health and Safety Laboratory).

Global Inventory of Strontium-90

Essentially all of the ^{90}Sr injected into the upper atmosphere during the period of weapons testing prior to the test-ban agreement was deposited on the earth's surface by 1970. This is seen in Fig. 14-5 in which monthly deposition is plotted for the period 1954 to early 1971. The global accumulation of ^{90}Sr from 1958 through 1970 is shown in Fig. 14-6 (Volchok and

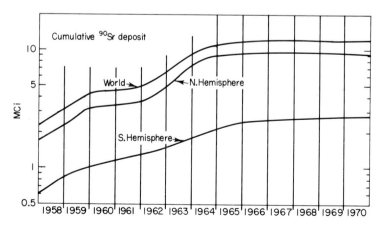

Fig. 14-6. Accumulation of ^{90}Sr on the earth's surface from 1958 through 1970 (Volchok and Kleinman, 1971).

Kleinman, 1971), in which it is seen that worldwide deposition reached a peak of about 12.5 Ci by late 1967. Deposition in the Southern Hemisphere was at this time less than one-third that in the Northern Hemisphere. The ground deposition should have diminished at a rate of 2.5%/year because of radioactive decay, but Volchok (1970) has shown that tests by the Chinese and French in 1968–1969 resulted in ^{90}Sr deposition about equivalent to decay of the cumulative ground deposit.

Worldwide deposition of ^{90}Sr reflects the stratospheric inventory, which has been estimated by Krey *et al.* (1970) using data from various high-altitude sampling programs. Krey's estimates are shown in Fig. 14-7 in which the exponential diminution following cessation of tests by the United States and Soviet Union in 1963 is clearly evident, with a half-life of about 1 year. The interruption in this decline owing to additions from atmospheric tests between 1967 and 1970 is also apparent.

Within the hemispheres, the deposition of ^{90}Sr was far from uniform, as can be seen from Fig. 14-8. The band of relatively heavy fallout (60–80 mCi km^{-2}) in the northern midlatitudes is conspicuous and is believed to be due to meteorological factors that result in increased stratospheric–

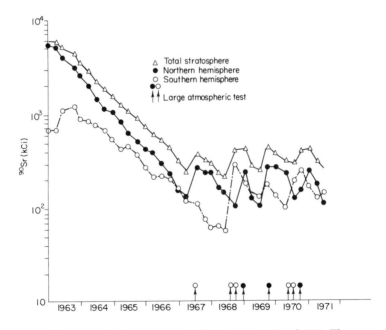

Fig. 14-7. The stratospheric inventory of ^{90}Sr between 1963 and 1970. The exponential decline following the 1963 test-ban agreement was interrupted by large-scale tests undertaken by France and China from 1967–1970 (Krey *et al.*, 1970).

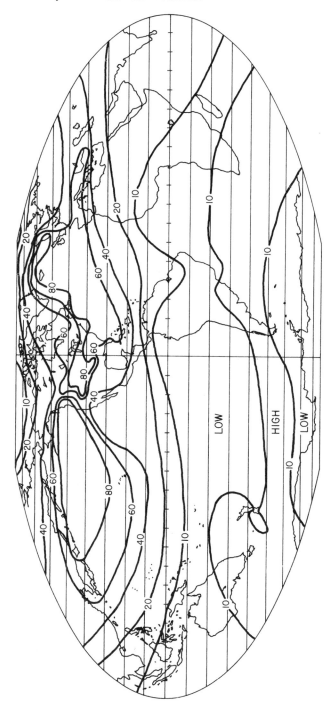

Fig. 14-8. Isolines of cumulative ⁹⁰Sr deposits based on analyses of soils collected 1965–1967 (mCi km⁻²) (United Nations Scientific Committee on the Effects of Atomic Radiation, 1969).

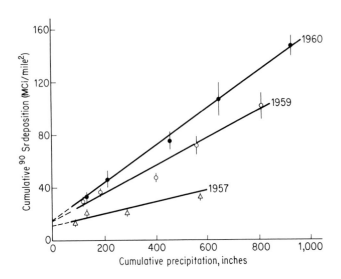

Fig. 14-9. Dependence of fallout on precipitation on the Olympic Peninsula (Hardy and Alexander, 1962).

tropospheric transfer in the spring months. This phenomenon was discussed in Chapter 5.

At any given latitude the deposition is affected by the amount of rainfall. For example, analysis of soil samples (Alexander, 1959) from the band 20 to 30°S latitude showed a range of ^{90}Sr values from 0.4 mCi/square mile at Antofagasta, Chile to 11.6 mCi/square mile in Brisbane, Australia. The mean of samples collected at seven localities was 6.1 mCi/square mile. The low value at Antofagasta is explained by the fact that no precipitation was recorded from January 1, 1953 to the date of sample collection several years later. The dependence of fallout on the amount of precipitation is illustrated elegantly in Fig. 14-9, in which the cumulative fallout is plotted against rainfall at five sites on the Olympic Peninsula in Washington State (Hardy and Alexander, 1962).

There has been attention given to the possibility that deposition of ^{90}Sr (and other dust-borne substances) may be greater on the ocean surfaces than on land, owing possibly to the effect of ocean spray as a scavenging mechanism (UNSCEAR, 1969). This possibility was initially suggested as the result of ocean water analyses that indicated the oceanic inventories might be higher than could be explained on the basis of fallout measurements made on land (Belyaev *et al.*, 1965). However, field and laboratory studies by Freudenthal (1970a, b) seem to indicate that the phenomenon of spray scavenging is not responsible for a significant increase of oceanic fallout.

Radiostrontium in Foods

It was seen earlier that the ^{90}Sr content of plants is due in part to uptake from soil and in part to foliar deposition. The relative proportions of the two components depend on the rate of fallout in relation to the cumulative soil deposit.

The dietary sources of ^{90}Sr depend in part on the food consumption habits of the population, including differences in the kinds of foods eaten and the manner in which the food is processed or prepared. In the United States the per capita daily calcium intake is 1 g, but in other countries this may vary somewhat and, moreover, may be derived from different foods, as was discussed in Chapter 5 and summarized in Table 5-1.

For many years, the AEC Health and Safety Laboratory has reported on the whole diet ^{90}Sr content in two cities, New York and San Francisco. Chicago was included until 1967. There have been significant differences in the daily amounts of ^{90}Sr ingested by the inhabitants of these cities, as can be seen from Fig. 14-10, in which are given the daily intakes in each of these cities for the period from 1960 through 1968 (Harley, 1969). It is seen that New York has been consistently higher than Chicago and San Francisco. A comparison of typical diets in New York and San Francisco is given in Table 14-1, which shows that on an annual basis the average New Yorker was ingesting more than 3.7 times amounts of ^{90}Sr ingested by inhabitants of San Francisco. The origin of this difference is not understood but may be due in part to the somewhat lower annual rainfall in regions that supply food to the San Francisco area. Bennett (1971) estimated that based on data from the food sampling program, average daily intakes for 1971 would be 11.8 pCi/day in New York and 4.1 pCi/day in San Francisco.

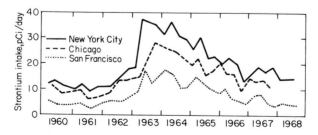

Fig. 14-10. Estimated daily intake of ^{90}Sr in three United States cities 1960–1968 (Harley, 1969).

TABLE 14-1

STRONTIUM-90 IN DIET[a]

90Diet category	Intake (kg/year)	Calcium (g/year)	Percent of total intake	90Sr (pCi/kg)	90Sr (pCi/year)	Percent of total intake	90Sr (pCi/kg)	90Sr (pCi/year)	Percent of yearly intake
Dairy products	200	216.0	58	10.4	2080	38	1.7	340	23
Fresh vegetables	48	18.7	9	12.2	586	22	4.4	211	26
Canned vegetables	22	4.4		9.3	205		1.9	42	
Root vegetables	10	3.8		4.9	49		4.5	45	
Potatoes	38	3.8		9.5	361		1.5	57	
Dried beans	3	2.1		3.6	11		5.4	16	
Fresh fruit	59	9.4	3	18.2	1074	22	1.4	83	9
Canned fruit	11	0.6		2.4	26		1.7	19	
Canned fruit juices	28	2.5		3.3	92		1.1	31	
Bakery products	44	53.7	20	5.4	238	11	4.7	207	29
Flour	34	6.5		6.9	235		3.4	116	
Whole grain products	11	10.3		8.5	94		6.3	69	
Macaroni	3	0.6		4.7	14		2.7	8	
Rice	3	1.1		2.3	7		9.7	29	
Meat	79	12.6	8	10.0	79	3	0.7	55	7
Poultry	20	6.0		2.4	48		1.1	22	
Eggs	15	8.7		3.4	51		5.1	23	
Fresh fish	8	7.6	2	0.5	4		0.0	0	0
Shellfish	1	1.6		1.4	1		0.3	0	
Water	400	2–10		0.5	200	4	(0.2)	80	6
Yearly intake (rounded)					5450			1450	

[a] Harley (1969).

Fig. 14-11. [90]Sr in whole milk in New York City, 1954–1971. One liter of milk contains ~1 g of calcium (U.S. Atomic Energy Commission, Health and Safety Laboratory).

The difference in the contribution from milk is particularly significant, the New York:San Francisco ratio being 6.1 on an annual basis. The reason why fresh fruit should contribute so significantly to the ^{90}Sr intake of New Yorkers is not explained.

So much for the differences within the United States; Harley (1969) has also compared the New York diet with similar data reported for the Soviet Union by Petukhova and Knizhnikov (1969) (Table 14-2). During 1968, the Soviet Union diet contributed 12,990 pCi ^{90}Sr compared to 7170 pCi ^{90}Sr for a representative New York diet. Since the daily calcium intake is higher in New York than in the Soviet Union (1 g/day compared to 0.6 to 0.8 g/day), the difference when expressed as strontium units is even greater, about 19 SU for the New York diet and 54 SU for the USSR.

The data of Fig. 14-10 show that the ^{90}Sr content of the total diets diminished with a half-time of $3\frac{1}{2}$ to 4 years, following the peak values of 1963–1964 (Harley, 1969).

Fig. 14-12. [90]Sr in tap water in New York City, 1954–1971 (U.S. Atomic Energy Commission, Health and Safety Laboratory).

TABLE 14-2

DIETARY ^{90}Sr IN THE UNITED STATES AND SOVIET UNION, 1966[a]

Diet category	United States diet (New York City)				Soviet Union diet (country)[b]			
	Intake (kg/year)	^{90}Sr (pCi/kg)	^{90}Sr (pCi/year)	Percent of total intake	Intake (kg/year)	^{90}Sr (pCi/kg)	^{90}Sr (pCi/year)	Percent of total intake
Milk	200	13.4	2,970	42	110	14	1,540	12
Bread	89	17.4	1,580	22	220	33	7,250	55
Meat	114	2.8	320	4	60	10	600	5
Cereals	6	6.3	40	1	20	18	360	3
Fish	9	1.8	20	—	4	30	120	1
Potatoes and vegetables	219	8.3	1,800	25	220	12	2,640	20
Water	400	1.1	440	6	440	1.1	480	4
Yearly intake			7,170				12,990	

[a] Harley (1969).
[b] Based on data given by Petukhova and Knizhnikov (1969).

Milk is an important source of ^{90}Sr and a good index of exposure of growing children in many countries where dairy products are the main source of calcium. The United States Public Health Service operates an extensive milk sampling network in cooperation with many local communities. The number of stations has varied from time to time, but in April, 1971 there were 156 stations located in 50 states.

The longest series of monthly samples is from the program of the AEC Health and Safety Laboratory, which has been sampling both powdered and fresh milk from New York State since early 1954. The ^{90}Sr content of liquid whole milk in New York City through 1970 is shown in Fig. 14-11, in which it is seen that the milk contained as much as 40 pCi ^{90}Sr/g Ca in 1963. The tap water in New York City reached a peak of 2.12 pCi/liter during this period, as can be seen in Fig. 14-12.

The relationship of the ^{90}Sr content of cows' milk, the rate of fallout, and cumulative deposition has been described by Knapp (1961) in the following model [Eq. (14-1)]:

$$\begin{pmatrix} \text{Average } ^{90}\text{Sr level in} \\ \text{United States milk} \\ \text{supplies, expressed} \\ \text{in pCi/liter} \end{pmatrix} = A \begin{pmatrix} \text{average cumulative} \\ ^{90}\text{Sr level in United} \\ \text{States soil, expressed} \\ \text{in mCi/square mile} \end{pmatrix}$$

$$+ B \begin{pmatrix} \text{average } ^{90}\text{Sr deposition} \\ \text{for the preceding} \\ \text{month, expressed in} \\ \text{mCi/square mile} \end{pmatrix} \quad (14\text{-}1)$$

Solution of these equations gives

$$A = 0.11 \text{ pCi/(liter)(mCi)(square mile)}$$

$$B = 2.6 \text{ pCi/(liter)(mCi)(square mile)}$$

and Eq. (14-1) becomes

$$\begin{pmatrix} \text{Average } ^{90}\text{Sr level in} \\ \text{United States milk} \\ \text{supplies, expressed} \\ \text{in pCi/liter} \end{pmatrix} = 0.11 \begin{pmatrix} \text{average cumulative} \\ ^{90}\text{Sr level in United} \\ \text{States soil, expressed} \\ \text{in mCi/square mile} \end{pmatrix}$$

$$+ 2.6 \begin{pmatrix} \text{average } ^{90}\text{Sr deposition} \\ \text{for the preceding} \\ \text{month, expressed in} \\ \text{mCi/square mile/month} \end{pmatrix}$$

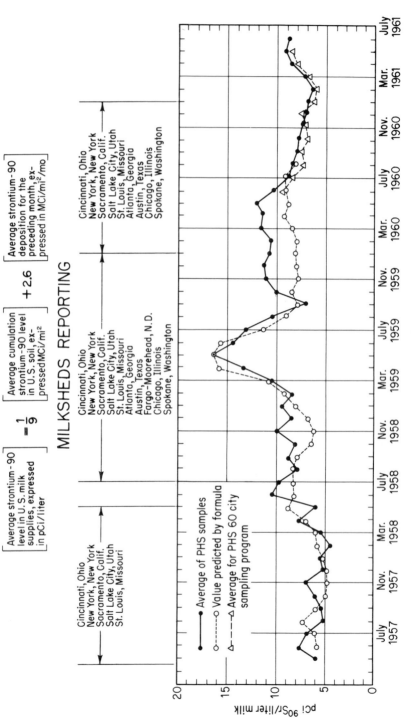

Fig. 14-13. Comparison of predicted and observed concentrations of ^{90}Sr in milk from 1957 until 1961 (Knapp, 1961).

This model, when applied to data from monitoring programs in the United States between 1957 and 1961, gives the data in Fig. 14-13, which shows reasonably good agreement between the predicted ^{90}Sr content of milk and the *average* concentrations actually observed in the milk sampling network.

The amount of ^{90}Sr contributed by grains depends on the milling practices. In the United States, Canada, and the United Kingdom, it has been shown that the ^{90}Sr:calcium ratio in flour was approximately one-third to one-half of that in the whole grain and one-quarter of that in bran (UNSCEAR, 1962). Milling reduces the ratio of ^{90}Sr:calcium in rice grain to one-fifth to one-tenth of the value in whole rice. Harley (1969) explained the relatively large contribution of bread to the daily intake of ^{90}Sr in the Soviet Union as being due to the use of black bread that includes ^{90}Sr deposited on the surfaces of wheat and rye.

In parallel with the food sampling network, many countries have designed sampling systems to provide samples of human bone for ^{90}Sr analyses. It has been found that a relatively simple model for the uptake of ^{90}Sr can account for the observed distribution of this radionuclide in human bones, if: (1) the rate of exchange of new strontium with strontium already in the bone is constant with age; (2) the ^{90}Sr content of the diet is known for each year of the person's life; (3) the annual accretion of calcium in the human skeleton is known; and (4) the discrimination factor between strontium and calcium is known and it is constant with age. An equation proposed originally by Kulp and Schulert (1962) [Eq. (14-2)] which predicts the concentration of ^{90}Sr in individuals of a given age is as follows:

$$Q_B = \frac{\sum_{i=1}^{P} [Ca_{n_i}A_{n_i} + Ca_{ex_i}(A_{n_i} - A_{s_{i-1}}) + 0.975\ Ca_{s_{i-1}}A_{s_{i-1}}]}{Ca_{s_p}}$$

(14-2)

where

Ca_n = grams of calcium added in a given 12-month period in the formation of new bone (depends on the age of the individual)
Ca_{ex} = grams of calcium exchanged per year
Ca_s = grams of calcium in the skeleton
Ca_{s_p} = total grams of calcium in present skeleton
A_{n_i} = specific activity of newly depositing bone (pCi ^{90}Sr/g Ca)
$A_{s_{i-1}}$ = specific activity of total skeleton in preceding year (pCi ^{90}Sr/g Ca)
Q_B = specific activity of total skeleton at midpoint of Pth year of ingestion (pCi ^{90}Sr/g Ca)

In the above equation the summation is over the years for which there

TABLE 14–3

METABOLIC PARAMETERS USED TO PREDICT ^{90}SR CONCENTRATIONS IN HUMAN BONE[a]

Age	Bone/diet discrimination factor (K)	Annual bone turnover rate (f)	Net calcium accretion rate (g/year)
0	0.10[b]	—	28
1	0.35	0.75	72
2	0.25	0.25	47
3	0.19	0.10	32
4	0.18	0.05	22
5	0.17	0.04	18
6	0.16	0.04	20
7	0.15	0.04	25
8	0.13	0.04	33
9	0.13	0.04	44
10	0.13	0.07	55
11	0.14	0.08	67
12	0.15	0.08	76
13	0.16	0.09	85
14	0.17	0.09	91
15	0.18	0.08	91
16	0.19	0.08	88
17	0.20	0.07	79
18	0.21	0.06	62
19	0.23	0.05	38
20	0.25	0.035	5
21	0.25[c]	0.035[c]	0[c]

[a] Rivera (1965).
[b] Fetal bone/mother's diet.
[c] These values are assumed to persist.

is measurable ^{90}Sr in the diet, and the factor 0.975 corrects for the radioactive decay of ^{90}Sr in the skeleton. The parameters of this equation were refined by Rivera (1965) using data from various sources, as shown in Table 14-3. The skeletal ^{90}Sr burdens of New York City residents of various ages when calculated on this basis are in good agreement with the results obtained by radiochemical analysis, as shown in Fig. 14-14.

As noted earlier, the ^{90}Sr content of Soviet Union diets was shown to be higher than in New York. The skeletal burden of ^{90}Sr is also higher in the Soviet Union, but not proportionally so. The reason for the lack of proportionality, implying different observed ratios in the Soviet Union compared to the United States, is not understood (UNSCEAR, 1969).

The next step in evaluating the significance of generalized ^{90}Sr contamination is to translate the estimated skeletal burdens into dose. Because of the complexity of the structure and physiology of the skeleton, this is not a simple matter (Björnerstedt and Engstrom, 1960; Spiers, 1966; UNSCEAR, 1969). The structure and function of bone are such that the dose to two tissues are of interest: the marrow and the endosteum. The dose to each of these tissues is contributed in part by ^{90}Sr deposited in cortical bone and in part from the deposit in trabecular bone. The dose estimate is not simple if one assumes uniform skeletal deposition of ^{90}Sr, and it is even more difficult in the case of nonuniform deposition. The latter case is encountered in practice owing to the fact that the pattern of skeletal deposition is dependent on the variations of dietary ^{90}Sr with time and variations in the rate of skeletal calcium accretion and turnover with age.

Taking all available information into consideration and using the model proposed by Spiers, UNSCEAR (1969) estimated that for the case of nonuniform skeletal deposition the dose to bone marrow is 0.43 mrem/year/SU and that the dose to endosteal tissue is 0.82 mrem/year/SU. The estimates are based on the ^{90}Sr content of samples of vertebral bone.

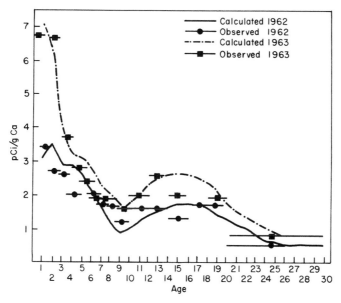

Fig. 14-14. [90]Sr in bone of New York City residents of various ages (Rivera, 1965).

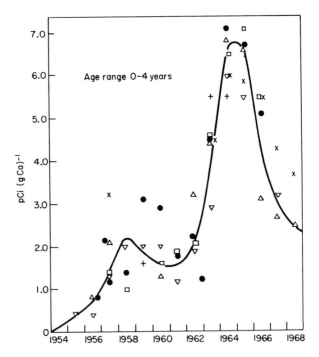

Fig. 14-15. ^{90}Sr/Ca ratios in human bone samples in North Temperate Zone (age range 0–4 years). Legend: □, Canada; △, Denmark; ▽, Federal Republic of Germany; ◇, France; +, Poland; ×, Soviet Union (Moscow); ●, United Kingdom (adults, West London; other ages, national); ○, United States (New York City)(United Nations Scientific Committee on the Effects of Atomic Radiation, 1969).

The ^{90}Sr content of human bone in the North Temperate Zones from 1955 through 1968 is shown in Figs. 14-15 and 14-16 for two age groups, 0–4 and 5–19 years, respectively (UNSCEAR, 1969). The peak concentrations occurred in 1965–1966, reaching a maximum of about 7 SU in children up to 4 years of age. The lifetime dose commitment to endosteal cells from ^{90}Sr for residents of the North Temperate Zone was calculated by UNSCEAR to be 130 mrem on the average, compared to an average dose of 28 mrem in the South Temperate Zone. The bone-marrow dose commitment was calculated to be 64 and 14 mrem in the North and South Temperate Zones, respectively.

STRONTIUM-89

^{89}Sr behaves like ^{90}Sr in all respects, except that it is relatively more important as a foliar contaminant because of its short half-life. During periods of active testing, the ^{89}Sr component of the total radiostrontium dose to the skeletons of very young children may be predominant, but in older individuals the ^{89}Sr component tends to be minor because of the short half-life of this isotope. Thus, it has been shown (Dunning, 1962) that for any dose of ^{90}Sr incorporated into bone during the first year of a child's life the ^{89}Sr may deliver as much as 3.4 times the dose delivered by [90]Sr. However, unless fresh ^{89}Sr is added during the second year, this isotope will have decayed almost completely, so that the ^{89}Sr component would become relatively unimportant over the whole lifetime of the individual. The dose commitment to skeletal tissues from ^{89}Sr deposited from weapons tests has been calculated to be less than 1 mrem (UNSCEAR, 1969).

CESIUM-137

[137]Cs has a 30-year half-life compared to 28 years for ^{90}Sr, and it is produced somewhat more abundantly, in a ratio of about 1.6 on an activity basis. However, as noted in Chapter 5, ^{137}Cs is tightly bound by soil and thus is not readily incorporated metabolically into land plants, but does contaminate human foods by foliar absorption.

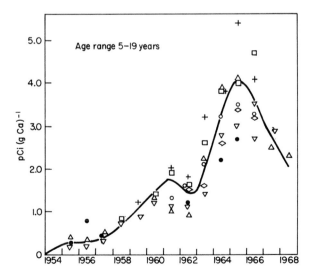

Fig. 14-16. [90]Sr/Ca ratios in human bone samples in North Temperate Zone (age range 5–19 years). See Fig. 14-15 for legend (United Nations Scientific Committee on the Effects of Atomic Radiation, 1969).

Measurement of ^{137}Cs in both food and people has been facilitated by the readiness with which this nuclide can be quantitated by γ spectrometry. It has proved convenient to report the ^{137}Cs in units of picocuries of ^{137}Cs per gram of potassium, although it is known that the metabolism of cesium and potassium is, in fact, somewhat different from food to food and organ to organ (Yamagata and Yamagata, 1960). For purposes of dose calculation, ^{137}Cs is usually assumed to be distributed uniformly throughout the body.

The distribution of ^{137}Cs in the stratosphere and the pattern of terrestrial deposition are similar to that for [90]Sr. A number of investigators have shown that fractionation of ^{137}Cs with respect to ^{90}Sr does not occur (UNSCEAR, 1969).

Gustafson (1969) used ^{90}Sr deposition measurements at various stations in the United States to estimate the ^{137}Cs deposition and concluded that over most of the country the accumulation by the end of 1965 ranged between 60 and 100 mCi/km², but that there was even less variation in the ^{137}Cs body burdens of the American population, owing presumably to the fact that many important items of food originate from many different areas of the country so that some blending takes place.

The relative ^{137}Cs content of various foods is given in Table 14-4, which shows that in a typical Chicago diet dairy products, grains, and meat products are the most important sources of this nuclide. Because cesium is bound so tightly in soil, the ^{137}Cs content of land-grown food is generally dependent on the rate of fallout rather than cumulative deposition. However, Gustafson (1969) has shown that this is not so for freshwater fish. In a lake system he studied, the half-life of the ^{137}Cs concentration in fish was about $2\frac{1}{2}$ years.

Gustafson (1969; Gustafson et al., 1970) has for many years maintained careful observations of ^{137}Cs in air, fallout, food, and humans in the Chicago area. His data from 1961 through 1970 are summarized in Fig. 14-17, which shows that deposition in the Chicago area reached a peak of about 130 mCi/km² during 1965 to 1966, and that this was reflected by a maximum annual dose of about 3 mrem owing to internally deposited [137]Cs.

[137]Cs delivers a somewhat higher dose through external radiation. By 1969, when the shorter-lived γ-emitting fallout radionuclides had decayed, the external radiation exposure owing to fallout on an open field in Illinois was almost entirely because of ^{137}Cs and was 13 mR/year. Gustafson et al. (1970) estimated that the indoor exposure, assuming 40% transmission, was, therefore, about 5 mR/year, and if one applies a factor of 0.6 to estimate gonadal exposure, the absorbed gonadal dose would be about 3 mrem/year. UNSCEAR (1969) estimated that as of 1968 the mean 50-year dose

TABLE 14–4

^{137}Cs in Chicago Diet (Argonne Data)

Diet category	Intake (kg/year)	Potassium g/kg	Potassium g/year	Percent of total intake	^{137}Cs (10/68) pCi/kg	^{137}Cs (10/68) pCi/year	Percent of total intake
Dairy products	200	1.4	280	21	18	3,600	29
Fresh vegetables	48	2.3	110		2	100	
Canned vegetables	22	1.3	29		9	200	
Root vegetables	10	2.9	29	29	7	70	8
Potatoes	38	4.5	171		17	650	
Dried beans	3	13.9	42		5	10	
Fresh fruit	59	1.9	112		4	240	
Canned fruit	11	1.2	13	13	18	200	10
Canned fruit juices	28	1.9	53		26	730	
Bakery products	44	1.2	53		21	920	
Flour	34	1.0	34		33	1,120	
Whole grain products	11	3.5	38	10	30	330	20
Macaroni	3	1.8	5		19	60	
Rice	3	N.d.[b]	—		N.d.[b]	—	
Meat	79	3.3	261		26	2,060	
Poultry	20	2.7	54	25	15	300	20
Eggs	15	1.5	22		7	100	
Fresh fish	8	3.4[c]	27		194[c]	1,550	
Shellfish	1	N.d.[b]	—	2	N.d.[b]	—	13
Water	400	N.d.[b]	—		(0.05)[d]	20	
Yearly intake (rounded)			1330			12,300	

[a] Based on data given in E. P. Hardy, Jr., and J. Rivera (Eds.), Fallout Program Quarterly Summary Report, September 1, 1968, USAEC Report HASL-204, January 1969, and P. F. Gustafson and J. E. Miller, Significance of Cs-137 in Man and His Diet *Health Phys.*, 16: 167–183 (February 1969). From Harley (1969).

[b] Not determined.

[c] Based on 90% ocean, 10% freshwater fish.

[d] Number in parentheses represents estimated value.

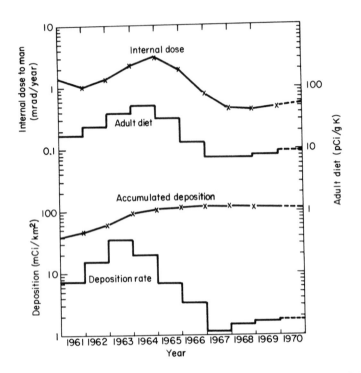

Fig. 14-17. Fallout ^{137}Cs deposition rate, accumulation, dietary levels, and dose to man in the Chicago area 1961 to mid-1970 (Gustafson *et al.*, 1970).

commitment from ^{137}Cs already deposited in the North Temperate Zone was 36 mrem from external radiation and about 21 mrem from internal radiation. The estimates for the South Temperate Zone were 8 mrem from external radiation and 4 mrem from internally absorbed [137]Cs.

As was discussed in Chapter 5, the ^{137}Cs absorption by man can be increased markedly by anomalous ecological factors. One example was in potassium-deficient soils on the island of Jamaica, where the ^{137}Cs content of milk was 10 to 100 times that of milk from farms with normal soil (Broseus, 1970). The most remarkable anomaly is in the sub-Arctic, where the body burdens of ^{137}Cs in individuals eating large quantities of moose or caribou meat have been more than 10 times higher than the local population average (UNSCEAR, 1969). In studies of Alaskan Eskimos, Eckert *et al.* (1970) found less ^{137}Cs in inhabitants of villages subsisting on reindeer meat than in villages where the main source of meat is moose on caribou (Chapter 5).

CARBON-14

We have seen in Chapter 7 that cosmic-ray reactions in the upper atmosphere result in the transmutation of atmospheric nitrogen to ^{14}C and that this isotope has been found to be in secular equilibrium throughout the biosphere in a concentration of 7.5 ± 2.7 pCi $^{14}C/g$ C. This equilibrium concentration is believed to have been unchanged for at least 15,000 years prior to 1954 when the advent of large thermonuclear explosions resulted in the production of additional ^{14}C in sufficient amounts to perturb the natural equilibrium. The ^{14}C content of the atmosphere is thought to exist as CO_2.

Because the average life of ^{14}C is 8000 years, the dose from ^{14}C introduced into the environment will be delivered for many generations. Since the genetic effects may be cumulative even in such small doses, there has been some concern over the genetic significance of adding additional quantities of this nuclide to the already existing pool of ^{14}C in the biosphere (Totter *et al.*, 1958; Leipunsky, 1957). A special reason for concern is that this isotope of carbon will enter the pool of biological carbon and may be incorporated into the molecules of which the genes are formed. Thus, in addition to the calculable mutations from ionization produced by the ^{14}C dose, a possible additional source of mutations may be the transmutation of the carbon within the genetic material itself (Zelle, 1960; Totter *et al.*, 1958).

Various estimates of the amount of ^{14}C produced in the nuclear explosions have been based on estimates of the total yields of the explosions, the ratio of fission to fusion, and the interaction of neutrons of various energies with atmospheric nitrogen. However, the best estimates are those made by actual measurement of the ^{14}C activity in stratospheric air using high-flying aircraft and balloons. The excess of the ^{14}C global inventory can be taken to have been 63.4×10^{27} atoms or about 4.2×10^5 Ci in January, 1964, by which time the major injections of ^{14}C to the stratosphere had terminated (UNSCEAR, 1964). Carbon distributes itself quickly among the major environmental compartments—the stratosphere, troposphere, biosphere, and surface ocean waters. Transfer among these compartments takes place with time constants on the order of a few years at most (UNSCEAR, 1969), but transfer to the deep oceans proceeds more slowly. After relatively few years, the biospheric ^{14}C response to a stratospheric injection will reach equilibrium and decrease slowly at a rate determined by transfer of the ^{14}C to deep ocean water and possibly humus (Young and Fairhall, 1968).

By the end of 1967, the tropospheric ^{14}C content had increased about 510% above natural levels in the Northern Hemisphere and about 390% in the Southern Hemisphere (Nydal, 1968). Because of the short time constants involved in transfer from the atmosphere to biosphere, the ^{14}C content of human tissues and foods increased rapidly following the heavy testing schedules of 1961 and 1962. By 1964, the radiocarbon levels in human tissue had increased from 20 to 30% above the 1950 values (Drobinski, 1966). Based on these and other observations, the dose commitment up to the year 2000 from ^{14}C produced up to the time of the nuclear weapons test-ban agreement is estimated to be 13 mrem in both the North Temperate and the South Temperate Zones (UNSCEAR, 1969).

PLUTONIUM-239 AND -238

[239]Pu and [238]Pu are both produced in nuclear and thermonuclear bombs. It has been estimated that about 300 kCi of [239]Pu has been distributed globally, mainly from explosions of megaton range weapons that injected debris into the stratosphere (Harley, 1971). Transfer from the stratosphere, and deposition from the troposphere to earth's surface, proceeds at the same rates as ^{90}Sr, with the result that the ratio of [239]Pu/^{90}Sr has been found to be remarkably constant since the cessation of large-scale tests in 1963. Plutonium is tightly bound by soils and is present in biota in only minute amounts (Romney and Davis, 1972).

Although there have been relatively few soil or fallout measurements, surface deposition has been estimated by Harley using the known ^{90}Sr deposition and the ratio [239]Pu/^{90}Sr in stratospheric air, which has been observed to be about 0.017, corrected for decay to the time of production. Since deposition of ^{90}Sr in the North Temperate latitudes was about 80 mCi/km^2, by using the factor 0.017, one can estimate the deposition of [239]Pu to be about 1.4 mCi/km^2.

Magno et al. (1967) undertook the measurement of the [239]Pu content of surface air, total diets, and human tissues during the period 1964–1967. The concentration in surface air in Winchester, Massachusetts was found to average 0.15 × 10^{-3} pCi/m^3 with a range from 0.047 × 10^{-3} to 0.44 × 10^{-3} pCi/m^3. The mean concentration in 33 samples of milk was (0.2 ± 0.5) × 10^{-3} pCi/liter, or below the level of significance. Total diets collected from six regions of the country ranged from (2.7 ± 2.3) × 10^{-3} to (5.8 ± 1.7) × 10^{-3} pCi/kg with a mean value of (3.8 ± 0.9) × 10^{-3} pCi/kg. The average plutonium dietary intake was estimated to be 7.0 × 10^{-3} pCi/day from which Magno et al. calculated that 2 × 10^{-7}

pCi/day reached the blood. Using the surface air data from Winchester, it was estimated that more than 100 times this amount reaches the blood via inhalation. Analyses of human tissues revealed the liver to have the highest concentration of ^{239}Pu, ranging from 0.21 to 2.52 pCi/kg with a mean value of 0.78 pCi/kg. The lung tissue ranged from 0.14 to 1.13 pCi/kg with a mean concentration of 0.45 pCi/kg. The ratio of liver-to-lung concentration was 1.6 ± 0.5. Samples of human bone composited according to age ranged from 0.04 to 0.12 pCi/kg with the highest values in the 5- to 8-year age group.

^{238}Pu is produced in much smaller quantities than ^{239}Pu. Nevertheless, production in weapons testing was sufficient, prior to the 1963 test-ban agreement, to mask the effect of 17 kCi ^{238}Pu injected into the upper atmosphere by an aborted space vehicle (see Chapter 16).

The critical organ for plutonium when present as an oxide aerosol is the lung. Based on the measurements noted above, it has been calculated that as of 1963 the 50-year dose commitment from ^{239}Pu, then present in the atmosphere, was 4.5 mrem to the whole lung and 1000 mrem to the pulmonary lymph nodes. The 50-year dose commitment from ^{238}Pu was less than 10% of the dose from ^{239}Pu (Environmental Protection Agency, 1972).

RADIOIODINE

Although several species of radioiodine are produced in fission, only one, ^{131}I, is of major significance so far as worldwide fallout is concerned (Holland, 1963; Eisenbud and Wrenn, 1963). This nuclide is produced copiously, about 64 MCi/megaton of fission. We have seen earlier that prior to the test-ban agreement the fission yields of all nuclear weapons tests conducted in the atmosphere totalled about 193 megatons. Thus, in excess of 12 billion Ci of ^{131}I were released to the atmosphere, most of which was produced by explosions in the megaton range of yield. The bulk of the ^{131}I was thus injected into the stratosphere where it decayed substantially before transferring to the troposphere and depositing on the earth's surface. However, the ^{131}I produced by explosions having yields less than 100 kilotons remained in the troposphere, together with a small fraction of the debris from megaton yield weapons. Substantial amounts of radioiodine were distributed throughout the world in this manner.

Van Middlesworth (1954) demonstrated the presence of ^{131}I in cattle thyroids throughout the United States during the period of nuclear weapons testing in Nevada in 1953. This was the first indication that this radionuclide might be present in the human food chain during and immediately following weapons testing. Comar *et al.* (1957) subsequently undertook measurements of the ^{131}I content of human and cattle thyroids during a 23-month period from January, 1955 to December, 1956.

The plaucity of data during the 1950's, when extensive open-air testing of nuclear weapons was taking place within continental United States, was due to the fact that radioiodine was relatively difficult to measure with the instrumentation then available. Measurement soon was to become a relatively simple procedure by means of γ spectrometry, but this technique was not generally available prior to the first test moratorium of 1958, and no systematic measurements of radioiodine in human tissues were made during the period from 1950 to 1958, when copious amounts of radioiodine were being released to the atmosphere from tests in Nevada. Radioiodine contamination of milk during that period may have reached higher levels in many parts of the country than at any time since. Lewis (1959), based on study of the published values of ^{131}I in cows' milk, concluded that over the 5-year period prior to 1958 the average accumulated dose to the thyroids among children in the United States was 0.2 to 0.4 rad. This estimate was somewhat higher than that of Comar et al., but it is not unreasonable in view of the fact that Comar's thyroid samples were collected from human beings who died in hospitals and may not have been representative of people on a normal diet.

Following weapons tests in Nevada in the early 1950's, many of which were near-surface tests, it was known that relatively heavy fallout occurred at distances of a few hundred miles from the test explosions (Eisenbud and Harley, 1953, 1956). More than a decade later, when it was realized that radioiodine was a significant source of exposure at some distance from the tests, efforts were made to reconstruct the thyroid doses in several communities where relatively heavy fallout was known to have occurred (Knapp, 1963; Pendleton et al., 1963, 1964). Pendleton and his associates used 24-hr air samples made in a number of communities to deduce the ^{131}I content of cows' milk and the thyroid dose to infants consuming the milk. It was concluded that the exposures ranged from about 1 rad to as high as 84 rad in the community of St. George, Utah following a test on May 19, 1953. These estimates appear highly credible when all available information is taken into consideration and are supported by more recent analyses of existing data by Tamplin and Fisher (1966), who concluded that the thyroid doses to children in Washington County, Utah exceeded 100 rad.*

When the Soviet Union resumed weapons testing in 1961, γ spectrometry

---

* Studies of 4000 Utah and Arizona children by state and federal health authorities (Weiss et al., 1971) led to the conclusion that thyroid nodularity during examinations in 1965–1968 occurred no more frequently among exposed children than to those not exposed.

had progressed to the point where *in vivo* measurements of ^{131}I in human thyroids were feasible (Laurer and Eisenbud, 1963). Measurements could also be made of human thyroid tissue that became available at autopsy, as well as of the ^{131}I content of cows' milk and other foods. Measurements made in New York City during weapons tests in the Marshall Islands and Siberia showed that during two 3-month periods in 1961 and 1962 the thyroid doses in milk-drinking children was 50 and 140 mrem, respectively (Eisenbud *et al.*, 1962, 1963a).

Calculation of the population thyroid dose from the [131]I (Eisenbud *et al.*, 1963b) proved to be greatly facilitated by the remarkable uniformity of the ^{131}I concentration in the milk distribution system of New York City, where the above studies were made (Eisenbud *et al.*, 1963c). The variations from dairy to dairy were so random in nature that at the end of an extended period of sampling, such as 1 month, the mean concentrations of the milk from the various dairies were not significantly different from each other. It was concluded that, at least during the period when the study was conducted, the practice of analyzing single, daily random samples of the milk in New York City provided a satisfactory estimate for the milk supply as a whole. However, the fallout then occurring was from weapons tests conducted nearly halfway around the world, and the levels of contamination are understandably relatively uniform at such distances.

The excellent correlation between thyroid radioiodine and concentration of radioiodine in milk is shown in Fig. 14-18, in which are plotted the total thyroid ^{131}I content in a number of individuals during a 3-month period in 1961 when Soviet Union testing was in progress. On the same chart are the thyroid concentrations (picocuries per gram of thyroid) and the milk concentration of radioiodine. The thyroid dose during this period was estimated from the area under the solid lines as explained in the legend. In this case, integrated thyroid dose to maximum exposed persons was estimated from the integral under the solid line, which is the envelope of the plotted concentration of thyroidal [131]I. The area under the envelope can be shown to be 3400 pCi days/g of thyroid tissue, from which the thyroid dose can be determined as follows:

$$1 \text{ pCi day/g} = (2.22 \text{ disintegrations/min})(0.189 \text{ MeV/disintegration})$$

$$\times (1.6 \times 10^{-6} \text{ erg/MeV})(1440 \text{ min/day}) (100 \text{ erg/g})^{-1}$$

$$= 9.7 \text{ } \mu\text{rem/day}$$

and the integrated thyroid dose is, therefore,

$$(3400 \text{ pCi days/g})(9.7 \text{ } \mu\text{rad/day g})(10^{-3} \text{ mrem}/\mu\text{rem}) = 40.7 \text{ mrem}$$

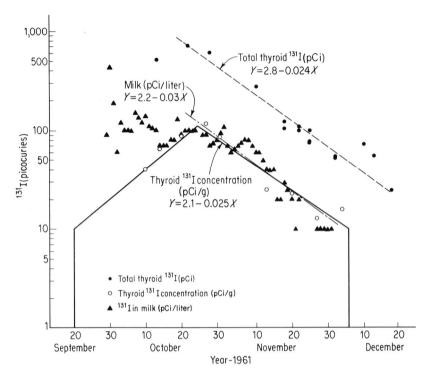

Fig. 14-18. Relationships between ¹³¹I content of fresh cows' milk and human thyroids in New York City during the Soviet Union tests of 1961 (Eisenbud *et al.*, 1962).

These data are for a large metropolitan area located thousands of miles from the test site. Unfortunately, there are relatively few data at significantly closer distances. Pendleton *et al.* (1963) estimated the doses to children in Utah, a few hundred miles from the scene of nuclear explosions in Nevada, to average 1 rem, with peak doses as high as 14 rem. However, there have been few studies such as these.

LUNG DOSE DUE TO INHALATION OF DUST

Radioactive dust particles that are too large to be respirable fall out quickly, and at distances beyond 200 miles from near-surface explosions the mass median diameter has been found to be 2 μm, which is close to the optimum particle size for lung retention (Eisenbud and Harley, 1953). Shleien *et al.* (1965) found that 88% of the total activity of airborne dust was contained in particles less than 1.75 μm in diameter. The dust concentrations to which people were exposed during the period of weapons testing were highly variable, depending on the age of the debris and distance from the test site. In one city, a few hundred miles from the Nevada test site, a 24-hr average concentration of 24 nCi/m³ was recorded following a test in 1952 (Eisenbud and Harley, 1953). The extent to which radioactive dust was then present in the atmosphere is illustrated in Figs. 14-2 and 14-3.

The thermonuclear tests between 1952 and 1962 were conducted at some distance from major centers of habitation and did not expose the public to the intense short half-lived radioactivity characteristic of the relatively fresh debris that blanketed much of the United States from tests in Nevada.

There have been very few estimates of the dose to the human lung during the period of active testing. Wrenn *et al.* (1964) undertook a series of measurements which showed that the principal dose was from the ^{95}Zr–^{95}Nb pair, it being estimated that the lung dose was about 3 μrad/day from lung burdens that ranged from 210 to 450 pCi at the time of death. These lung burdens were associated with average air concentrations over a 6-month period of 2 to 4 pCi/m³, several orders of magnitude lower than the concentrations reported above.

Wegst *et al.* (1964) autoradiographed the ashes from one human lung obtained in 1963 and ascertained the presence of 264 radioactive particles having a total activity of 436 pCi. As noted in Chapter 2, there is great uncertainty as to the method of calculating absorbed dose from β particles deposited in the lung, and the significance of the high doses in the vicinity of a β-radiation point source is not understood. Wegst *et al.* calculated that the tissue within 10 μm from the particle would receive 2×10^3 rem if the particle had a 120-day half-life in the lung. This volume would contain 16 cells. It is likely that the particle is in motion during much of the time it is in the lung, in which case the energy would be deposited in a larger volume of tissue with a corresponding reduction in the absorbed dose.

EXTERNAL RADIATION

It is very difficult to estimate the absorbed dose delivered externally by the radionuclides in intermediate and delayed fallout. In all but exceptional cases the ambient γ-radiation intensity is increased by too small a factor to permit convenient measurements with the instruments available to most laboratories. In lieu of continuous records of the observed change in the γ-radiation background, many investigators have calculated the dose using data on the quantities of the various radionuclides known to be present in fallout. These calculations give estimates of the γ dose, assuming a deposition of a given mixture of radionuclides distributed uniformly on an infinite smooth plane. Among the difficulties that present themselves in calculations of this kind are the following.

1. The dose integral is sensitive to the time of fallout, particularly if the fallout occurs within 2 or 3 days after the explosion. The time of arrival of the fallout is not always known.

2. The surface of the earth is not an infinitely smooth plane but is highly irregular, particularly in inhabited areas. The natural and man-made irregularities tend to absorb γ radiation and may modify the distribution of fallout to the extent that it is not correct to assume that the deposit is uniform.

3. The effect of weathering is difficult to evaluate. Rain will in most cases lessen the dose received by washing the fallout from places of habitation, but in some cases the rain may tend to concentrate deposits and thereby increase the dose.

4. Buildings provide an uncertain degree of shielding that depends on structural factors as well as the fraction of the day spent indoors.

Nevertheless, Beck (1966) has shown that estimates of the γ-radiation exposure rates based on rainfall deposition data are reasonably consistent with measurements made *in situ*, using γ-spectrometric techniques.

Gustafson (1971) has made careful measurements of the external radiation dose from fallout near Argonne National Laboratory since 1960. He estimated that the maximum open-field exposure was 56 mR/year in 1963, which diminished to 13 mR/year by 1969. Gustafson estimated that a more realistic estimate of the exposure would take into consideration the time a person spent indoors, in which case he estimates the exposure rates to have been 22 and 5 mR/year, respectively. However, these exposures, as noted earlier, are to dust from explosions many thousands of miles away. Localities closer to the test sites undoubtedly received very much higher doses, but only sporadic direct measurements have been made. The *average* dose commitment to inhabitants of the North Temperate Zone from external radiation owing to tests conducted prior to 1968 has been estimated to be 72 mrem (UNSCEAR, 1969).

The measurements of Beck included estimates of the contribution from the various nuclides present in fallout. In a total exposure of 76 mR during the period 1960–1964, the pair ^{95}Zr–^{95}Nb contributed about 37 mR, and ^{137}Cs accounted for 13 mR. The long life of ^{137}Cs results in a larger dose commitment from this nuclide.

MISCELLANEOUS NUCLIDES

Tritium is produced copiously in thermonuclear explosions, and the normal tritium content of surface waters in the North Temperate Zone reached several thousand picocuries per liter in the mid-1960's (Wyerman *et al.*, 1970). However, the physical and biological properties of tritium are such that the dose to humans was of a low order. Moghissi and Lieberman (1970) measured the tritium content of body water in children by means of tritium analyses of urine samples during the 2-year period 1967–1968, during which time the ³H content of urine decreased from 1.5 to 0.2 pCi/ml. The dose equivalent for a standard man was calculated to be 0.15 and 0.13 mrem for the 2 years, respectively.

⁵⁴Mn is produced as an activation product in thermonuclear explosions, as is ⁵⁵Fe, ⁵⁹Fe, ⁶⁵Zn, and a number of other nuclides. These are relatively short-lived nuclides that are readily detectable in the environment during and immediately following nuclear weapons testing, but they have not contributed significantly to human exposure. A possible exception is ⁵⁵Fe, which was first detected by Palmer and Beasley (1965) and has been shown by Wrenn and Cohen (1967) to have delivered doses of about 1 mrem to the erythrocytes of inhabitants of the New York City area. The dose to ferritin aggregates was calculated to be 235 mrem by including the energy of the short-ranged Auger electrons from ⁵⁵Fe (Wrenn, 1968).

Chapter 15

Peaceful Uses of Nuclear Explosives

The enormous amount of energy available from nuclear explosions, the compactness of the devices, and the relative ease with which the energy equivalent of millions of tons of TNT can be emplaced has resulted in widespread interest in the United States and the Soviet Union concerning the possibility of using nuclear explosives for civilian purposes. In the United States since 1957, this has been the subject of extensive investigations in a program that has been known allegorically as Plowshare. An extensive theoretical, laboratory, and field program that has been supported by the AEC has included a number of explosions at the Nevada test site designed to confirm calculations of the blast and radiation effects. In addition, two tests have been conducted to demonstrate the feasibility of stimulating the yields of natural-gas deposits, an application of the Plowshare program that has proved to be the most popular of all the various applications that have thus far been proposed.

Interest in natural-gas stimulation is based on the projected acute shortages in natural gas that are expected to develop in the next 15 to 20 years (Smith, 1971). To increase supplies, it is reasoned that nuclear explosives can be used to fracture rocks containing tightly bound natural gas that would otherwise not be available to wells. Two test explosions, called Gasbuggy and Rulison, have been conducted in New Mexico and Colorado and will be discussed briefly later in this chapter.

Other applications of nuclear explosives have not fared as well. Extensive studies were made of the feasibility of constructing a second canal across the Panama isthmus using a series of cratering explosions (Hughes, 1969), but after careful evaluation an Atlantic–Pacific Interoceanic Canal Study Commission concluded in 1970 that ". . .Although we are confident that some day nuclear explosions will be used in a variety of massive earth-moving projects, no current decision on U.S. canal policy should be made in the expectation that nuclear excavation technology will be available for canal construction. . ." (AEC, 1971a). Other potential applications such as harbor excavations, construction of dam embankments (Knox, 1969), and various other cratering and mounding applications have not as yet proved feasible for practical application.

In addition to gas stimulation and earth moving, several other applications have been or are being considered, including production of trans-uranic elements (Dorn, 1964) and power production by exploitation of the heat deposited by deep underground nuclear explosions (Kennedy, 1964; University of California *et al.*, 1964). In addition to its application to nat-ural-gas stimulation, the principle of underground fracturing has been proposed for recovery of oil from shale, for leaching copper ores, and for storage of natural gas (Nordyke, 1969).

An interesting Plowshare application has been reported by the Soviet Union (Werth, 1971), which has reported the use of a 30 kiloton deep underground nuclear explosion to extinguish an oil-well fire that had burned out of control for several years.

The overriding advantage of Plowshare is economy in the cost and em-placement of the explosives. The obvious major disadvantage is the po-tential for environmental contamination.

The cost of nuclear explosives ranges from 35 dollars per ton for a 10 kilotons explosive down to 30 cents for a 2 megaton explosive. Conven-tional chemical explosives cost 400–500 dollars per ton, and their sheer mass makes emplacement impractical and excessively expensive (Williamson, 1969). A nuclear explosion equivalent to 10 million tons of TNT can be achieved with a canister 65 in. in diameter and 30 ft long. (A 10 megaton charge of dynamite would occupy a volume approximately 10 ft $\times$ 10 ft $\times$ 400 miles long!) Whether or not the nonradiological problems associated with excavation undertakings involving such enormous amounts of energy can be solved is beyond the scope of this text, but an abundance of informa-tion in this regard has been accumulated as a result of laboratory and field studies (American Nuclear Society, 1969; University of California *et al.*, 1964; Joint Committee on Atomic Energy, 1968; U.S. Public Health Service, 1969a).

Objective evaluation of the radiological problems associated with the Plowshare program is made difficult by the fact that certain relevant characteristics of the devices used are classified, and the yields per ton of TNT equivalent for many of the key nuclides have not been published. In particular, one would wish to know the fission to fusion ratios of the devices and the amounts of tritium and activation products produced in a given Plowshare application.

In discussing the characteristics of radioactivity produced by nuclear explosives, Miskel (1964) assumes, for purposes of calculation, that the explosive yield of a 1 megaton device is 99% fusion and 1% fission, and that the fusion component (0.99 megatons) would produce 10^7 Ci of tritium and 1.4×10^{27} neutrons. The neutron flux would be produced almost entirely by the fusion component, compared to which the yield of the fission component (10 kilotons) would be less by about three orders of magnitude. A reduction in neutron flux by a factor of 100 is assumed by providing a boron absorber surrounding the device. This would result in a net neutron yield of about 1.4×10^{25}. The various radionuclides with half-lives greater than a few minutes that would be produced if such a 1 megaton device were exploded in basalt are given in Table 15-1.

TABLE 15-1

RESIDUAL RADIOACTIVITY FOR A 1 MEGATON DEVICE EXPLODED IN BASALT[a]

Nuclide	Half-life	Curies at t_0
Activation Products		
^{3}H	12 years	1×10^7
^{24}Na	15 hr	1.9×10^8
^{31}Si	2.62 hr	7×10^7
^{32}P	14.3 days	1.8×10^5
^{42}K	12.4 hr	2.8×10^6
^{45}Ca	165 days	4.3×10^4
^{55}Fe	2.7 years	7.5×10^5
^{56}Mn	2.58 hr	8.4×10^8
^{59}Fe	45 days	3.8×10^4
Fission Products		
^{90}Sr	28 years	1.8×10^3
^{99}Mo	66 hr	6.7×10^6
^{131}I	8 days	1.28×10^6
^{137}Cs	30 years	2.9×10^3
^{147}Nd	11 days	7.5×10^5

[a] Miskel (1964).

With these figures as background, it is interesting to note that two alternative proposals for design of the transisthmian canal (Klement, 1969) would have required 292 and 245 megatons of yield, respectively. For the larger requirement, tritium production as a by-product of canal construction would have been about 3×10^9 Ci.

The amounts of radioactivity that will remain essentially *in situ* following a Plowshare explosion and the quantities that will be dispersed to the environment would depend on the depth and size of the explosion, the underground hydrology, and the amount of radioactivity that vents to the atmosphere. From this point of view, three basic types of Plowshare explosions can be visualized. (1) The energy and debris is contained totally below ground, in which case there is no gross venting to the atmsophere, although some gradual seepage of the volatile constituents takes place. (2) The explosion, in relation to its size, is so close to the surface that debris is ejected to the atmosphere with partial fall-back in or near the crater thus produced. (3) The explosion vents to the atmosphere but is arranged so that there is almost total fall-back of debris, producing what is called a "retarc" in Plowshare nomenclature.

The underground explosion produces a chimney of rubble surrounded by fractured rock. The use of retarcs has been proposed for *in situ* leaching of low-grade ores, and cratering explosions would be used mainly for civil engineering purposes such as construction of canals or harbors.

Although plans for the construction of the transisthmian canal using nuclear explosives have now been canceled, it will be instructive to discuss briefly the dose estimates that had been projected for inhabitants along one of the projected routes. The route to be discussed, number 17, was one of the many that had been investigated and is shown in Fig. 15-1, which also presents the limits of the exclusion area, based on the criteria that the inhabitants would be evacuated at the start of the program and for some time afterwards, in order to limit their whole-body exposure to 0.1 R in 70 years.

Construction of the canal along route 17 would have required detonation of 437 devices, with a cumulative yield of 274.8 megatons exploded in 22 events over a period of 938 days. (Hughes, 1969). Cowser *et al.* (1967) and Kaye and Rohwer (1970) have undertaken detailed critical-pathway analysis of the projected transisthmian project, based on fallout estimates calculated by others using experience gained at the Nevada test site, where a number of explosions were undertaken to demonstrate cratering techniques prior to the 1962 test-ban agreement. The analysis was based on the living habits of 84,000 inhabitants along the projected course of the canal, and Table 15-2 gives the estimated dose for various modes of exposure to in-

TABLE 15–2

SUMMATION OF DOSE ESTIMATES FOR MODES OF EXPOSURE CONSIDERED, WATERSHED 5, ROUTE 17. EXPOSURE STARTING 2.652 YEARS AFTER THE FIRST DETONATION (30 DAYS AFTER THE LAST DETONATION)[a]

Body organ	Critical age group (years)	Estimated dose (rem) to age 70 years (internal + external)	Estimated maximum annual dose rate (rem/year) (internal + external)
Total body	1–5	1.00 + 4.25 = 5.25	0.968 + 1.96 = 2.93
Bone	0–1	5.75 + 4.25 = 10.0	5.21 + 1.96 = 7.17
Liver	1–5	1.27 + 4.25 = 5.52	1.17 + 1.96 = 3.13
Kidneys	1–5	4.17 + 4.25 = 8.42	3.97 + 1.96 = 5.93
Thyroid	1–5	5.37 + 4.25 = 9.62	5.34 + 1.96 = 7.30
Testes	1–5	1.15 + 4.25 = 5.40	1.12 + 1.96 = 3.08
Ovaries	1–5	1.00 + 4.25 = 5.25	0.968 + 1.96 = 2.93
Lungs	1–5	1.00 + 4.25 = 5.25	0.968 + 1.96 = 2.93
GI Tract	1–5	1.04 + 4.25 = 5.29	1.01 + 1.96 = 2.97

[a] Kaye and Rohwer (1970).

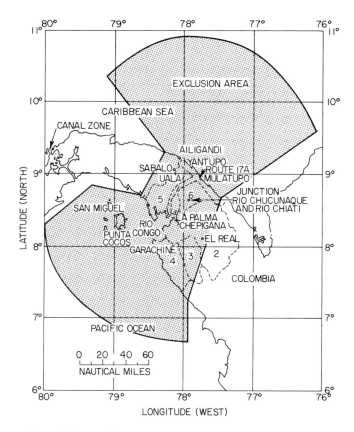

Fig. 15-1. Map of Panama illustrating the route 17 alignment, watersheds (numbered 1 through 7), and locations of towns and villages for estimates of external dose (Kaye and Rohwer, 1970).

digenes on water shed 5 of Fig. 15-1, who are evaculated from the the area prior to the first detonation and returned 30 days after the last detonation. The influence of reentry time on the total body dose is given in Fig. 15-2. If these projections are realistic, and considered by themselves, one would conclude that from the radiological point of view the project is feasible if the benefits from it can justify excluding several thousand inhabitants from their homeland for a period of several years. However, most projections of this kind must be accepted with the greatest of caution because of uncertainties in the ecological models on which the dosimetry is based. Moreover, this single project would make a substantial contribution to the world's inventory of tritium. From the tritium production rate previously

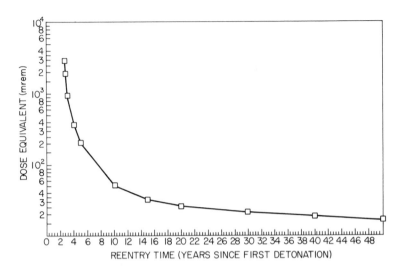

Fig. 15-2. Total estimated dose to the total body for reentry times up to 50 years after the first detonation, watershed 5, route 17 (Kaye and Rohwer, 1970).

given, it can be calculated that about 2.7×10^9 Ci of tritium would be produced from this one project, which would be approximately equal to the amount of tritium produced in all test explosions to date.

The practical applications of Plowshare have been limited until now in the United States to natural-gas stimulation and, as noted earlier, there have been two demonstrations, Gasbuggy and Rulison. The Rulison test, which was conducted in Colorado, attracted widespread attention because of the attempt of interveners to block the explosion because of potential hazards. The specific objectives of the test, which was conducted cooperatively by the AEC, Department of Interior, and two private industrial companies, were to measure the changes in gas production produced by the nuclear explosion, the extent of radioactive contamination of the gas, the seismic effects, and the various phenomena associated with underground blasting technology (Haas, 1971). The experience gained from projects Gasbuggy and Rulison have been examined by Barton *et al.* (1971) with a view toward estimating the potential risks to the general population from this method of natural-gas stimulation.

Three nuclides—tritium, ^{14}C, and ^{85}Kr—are the principal isotopes in the gas a few months after detonation. A comparison of the radioactivity in gas produced by the Gasbuggy and Rulison wells is given in Table 15-3. Project Gasbuggy (El Paso Natural Gas Co., 1965), had a yield of 29 kilotons and was exploded 3900 ft below the surface near Farmington, New

TABLE 15-3

COMPARISON OF RADIOACTIVITY IN GASBUGGY AND RULISON[a]

	Gasbuggy	Rulison
Explosive yield (kilotons)	29	40
Tritium		
Total in gaseous products (Ci)	2400	1100
Initial concentration (pCi/cm³)	700	175
⁸⁵Kr		
Total (Ci)	370	960
Initial concentration (pCi/cm³)	110	150

[a] Barton *et al.* (1971).

Mexico on December 10, 1967. The explosion created a chimney of broken rock about 333 ft high and a void volume of about 2 million ft³ (U.S. Atomic Energy Commission, 1971a). The Rulison experiment had a 40 kiloton yield and was exploded 8431 ft below the surface near Grand Valley, Colorado on September 10, 1969. It is seen from Table 15-3 that although the yield of the Rulison event was about one-third larger than Gasbuggy, tritium production was less than half. One can speculate that the reason for this is that the fission to fusion ratio of the Rulison device was higher, as evidence by the considerably greater ⁸⁵Kr production.

Exposure of the public might occur in the vicinity of a natural-gas processing plant or within a community in which natural gas is used for space heating or electrical power generation. Barton *et al.* (1971) have concluded that tritium is the critical nuclide and have calculated the population dose in a metropolitan area based on the assumption that the tritium content of the natural gas decreases exponentially with time owing to replacement of the radioactive gas by nonradioactive gas from surrounding rock. The assumption is also made that the average concentration of tritium in the natural gas over the lifetime of the well will be 1 pCi/cm³, an assumption the validity of which could only be accepted on the basis of considerably more data than now exists. After examining the use of natural gas in the Los Angeles basin and the San Francisco Bay area, Jacobs *et al.* (1972) concluded that the maximum exposure from tritium, if the average concentration is 1 pCi/cm³, would be 2.5 mrem/year with an average exposure of 0.50 mrem/year. Future studies of the gas produced by the Gasbuggy and Rulison wells will undoubtedly make it possible to make projections of this kind with more assurance than at present.

Chapter 16

Environmental Contamination from Accidents

When an accident occurs in which radioactive materials are released to the environment, an opportunity is presented to obtain useful information about the manner in which the release took place, the mechanisms by which the radioactivity was dispersed into the environment, and the ecological relationships that developed. Although every effort must be made to keep such accidents from happening, it is inevitable that they will occur from time to time, and it is important that the lessons learned be documented carefully.

This chapter reviews the information that has been obtained by studying the effects of certain accidents that have occurred up to the present time. Not included are a number of accidents of types that would not involve a potential for environmental contamination by radioactive substances. These include a criticality incident at a privately operated industrial facility in Rhode Island (Auxier, 1965; Shipman, 1961); a fuel meltdown at St. Laurent (Corbett, 1971); a criticality accident involving a Belgian reactor (Penelle, 1968); and a nuclear excursion at Livermore Research Laboratory (Kathren *et al.*, 1964). Descriptions of these and other nuclear accidents not described in this chapter are summarized by the United States Atomic Energy Commission (USAEC) (1971d), by Thompson and

Beckerly (1964), and in the quarterly publication *Nuclear Safety*. The general subject of the management of radiation accidents has been reviewed by the International Atomic Energy Agency (IAEA, 1967c, 1969). A common feature of all the accidents reviewed in this chapter is that radioactive environmental contamination resulted to some extent, and it is this aspect primarily that is considered.

Fallout from the Thermonuclear Detonation of March 1, 1954

Prior to March 1, 1954, there had been some speculation among the very few people who were then informed about the subject of fallout as to whether surface and near-surface bursts in the megaton range could produce lethal amounts of fallout over large areas. The many doubts that existed even as late as March 1, 1954 were quickly dispelled when extensive fallout occurred following detonation of a multimegaton device mounted on a barge over the reef of Bikini Atoll in the mid-Pacific Ocean.

A detonation of comparable size had occurred at Eniwetok in November, 1952, but very little fallout was observed, because the debris fell into the oceans and missed the relatively few scattered atolls in the Marshall Islands.

The barge on which the 1954 detonation occurred was located in shallow water so that a large amount of coral was incorporated into the fireball. The meteorological conditions were such as to produce a minimum amount of wind shear, and it is estimated that from 50 to 80% of the radioactivity from the detonation fell out in the idealized cucumber-shaped pattern in Fig. 16-1.

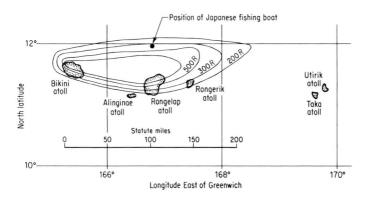

Fig. 16-1. The fallout pattern from the thermonuclear explosion of March 1, 1954 as reconstructed from various sources.

The first indication that radioactive fallout was occurring was an increase in a γ-ray detector located on Rongerik Island about 160 miles east of Bikini. This island was inhabited by 28 American servicemen who were operating a weather station. The fallout began about 7 hr after the detonation, and 30 min later the recording γ detector went off-scale at 100 mR/hr. The test headquarters was immediately notified, and aerial reconnaissance subsequently confirmed that fallout had occurred on Rongerik and that even heavier fallout had occurred on the inhabited Rongelap Atoll, about 105 miles east of Bikini, and Alinginae, located about 75 miles to the east southeast. Evacuation procedures were put into effect, beginning about 30 hr after the detonation, with the removal of the 28 Americans by air. Rongelap, Utirik, and Alinginae were evacuated subsequently. Figure 16-1 shows the relative locations of these atolls.

It was later found that Rongelap Island was so heavily contaminated that the natives would be unable to return for an extended period, and they remained on another Marshall Island atoll for 38 months until July, 1957. The 18 natives evacuated from Alinginae were actually natives of Rongelap who had been visiting the normally uninhabited island. These 18 natives ultimately returned to Rongelap with their kin.

The fallout on Utirik was insufficient to require extensive evacuation, and the natives of this atoll were returned to their homes shortly after the incident.

It was not known until some days later that a Japanese fishing vessel, the Fukuru Maru, was located about 80 miles to the east of Bikini at the time of the fallout. The boat had somehow not been observed in the customary aerial searches that preceded each detonation. When the explosion occurred, the men hauled in their lines and proceeded away from the detonation, but visible fallout began to occur about 4 hr later. These men lived on the vessel for 13 days until they returned to their home port of Yaizu.

NATURE OF THE RADIATION EXPOSURES

Although the fallout was visible on both Rongelap and the Japanese fishing boat, neither the Marshallese nor the Japanese took any precautions to minimize the amount of their exposure. This is understandable in view of their relative ignorance of the subject. Thus, the 64 natives of Rongelap and the 23 fishermen of the Japanese fishing vessel lived for 50 hr and 13 days, respectively, in intimate contact with the most radioactive inhabited environment that is known to have ever existed. Large particles of fallout fell into their hair and came into direct contact with the skin. The Japanese fishermen described the dust deposit as so heavy as to seem like a coating of snow (Japan Society for the Promotion of Science, 1956).

TABLE 16-1

SUMMARY OF DOSES RECEIVED FOLLOWING TEST EXPLOSION OF MARCH 1, 1954

Group	Number exposed	Time fallout started (hr after detonation)	Exposure duration	Whole-body dose (estimated rem)	Iodine dose to thyroid (rem)
Rongelap[a]	64	4–6	Evacuated in about 50 hr, returned in 38 months	175	100–150
Rongerik[a]	28	7	Permanently evacuated in about 30 hr	78	50
Alinginae[a]	18	4–6	Evacuated in about 50 hr, returned to Rongelap in 38 months	69	—
Utirik[a]	157	22	Evacuated in 55 to 78 hr, returned to Utirik	14	—
Fukuru Maru[b]	23	4	Remained on boat for 13 days	170–700	20–120

[a] Cronkite *et al.* (1956); Conard *et al.* (1960).
[b] Kumatori *et al.* (1970).

The whole-body and thyroid doses received by the several groups, where estimates are available, are given in Table 16-1. The dust filled the air and fell on the bodies of the victims as well as into their food and water. The Marshallese sat on the contaminated ground, and the Japanese sat on the contaminated deck. No special precautions were taken in the consumption of food, and no special hygenic procedures were followed except that a number of the Marshallese swam in the surf during the period of exposure.

MEDICAL INVESTIGATIONS

Excellent reports of the American and Japanese studies of this accident are available (Conrad *et al.*, 1970a,b; Japan Society for the Promotion of Science, 1956; Kumatori *et al.*, 1965). The Marshallese were evacuated to the island of Kwajalein, where an emergency hospital was established and teams of specialists from the United States made available to conduct examinations of the natives. Survey teams were sent to the islands for radiological measurements. The Japanese did not enter their home port of Yaizu until March 14, by which time they were already suffering from the effects of exposure, but it was another 2 days until it became known that their maladies were due to their having been subjected to radioactive fallout. The 23 men were hospitalized in Yaizu and Tokyo.

The early effects of their exposure may be described under (1) effects on the skin, (2) hematological effects, and (3) effects due to internal emitters.

1. *Skin Effects.* Itching and burning sensations were experienced from 1 to 2 days after the fallout by all groups except the Utiriks. After various time intervals, ranging from 3 days in the case of the Japanese fishermen to 21 days at Rongerik and Alinginae, skin lesions and epilation began to develop. The lesions became ulcerous in 70% of the Japanese fishermen and in about 25% of the Rongelaps. The investigators found that the severity of the injuries at Rongelap and on the Fukuru Maru was definitely related to obvious factors such as whether outer garments were worn or whether the invididuals went swimming. For example, Rongelap children who spend much time wading in the lagoon had fewer foot injuries, and the worst burns among the Japanese fishermen were on two men who did not wear hats.

2. *Hematological Effects.* The detailed hematological findings need not be reviewed here except to say that the blood changes were related to the severity of the whole-body γ dose. The diminution in white blood count was most marked in the Japanese, among whom the values diminished to from 15 to 50% of normal in about 28 days, at which time slow recovery

began. The values in the Marshallese varied from 55% of normal at 44 days in the case of the Rongelaps to about 84% of normal in the case of the Utiriks who were subjected to the smallest dose.

3. *Dose from Internal Emitters.* Until they were removed from the contaminated environment, the Marshallese and Japanese existed under conditions which tended to maximize the opportunity for inhalation and ingestion of radioactive substances. Life on a tropical atoll and on a fishing vessel in the tropics is largely an outdoor existence. The Marshallese drank from exposed cisterns and ate food which was exposed to the open air. The Japanese ate raw fish which were certainly in contact with the contaminated deck. Almost everything that came off the ship in Yaizu 13 days later was heavily contaminated, and this must have been even more true during the first few days after exposure.

It is, therefore, not surprising that on initial examination both the Marshallese and Japanese were found to be radioactive, not only because of superficial contamination but also because fission products were found to be present within the body. In the Marshall Islands, the first samples collected for radiochemical analysis were obtained 15 days after the detonation. The results of urinalysis are given in Table 16-2 (Cronkite *et al.*, 1956). It is seen that the principal absorbed isotope was ^{131}I. Two independent estimates of the amount initially deposited were 6.4 and 11.2 μCi. These data served as the basis for the estimate that the Rongelap natives received from 100 to 150 R to the thyroid in addition to the whole-body dose of 175 R.

TABLE 16-2

MEAN BODY BURDEN OF THE RONGELAP GROUP[a]

Radioisotope	Activity at 82 days (μCi) (USNRDL)[b]	Activity at 1 day (μCi) (USNRDL)	Activity at 1 day (μCi (LASL)[c]
^{89}Sr	0.19	1.6	2.2
^{140}Ba	0.021	2.7	0.34
Rare-earth group	0.03	1.2	—
^{131}I (in thyroid)	0	6.4	11.2
^{103}Ru	—	—	0.013
^{45}Ca	0	0	0.019
Fissile material	0	0	0.016 (μg)

[a] Cronkite *et al.* (1956).
[b] U.S. Naval Radiological Defense Laboratory.
[c] Los Alamos Scientific Laboratory.

Samples of urine were also collected from the Japanese fishermen and were sent to New York for radiochemical analysis at the USAEC Health and Safety Laboratory (Kobayashi and Nagai, 1956). Very little can be concluded from the reported data except the all-important finding that only minimal urinary excretion of fission products were reported, despite the fact that the men lived for 13 days on the contaminated fishing vessel. The excretion levels appeared to be less than those reported for the Marshall Islands. The β activity of oxalate precipitates from samples collected on April 21, 1954 ranged from 10 to 110 pCi/liter compared with values as high as 37 pCi/liter obtained on five unexposed Americans and analyzed at the same time.* The analyses were formed by precipitating the mixed fission products with calcium oxalate, thus excluding potassium which is normally present in urine. The total β activity of the urine of the Japanese was only about one-tenth of that of the Rongelap natives. Two samples collected on March 26 from two patients who were believed to be the most heavily exposed of the Japanese fishermen were found to have 320 to 230 pCi/liter. This is consistent with the somewhat lower values reported about 1 month later but inconsistent with samples analyzed from five other fishermen on March 29, which ranged from 590 to 2500 pCi/liter. These samples were so much higher than 49 others collected from the 23 men and submitted separately that it has been thought by the chemists that these samples may have become contaminated.

One of the fishermen died of a liver disease 6 months after the accident. The results of radiochemical analysis of his tissues as reported by Tsuzuki (1955) are given in Table 16-3. The amount of radionuclides in the tissues of this man was so low as to rule out the significance of the delayed radiation because of intermediate and long-lived radioisotopes in his body. However, as noted earlier, the dose from short-lived iodine was certainly significant in all these cases.

FOLLOW-UP STUDIES

The exposed Marshallese were evacuated by plane and ship about 2 days after the fallout and were taken to Kwajalein where they remained for about 3 months, during which time extensive medical studies were undertaken. The Rongelap natives were then taken to a temporary village on Majuro Atoll where they lived for $3\frac{1}{2}$ years, by which time the radiation levels on Rongelap had been reduced to a point where they could be returned. The Utirik and Ailinginae inhabitants were allowed to return to

---

* These data were obtained prior to the availability of γ spectrometry, which permits a more sophisticated analysis.

TABLE 16-3

DISTRIBUTION OF RADIOACTIVITY IN DECEASED FISHERMAN[a]

Fraction	Probable nuclide	Liver (pCi/g)	Fresh tissue			
			Kidney	Lung	Muscle	Bone
Ru + Te	^{106}Ru + ^{106}Rh ^{129}Te	<0.1	0.9	<0.1	0.2	2
Zr + Nb	^{95}Zr + ^{95}Nb	1	1	0.4	0.3	2
Rare-earth elements	^{144}Ce + ^{144}Pr	2	1	0.5	0.5	20
Sr	^{90}Sr + ^{90}Y	0.6	0.4	~0.1	<0.1	1

[a] Tsuzuki (1955).

their atolls following their 3-month stay on Kwajalein. Annual medical examinations of the Marshallese, undertaken under the direction of Dr. Robert A. Conrad of the Brookhaven National Laboratory, have revealed a number of late effects of irradiation. The findings of Conrad and his associates are thoroughly reviewed in a summary published 15 years after the fallout occurred (Conrad et al., 1970b). Over this period of time the original acute β burns sustained by about 20 Marshallese continued to show residual scarring and pigment changes, but none of the lesions developed into chronic radiation dermatitis nor was there evidence of malignant skin change. Peripheral blood studies indicated a slight depression in the white cell and platelet counts, implying a residual effect on the blood-forming mechanism that has been confirmed by bone-marrow biopsies on some of the exposed people. Changes have also been seen in the red cell precursors as well as in the peripheral blood chromosomes. Atypical lymphocytes are present in the exposed group and in some children born subsequently of exposed parents. General mortality among exposed group has been generally higher than the Marshallese, but none of the deaths could be attributed directly to radiation exposure, and the higher mortality is difficult to interpret in view of the small numbers of people involved.

During the first 4 years following exposure, the frequency of miscarriages and stillbirths was higher among exposed women, but no difference has been observed since that time.

Growth retardation has been observed in boys exposed as infants, 2 of whom showed marked hypothyroidism with atrophy of the thyroid gland. Thyroid pathology has been the most remarkable finding among these natives. Of 66 surviving Rongelap and Ailinginae inhabitants, 16 have developed benign thyroid nodules, 3 have developed thyroid cancer, and 2 have shown thyroid atrophy with associated hypothyroidism. Ninety percent of those showing thyroid injury were less than 10 years of age at the time of fallout.

Other findings include a possibly higher incidence of lenticular imperfections in the exposed group and some biochemical indications of more rapid aging.

Some low-level residual radioactivity remained on Rongelap at the time the natives returned in 1957. It was found that the coconut crab, a much-favored food among the Marshallese, tended to concentrate ^{90}Sr, and consumption of local crabs was banned accordingly. After 15 years, the concentration of ^{90}Sr remained high in the flesh of these crabs, about 700 pCi/g Ca. Whole-body counts of the natives indicated elevated levels of ^{137}Cs and ^{60}Co. By 1965, the body burdens of ^{137}Cs ranged from 10 to 20 nCi. A number of individuals then living on Rongelap were not present at

TABLE 16-4

ESTIMATE OF ^{90}Sr IN DIET OF RONGELAP ADULTS, 1958[a]

Diet	A Daily intake (g wet wt)[a]	B Ca content (mg/g)	A × B Daily Ca intake (mg)	D Fraction of total Ca intake[b]	E ^{90}Sr content (pCi/g Ca)	D × E Contribution to total daily ^{90}Sr intake (pCi/g Ca)
Meat from mature coconut	89	0.075	6.7	0.008	1200	9.6
Meat from drinking coconut	75	0.14	10.5	0.013	210	2.7
Milk from green coconut	116	0.15	17.4	0.022	1000	22.0
Pandanus, edible portion	79	0.15	11.9	0.015	930	14.0
Arrowroot	58	2.10	121.8	0.152	19	2.9
Breadfruit	45	0.60	27.0	0.034	260	8.8
Fish	139	0.13	18.1	0.023	280	6.4
Clams	45	4.00	180.0	0.225	5	1.1
Crabs, land	14	4.00	56.0	0.070	(4000)	(280.0)
	660	—	449	0.56	—	67.5

[a] Conard *et al.* (1960). Based on average daily diet of 14 Rongelap males.

[b] Based on total calcium intake of 0.8 g/day.

[c] The diet also included imported foods; rice, canned C rations, flour, tea, milk, salt, and sugar.

the time of fallout. The ^{137}Cs burdens of the two groups were indistinguishable, implying that the ^{137}Cs was absorbed after the natives returned or that the levels are generally higher throughout the Marshall Islands. It should be noted that extensive tests of nuclear weapons took place on the Marshall Islands and that the fallout from explosions in 1961 and 1962 may have added significantly to the exposure of these natives. It was estimated (Cronkite *et al.*, 1956) that the mean ^{90}Sr burden of the Rongelap people when they returned to the island in 1957 was 0.1 nCi. This had increased to 2 nCi by 1958 and 6 nCi by 1959. By 1962, it was estimated from urine analysis that the mean body burden of ^{90}Sr was 12 nCi. Analysis of bone samples from a deceased Rongelap woman in 1962 corroborated this estimate (Conrad *et al.*, 1970b). The body burdens of both ^{137}Cs and ^{90}Sr began to drop subsequent up to 1963. It is not known if this is due to lowered radioactive contamination of food from the island of Rongelap or whether the daily intake is being diluted by an increasing amount of imported food. The ^{90}Sr content of food on the Rongelap Atoll in 1958 is given in Table 16-4.

KNOWLEDGE GAINED

The events subsequent to the fallout of March 1, 1954 led directly and indirectly to extensive investigations, all which have produced important information about the behavior of radioactivity in man's environment. If one screens the findings of the various studies so as to exclude findings of detail, certain important conclusions remain.

1. The detonation of March 1, 1954 demonstrated conclusively that fallout is a major consequence of the use of nuclear weapons in the megaton range. From the observations that were made following this explosion, the magnitude of fallout was defined for the first time and its potential significance in time of war elaborated.

2. Severe exposure to residual radiation was associated with the fallout of visible dust. Although it cannot be concluded that this would always be the case regardless of the type or place of detonation, this observation must be taken into consideration in the indoctrination of populations who may someday be involved in nuclear war.

3. The β-ray burns owing to deposition of fallout on the skin may be avoided by simple procedures such as wearing clothes, staying indoors, and washing.

4. Inhalation and ingestion of radioactive substances appear to be very much less of a problem in fallout than had been previously thought. Both the Marshallese and Japanese lived intimately with their contaminated environment during and immediately after the period of exposure, but the

amounts of radionuclides deposited in their tissues did not contribute appreciably to the overall effects observed.

The Accident to the Windscale Reactor Number One of October, 1957

The Windscale works of the United Kingdom Atomic Energy Authority is located on a low-lying coastal strip in the northwest of England. At this site are two air-cooled graphite-moderated natural-uranium reactors employed in the production of plutonium. The core of one of these reactors was partially consumed by combustion in October, 1957, resulting in the release of fission products to the surrounding countryside.

Reference has been made in Chapter 9 to the fact that graphite tends to store energy as a result of neutron irradiation and that a controlled release of this energy can be achieved by annealing. The 1957 accident was caused by the release of this stored energy at an excessive rate during the regular annealing procedure, during which the temperature of the core was raised using nuclear heat. The release of stored energy was excessive in portions of the core, but the local increases in temperature that were occurring went undetected because of insufficient core instrumentation (United Kingdom Atomic Energy Office, 1957, 1958). Failure of a fuel cartridge evidentally resulted from this factor. The metallic uranium and graphite began to react with air, and from the time combustion began on the morning of October 8 until the fire was extinguished on the afternoon of October 12 a substantial portion of the core was destroyed. It is estimated that during this incident the following isotopes were released to the environment (Dunster, 1958b):

Isotope	Curies
^{131}I	20,000
^{137}Cs	600
^{89}Sr	80
^{90}Sr	9

The first evidence that a mishap had occurred was the observation of elevated β activity of atmospheric dust collected by an air sampler located in the open about 0.5 mile from the reactor stack. A concentration of about 3000 disintegrations per minute (dpm)/m^3 of air was observed, this being 10 times the level normally present from the atmospheric radon and thoron daughter products. Air samples collected elsewhere in the vicinity of the reactor confirmed that a release of radioactivity to the atmosphere was occurring.

Visual inspection through a plug hole in the face of the reactor revealed that the uranium cartridges were glowing at red heat in about 150 fuel channels. Because of distortion that had already occurred, these cartridges could not be discharged, but the fuel was removed from channels adjacent to the affected area, thereby creating a fire break which served to limit the extent of the mishap. For several hours, various schemes were devised for extinguishing the slowly burning core, but none was effective. On the following day, what must have been a most difficult decision was reached, and the graphite core was cooled by flooding the core with water. The reactor was cold by the afternoon of October 12.

ENVIRONMENTAL SURVEY PROCEDURES

When it was discovered that a mishap had occurred, procedures were implemented to determine the extent of exposure from the following sources: (1) external radiation, (2) inhalation of radioactive dust or vapor, and (3) contaminated food and water.

Vehicles equipped with radiation detection equipment were dispatched downwind from the stack. It was observed that the highest radiation level occurred directly under the plume at a point about 1 mile downwind, where the highest radiation level was 4 mR/hr compared to a maximum of about 0.20 mR/hr owing to deposited radioactivity about 3 miles south of the reactor. It was subsequently determined that the maximum dose of external radiation which would be received by a person remaining out-of-doors for 1 week following the accident would be in the range of 30 to 50 mR.

During the period of release, about 12,000 air samples were collected on the site and about 1000 were collected in the environs. As would be expected, they revealed wide variations, with concentrations ranging as high as 0.45 pCi/ml. The average concentration during the period of the incident was approximately 4.5×10^{-3} pCi/ml, which is about 50% greater than the ICRP standard for permissible continuous exposure to ^{131}I (International Commission on Radiation Protection, 1960). As described by Dunster, the atmospheric contamination from the incident "rose on occasion to worrying but not dangerous levels on the site, while the dilution resulting from wind variations considerably reduced the hazard in the district."

It was found that, beginning on the afternoon of the first day, milk from cows in the vicinity of Windscale was contaminated with ^{131}I. Up to that time no emergency level had been established for short-term permissible exposure to radioiodine in food but, on consideration of the problem, the Medical Research Council promptly recommended that the maximum permissible concentration (MPC) be 0.1 μCi/liter and that all milk containing more radioiodine than this be discarded. The manner in which the Council arrived at this figure is of some interest in illustrating the thinking of such groups and the ingenuity with which workable figures can be arrived at on short notice when the occasion demands.

The Council started with the knowledge that cancer of the thyroid in children had been known to occur following doses greater than 200 rad. Although no cases were known to have occurred following exposure to smaller doses, the data were insufficient to permit the conclusion that 200 rad was actually the threshold for tumor production. It was, therefore, decided to limit the dose to children to a maximum of 20 rad. The amount of radioiodine in milk that would produce this dose in children became the permissible level for the entire population, children and adults alike.

Constants furnished in the ICRP tables were used to relate the concentration of radioiodine in milk to thyroid dose. One microcurie of ^{131}I/g of thyroid was calculated to deliver an integrated dose of 130 rad. The mass of the child's thyroid was taken to be 5 g, and the thyroid was assumed to retain 45% of the ingested iodine. On this basis the limiting concentration in milk was computed to be 0.15 μCi/liter, which was rounded off to 0.1 μCi/liter.

Up to 300 samples/day were analyzed for their ^{131}I content by γ spectrometry. Milk measurements were originally hampered by the fact that radiochemical methods were used at Windscale in the absence of spectrometric equipment, but beginning on the fifth day, spectrometric analyses were undertaken using equipment located in various laboratories in the United Kingdom. The distribution of ^{131}I in milk on October 13 is shown in Fig. 16-2. Altogether the milk exceeded 0.1 μCi/liter in an area of approximately 200 square miles stretching in a southeasterly direction from Windscale. The irregularity of desposition in Fig. 16-3 is explained by the changing meteorological patterns that existed during the period of emission (Crabtree, 1959). The highest concentration of radioiodine in milk was 1.4 μCi/liter obtained from a farm located about 10 miles from Windscale in the direction of the main plume.

It was observed that a useful correlation existed between the amount of radioiodine in milk and the γ-radiation levels in the area. This is seen by comparing Fig. 16-2 to the γ-radiation levels shown in Fig. 16-3. The concentration of ^{131}I in milk exceeded 0.1 μCi/liter in pastures where the γ radiation exceeded 0.035 mR/hr.

The criteria for restricting the sale of milk remained unchanged throughout the episode. By November 4, the permissible level was exceeded only in a region extending about 12 miles southward from Windscale. This area remained under restriction until November 23.

Drinking-water samples were collected from reservoirs and streams during the period immediately following the emergency, and no concentrations of radioiodine or other radionuclides were found to exceed the MPC permitted by the ICRP.

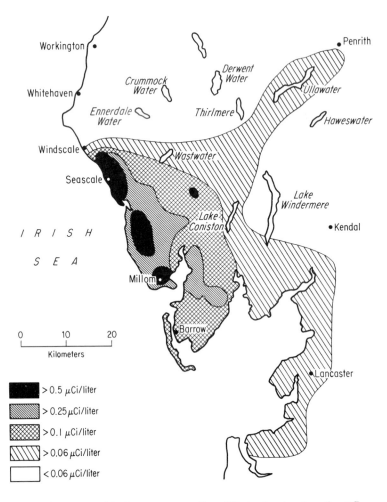

Fig. 16-2. The geographical area surrounding Windscale showing the ^{131}I concentration in the milk from various districts 5 days after the accident (Dunster *et al.*, 1958).

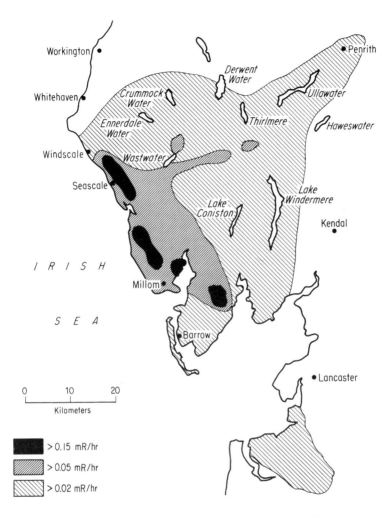

Fig. 16-3. Environmental γ-radiation levels in the vicinity of Windscale 5 days after the accident (Dunster *et al.*, 1958).

Children and adults living downwind as far as 24 miles from Windscale were scanned for iodine uptake with a scintillation counter. Among 19 children studied, the highest dose received was estimated to be 16 rad. The highest adult dose was estimated to be 9.5 rad.

Milk and other foods in the vicinity were studied for [89]Sr. It was found that the concentrations of these isotopes in foods did not exceed the levels known to exist in the area prior to the accident.

LESSONS LEARNED FROM THE WINDSCALE EXPERIENCE

The reactor designers and operators as well as the environmental specialist have much to learn from an accident of this kind. Presumably, this accident would not have occurred if the fuel had been fabricated from uranium oxide rather than metallic uranium, which is known to be pyrophoric. The incident emphasizes the need for great caution in the annealing of graphite reactors, but these and other aspects relating to design and operation of reactors are beyond the scope of this text. We are limiting our concern to the evaluation and management of the contaminated environment.

If we have learned much from this experience, it is due to the exemplary manner in which the British scientists conducted their investigations and published their reports. The accounts of this experience published by Dunster, the Medical Research Council, and others are to be admired for their thoroughness and promptness.

The following may be enumerated as being the significant conclusions derived from this accident.

a. In the slow oxidation of fuel of the Windscale type, the radioiodine was released preferentially from the core. In such an air-cooled reactor, if the stack effluents are filtered, all other isotopes can be ignored insofar as their public health consequences are concerned.

b. The dose to individuals from inhalation of the iodine or from direct exposure to the plume or the deposited radioiodine is negligible compared to the dose that would be received from eating crops or dairy products grown in the area.

c. The extent of iodine contamination can be determined most quickly by scanning the γ-radiation levels in the area. If weather permits, this may be done from aircraft; otherwise motor vehicles are practical.

d. When the ^{131}I concentration in milk exceeds 0.1 $\mu Ci/liter$, the dose to the thyroids of children may exceed 20 R. As a first approximation, it may be assumed that the concentration of radioiodine in milk will exceed 0.1 $\mu Ci/liter$ if it is produced by cows grazing in an area where the γ-radiation levels exceed 0.05 mR/hr.

e. γ spectrometry is far superior to radiochemical analysis for monitoring food supplies for contamination by radioiodine.

In addition to the above, one cannot read the accounts of the British experience without seeing the need for extensive advance planning. An incident of this kind taxes the health physics capacity of the local organization to the extent that only by pooling resources on a regional or national basis can the necessary technical assistance be brought to bear. It is of interest in this connection that at the height of the Windscale survey about

15 vehicles, each having a team of two men, were used for radiation survey-
ing and sampling. In addition, about 20 persons were employed in handling
and recording the samples, and about 150 radiochemists throughout the
United Kingdom assisted in the analysis of samples.

The Oak Ridge Plutonium Release of November, 1959

The radiochemical-processing pilot plant at Oak Ridge National Laboratory was built in 1943 and contains equipment for the chemical processing
of highly irradiated fuels. In November, 1959, a chemical explosion occurred in one of the shielded cells during a period when decontamination
of the process equipment was in progress. No one was injured, and the
monetary loss because of damaged equipment was relatively minor; however, the explosion resulted in extensive plutonium contamination of the
pilot-plant building, nearby streets, and building surfaces. The ensuing
cleanup operations were costly, and the contaminated areas could not be
usefully employed for many weeks while the cleanup was in progress.

The chemical explosion which occurred during decontamination of an
evaporator is thought to have resulted from the formation of compounds
such as picric acid when concentrated hot nitric acid was mixed with a
proprietary decontaminating agent containing phenol. A small quantity of
this solution had been left in the equipment because a normal water wash
was omitted, and the explosion occurred when nitric acid was later introduced into the evaporator and was brought to the boiling temperature
(King and McCarley, 1961).

The explosion breached a door leading from the cell directly to the outside of the building, and plutonium released through this doorway contaminated nearby streets and building surfaces. The adjacent air-cooled
graphite-reactor building became contaminated when plutonium was
drawn into the ventilation system. In addition, plutonium was forced
through penetrations in the concrete cell walls into the remainder of the
chemical-processing building. In all, an area of 1000 ft in diameter was
contaminated.

In addition to ^{239}Pu, ^{95}Zr and ^{95}Nb were also released, but the latter were
of secondary importance.

As reported by King and McCarley, the radiological safety procedures
consisted of the following: (i) immediate containment of the radioactivity
to prevent it from spreading to other laboratory areas; (ii) decontamination
of areas or building that were slightly contaminated and could easily be
made available for service; (iii) decontamination of streets and exterior
surfaces of buildings; and (iv) decontamination of the graphite-reactor
building and the radiochemical-processing pilot-plant building.

The extent of plutonium contamination is illustrated in Figs. 16-4 and 16-5, which give the distribution of contamination within the radiochemical-processing pilot plant and its immediate vicinity. The cell within which the explosion took place contained from 10^4 to 10^8 dpm/100 cm². A penthouse above the cell was contaminated to the extent of 5000 to 50,000 dpm/100 cm². Streets and buildings in the immediate vicinity were contaminated to levels which exceed 100,000 dpm/111 cm², but were for the most part below 100 dpm/100 cm².

The excellence of the decontamination procedures employed is shown by the fact that no employees were overexposed during the entire period of cleanup. Protective clothing consisted of two sets of coveralls, two pairs of shoe covers, two pairs of rubber gloves, an assault mask, and a hood. Wrists and ankles were sealed with masking tape. Employees leaving contaminated areas passed through monitoring stations where their protective clothing was removed. The men then showered and received a final examination for α contamination before being released.

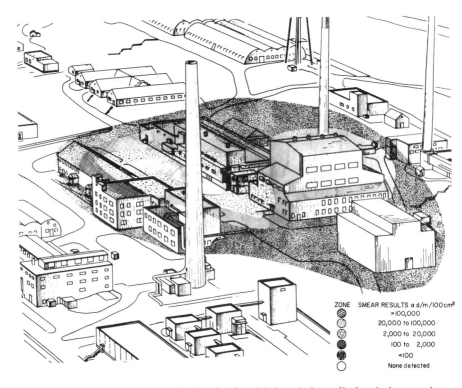

Fig. 16-4. Plutonium contamination in the vicinity of the radiochemical-processing pilot plant at Oak Ridge following the explosion (King and McCarley, 1961).

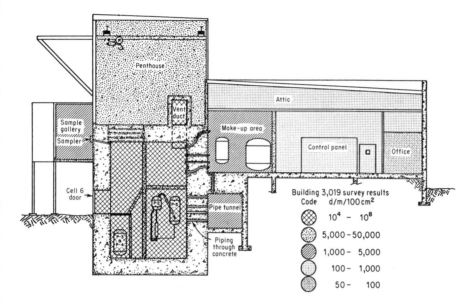

Fig. 16-5. Contamination levels within the chemical-processing pilot plant after the plutonium release (King and McCarley, 1961).

Containment of the radioactivity was achieved initially by fixing the surface contamination in a number of ways. Wash down was not attempted because of the possibility that the plutonium might wash into inaccessible places and because the accumulation of wash water would have created a problem.

Fixation was accomplished by resurfacing roads and by painting roofs, walls, and equipment. Paint was even sprayed on grass lawns and sidewalks that were found to be contaminated.

The decontamination procedures consisted of brushing and sponging, scraping, grinding, and various other techniques which are enumerated in Table 16-5. The target levels for decontamination are shown in the following tabulation:

	Direct reading	Transferrable
Maximum	300 dpm/100 cm²	30 dpm/100 cm²
Average*	30 dpm/100 cm²	3 dpm/100 cm²

* The number of samples considered in deriving an average included at least 10 samples, and there was at least one sample from each square meter of the projected surface area.

TABLE 16-5

SUMMARY OF DECONTAMINATION TREATMENTS FOR VARIOUS SURFACES[a]

Surface	Character of contamination	Primary decontamination treatment	Clean-up rate (ft²/man-hr)	Other decontamination treatment
All (walls, floor, ceiling)	Transferable	Scrubbed with detergent and water and brush or sponge	27	Dusty areas vacuumed
Painted metal (walls and ceiling)	Fixed	Paint removed with paint remover and scrapers, and surface scrubbed with soap and water	4	Outer layer of paint removed with sandpaper
Concrete (floor)	Fixed	Ground with terrazzo floor grinding machine	5	"Hot" spots chipped out and vertical surfaces washed with dilute hydrochloric acid
Bare metal (SS[b] piping and tanks)	Fixed	Rinsed with dilute nitric acid and scrubbed with steel wool	—	Surfaces abraded with emery paper
Bare metal (other than SS[b])	Fixed	Abraded with emery paper or ground to remove pits	—	—
Lead shielding	Fixed	Rinsed with dilute nitric acid	—	—
Oily metal (pumps)	Fixed	Washed with Gunk, a commercial solvent	—	—

[a] King and McCarley (1961).
[b] SS: stainless steel.

When decontamination to this extent was not practical, but where the levels were less than ten times the target levels, the surfaces were covered with brightly colored enamel and then covered with either paint or concrete. The brightly colored paint would serve as a warning in the future should the protective coating be removed.

The decontamination proceeded smoothly, but the incident resulted in costly interruptions to normal operations. The Oak Ridge National Laboratory graphite reactor was not in operation from November 20 to December 22. The processing cells directly involved in the incident could not be cleaned up for about 8 months until building modifications could be made to provide for improved containment of radioactivity.

It is estimated that only 15 g of plutonium were blown out of the evaporator subcell. This is a relatively small amount of material; yet it resulted in contamination that could only be removed after an expediture of hundreds of thousands of dollars and a costly interruption to an important research reactor and other activities in the immediate vicinity. All this occurred after 16 years of safe operation and as a result of the rather subtle error of introducing phenol into a decontaminating agent which was later to be mixed with nitric acid. The incident serves well to illustrate the extraordinary care which must be exercised in the design and implementation of operations such as these.

One cannot read the reports of the accident discussed above without being impressed with the excellent manner in which the emergency was managed and the high degree of coordination and competence which was demonstrated. One can imagine how different the situation might have been had the accident occurred anywhere but in a national laboratory which had had unique experience with many minor incidents of this type. A well-trained technical staff was available, backed up by a well-indoctrinated labor force and a host of material and services of a type which can only be found at large centers such as Oak Ridge.

One of the conclusions reached from investigations of this accident was that if costly interruptions to nearby facilities are to be avoided, double containment must be provided for operations of this kind. As a result, immediately following the explosion, high-level radiochemical operations at Oak Ridge were suspended pending studies of the need for secondary containment.

The Army Stationary Low-Power Reactor (SL-1)

An explosion of the Army Low-Power Reactor (SL-1) during the month of January, 1961 resulted in the deaths of three military personnel at the National Reactor Testing Station in Idaho. The reactor was a direct-cycle boiling-water unit designed to operate at a level of 3 MWt and was fueled with enriched uranium plates clad in aluminum. After a little more than 2 years of operation, the SL-1 reactor was shut down on December 23, 1960 having accumulated an operating history of approximately 950

MW days. A 12-day maintenance program was contemplated, and the reactor was scheduled to resume full power on January 4, 1961 (Buchanan, 1963; Horan and Gammill, 1963). On the night of the accident the crew consisted of three subprofessional military personnel of whom two were licensed reactor operators and one was a trainee.

The first indication of trouble at the reactor was given by remote radiation and thermal alarms which caused the fire department and health physicists to respond. On arriving at the SL-1, they encountered radiation

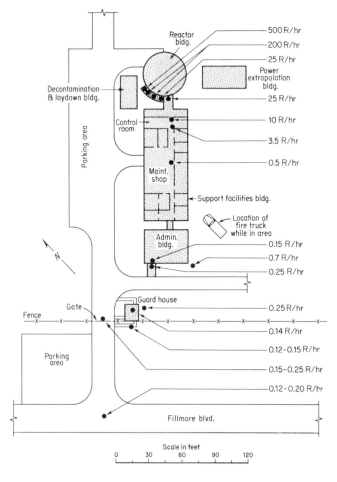

Fig. 16-6. General layout of the SL-1 area showing the radiation levels that were found after the explosion (Horan and Gammill, 1963).

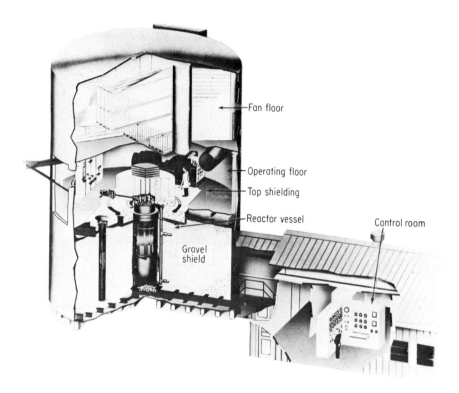

Fig. 16-7. Cutaway drawing of the SL-1 reactor building (Horan and Gammill, 1963).

fields of 200 mR/hr extending for a few hundred feet from the reactor building.

The general layout of the SL-1 area is shown in Fig. 16-6, and a cutaway of the reactor building is shown in Fig. 16-7. On entering the buildings, the emergency crews found neither a fire nor any of the three operators. High radiation levels were encountered, as can be seen in Fig. 16-6. A brief reconnaissance of the floor of the reactor building indicated radiation levels as high as 500 R/hr. A brief search of about $1\frac{1}{2}$-hr after the accident disclosed the whereabouts of the dead body of one of the personnel. A second man who was still alive was removed but died shortly thereafter. The third body was subsequently found, but the two dead men were not removed for several days, during which time several hundred men were engaged in recovery operations. Twenty-two personnel received radiation exposures in the range of 3 to 27 R (AEC, 1961). This is an excellent record, considering the complexity of the rescue and recovery operations and the fact that the radiation levels in some areas approached 1000 R/hr.

According to Horan, the bodies had been saturated with contaminated water and penetrated by particles of fuel. The radiation intensities at the bodies were 100 to 500 R/hr at 6 in.

The accident evidently resulted when withdrawal of a single control rod caused the reactor to go into the prompt critical condition. Why the rod was withdrawn may never be known.

Although the exact cause is not known, indications that a nuclear excursion took place were provided by the fact that metallic articles and jewelry on the bodies of the dead men showed signs of neutron activation (AEC, 1961). The excursion was evidently sufficient to result in a violent explosion, resulting in marked damage to the reactor core with expulsion of the core "innards."

Despite the fact that the excursion was a violent one and the reactor was not surrounded by a vapor container, the radiation levels outside the reactor building were minimal, being about 5 mR/hr at 1000 ft. Essentially all the radioactive material with the exception of ^{131}I was contained within

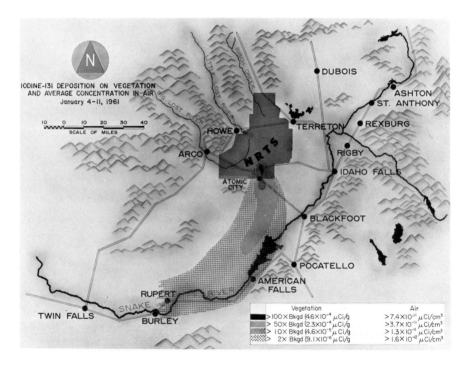

Fig. 16-8. The geographical vicinity of the SL-1 reactor showing the observed radiation levels (Horan and Gammill, 1963).

Fig. 16-9. The 13 ton core and pressure vessel of SL-1 being removed from the reactor building 11 months after the accident. The top and one side of the building have been removed. The reactor is to be lowered into the plastic-covered concrete cask on the flat-bed trailer (U.S. Atomic Energy Commission).

a 3 acre plot. Thus, despite the fact that the reactor core contained approximately 1 MCi of medium to long-lived isotopes, it is thought that less than 10 Ci of ^{131}I were released during the first 16 hr, resulting in contamination of the atmosphere and vegetation, as shown in Fig. 16-8.

Several months after the accident, it was decided to dismantle the reactor, an operation that presented many difficulties because of the high levels of radiation. Figure 16-9 shows the core being hoisted from the containment shell preparatory to removal to a hot cell for study.

The Houston Incident of March, 1957

A relatively minor accident which occurred in Houston, Texas in March, 1957 and which received extensive publicity (Look Magazine, 1960) is worth discussing. From this incident one can learn very little of a positive nature, but one can profit from an examination of the details of the accident and subsequent events, because mismanagement of the circumstances provides many illustrations of what not to do.

The company was licensed by the AEC to encapsulate sources for γ-ray cameras. In March, 1957, two men were opening a sealed can containing 10 pellets of ^{192}Ir, each pellet being a $\frac{1}{8} \times \frac{1}{8}$ in. right cylinder containing about 35 Ci.

It was necessary to remove a plug at the end of the can to have access to the pellets, and this was accomplished on a jeweler's lathe in a sealed Plexiglas box which in turn was located in a hot cell. The operation was performed with master-slave manipulators through 33 in. of concrete.

When the small can containing the pellets was removed from its larger container and opened, two of the pellets were found to be in a loose dusty form. Some of the dust escaped from the Plexiglas box and hot cell, as was evidenced by the fact that an air monitor in the laboratory indicated the presence of radioactive dust. One of the two employees involved in the operation was dressed in street clothes and left the laboratory. The second employee, who was dressed in work clothing and wearing a respirator, remained in the room for an unknown period of time.

The fact that the laboratory had become contaminated seems to have come to the attention of the plant management for the first time about a month after the incident occurred. It was then observed that contamination existed in the vicinity of the hot cell and that it had been spread to employees' street clothes, shoes, and even into their homes and automobiles. In the intervening weeks, various steps had been taken to encapsulate and store the sources, and items of equipment which became contaminated as a result of the incident had been stored away after unsuccessful attempts at decontamination.

The AEC was notified, somewhat belatedly, about 5 weeks after the accident, and shortly thereafter the company retained the services of a private company to survey and decontaminate the affected areas. Of 19 private homes that were checked, 8 showed evidence of contamination; 7 out of 53 automobiles were found to have traces of contamination.

Ninteen employees and members of their families were examined by a physician. One neighbor of an employee was also examined. Except for minor radiation burns on each of the two employees who were present at the time of the incident, the medical examinations were negative.

Very few quantitative data are available concerning the levels of contamination encountered. However, it is reported that one of the employees received an exposure of 1.7 rem during 1 week when decontamination of the laboratory was in progress. His 13-week exposure was 3.9 rem, which exceeds the permissible level by 25%.

This incident was widely publicized and its effect exaggerated. In addition to sensational newspaper accounts, the incident became the subject of nationwide televised programs and was featured in nationally distributed publications. One of the main themes of the publicity was that the affected individuals and their families became socially ostracized because of their "radioactivity" and that serious disabilities were developing not only in the two employees but also in their families.

The incident occurred very soon after the licensing procedure of the AEC was put into effect, and the licensee was apparently ignorant of the need to notify the AEC that an accident had occurred. Moreover, the regulatory apparatus of the AEC had not as yet been confronted with an incident of this type, and its handling of the incident indicated that there was as yet much to be learned about the role of this agency in incidents of this kind.

Many experts who might have been sent to Houston by the AEC to advise and assist its licensee to minimize the consequences of the incident were not made available, apparently because the regulatory function of the AEC in this particular instance forbade its participation in an advisory capacity. As a result, much potentially valuable information was never obtained. For example, urinalyses and whole-body γ-ray measurements would have certainly been indicated for the two men who were known to have been exposed. Other employees might also have been so examined. Depending on these findings, it might or might not have been desirable to have examined the families and neighbors.

The somewhat extensive decontamination procedures in the homes of the employees, including such drastic procedures as cutting out portions of rugs to remove observable contamination, were undertaken without any quantitative data that would indicate that such measures were necessary. In summary, a valuable study which might have shed useful quantitative information as to the consequences of an incident of this kind was not undertaken, and in the absence of reliable information, rumor and sensationalism swelled in uncontrolled fashion about the unfortunate principals for months afterward.

In 1961, after 4 years of exaggerated stories about the effects of the accident, some of the principals were examined at AEC expense by physicians at the Mayo Clinic (Atomic Industrial Forum, 1961). It was found that none was suffering from the radiation effects that were repeatedly alleged. However, as is so often true in incidents of this kind, the factual announcement of the findings at the Mayo Clinic did not receive nearly as much public notice as the more sensational claims of injury.

Fermi Reactor Fuel Melt-Down

The Enrico Fermi reactor is a sodium-cooled fast breeder located on the western shore of Lake Erie near Monroe, Michigan. The first core was loaded in 1963, and the full power of 200 mW [66 MWe (megawatt equivalents)] was authorized in December, 1965 (Atomic Power Development Associates, 1967; Scott, 1971).

The reactor core consists of 105 subassemblies containing uranium oxide fuel enriched to 25.6% and assembled in the form of pins clad with zirconium. The cooling system is basically that of the LMFBR, described in Chapter 9.

On October 5, 1966 at a power level of 27 MW, radiation monitors in the building's ventilation exhaust ducts sounded an alarm which resulted in automatic isolation of the building and, within a few minutes the reactor was manually shut down. In the few minutes prior to the automatic alarm, it was noted that the core exit temperatures were abnormally high. At the time of the radiation alarm, no personnel were in the reactor building.

The cause for the observed anomalies was not immediately apparent, but investigation soon revealed that about 10,000 Ci of fission products had been released to the primary sodium loop and the reactor cover gas. Radioactive noble gases, but no iodine, was found to be present. No particulate or gaseous contamination was observed outside the containment structure.

Many months of careful work, during which time the sodium loop was drained to permit inspection of the reactor vessel, disclosed that coolant flow had been blocked by pieces of zirconium that had become detached from a conical flow guide. As a result, partial melt-down of two fuel assemblies occurred. Repairs of the reactor were undertaken, and full power operation was achieved in October, 1970, 4 years after the incident.

This was an expensive accident for the reactor owners, but the radioactivity released from the core was properly contained and no exposure to the employees or the public resulted. The highest radiation level as a result of this incident was 9 mR/hr at one point on the outer surface of the reactor building.

Abortive Reentry of the SNAP 9-A

A navigational satellite launched on April 21, 1964 and carrying a SNAP generator failed to reach orbital velocity and reentered the atmosphere at about 150,000 ft over the Indian Ocean. The isotopic power unit was known as SNAP 9-A and contained about 17,000 Ci of ^{238}Pu (Krey, 1967).

The circumstances were such that the unit should have been volatilized by the heat of reentry, as was discussed in Chapter 11. ^{238}Pu was present in the upper atmosphere at the time as a residue from earlier nuclear weapon tests, but the system of high-altitude balloon sampling, described in the Chapter 14, served to demonstrate the sudden appearance of a new source of ^{238}Pu which was first detected at an altitude of 108,000 ft 4 months following the abort. An interesting subsequent finding was that the stratospheric distribution of the debris during the next several years followed the predictions that had been made on the basis of transport models of the stratosphere developed through studies of the behavior of debris from weapons testing (Kleinman, 1971). However, the concentration of ^{238}Pu of SNAP 9-A origin was somewhat lower in ground-level air than had been predicted (Shleien *et al.*, 1970). About 16 kCi of ^{238}Pu, representing 95% of the amount originally injected, has been estimated to have deposited by the end of 1970. The estimated stratospheric inventories are given in Fig. 16-10, in which it is seen that the debris apparently took about 2 years in which to diffuse to the sampling altitudes, following which depletion of the stratospheric inventory began to take place exponentially with a half-time of about 14 months (Krey *et al.*, 1970). Particle-size analysis of the debris

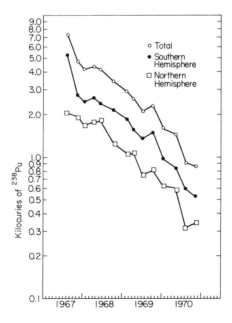

Fig. 16-10. ^{238}Pu stratospheric inventories, 1967–1970 (Kleinman, 1971).

indicates that the particles ranged from 5 to 58 mμ with an arithmetic mean of about 10 mμ.

The ^{238}Pu content of ground-level air has been monitored by de Bartoli and Gaglione (1969) and by Shleien et al. (1970), on the basis of which it has been estimated (Schleien et al., 1970) that the dose to the lung from the SNAP debris is small compared to the dose from ^{238}Pu in fallout from nuclear weapons testing. The 50-year dose commitment to the lung as a whole was estimated to be 0.35 mrem for exposure in 1963, prior to the SNAP 9-A abort, compared to 0.06 and 0.08 mrem for exposure in 1965 and 1969 (Environmental Protection Agency, 1971). The 50-year dose commitment to the respiratory lymph node from ^{238}Pu exposure between 1961 and 1968 was estimated to have been 36 mrem (Shleien, et al, 1970).

Plutonium Fire at Rocky Flats

A major fire occurred at the AEC's plutonium-processing plant operated by Dow Chemical Co. at Rocky Flats, Colorado on May 11, 1969 (U.S. Atomic Energy Commission, 1969b). The extent of off-site plutonium contamination received national press attention several months later as a result of measurements made by a group of independent scientists whose findings were reported to the public (Colorado Committee for Environmental Information, 1970). A subsequent study of the plutonium content of soil in the vicinity of the plant was undertaken by the AEC (Krey and Hardy, 1970) and confirmed that plutonium was indeed present but that the distribution in the vicinity of the plant was incompatible with the known meteorological circumstances at the time of the fire. The off-site inventory of ^{239}Pu, integrated within the 3 mCi/km^2 contour, was found to be 2.6 Ci and extended for a distance of more than 7 miles in a southeastern direction. However, the winds were generally blowing toward the southwest at the time of the 1969 fire. It was concluded that the observed plutonium deposition, although clearly from the Rocky Flats plant, was compatible with a gradual accumulation of plutonium owing to many small emissions from the plant. This conclusion followed from the observation that the pattern of plutonium deposition conformed to the annual distribution of wind direction at the site. The data also suggested that a very considerable area, up to about 40 miles to the east and north of the plant, was contaminated to the extent of an additional 1 mCi/km^2, accounting for an additional 3.2 Ci of Rocky Flats plutonium. The "background" distribution of ^{239}Pu owing to fallout from weapons testing was estimated to be about 1.6 mCi/km^2 in this general area. Subsequent measurements (Volchok et al., 1972) have indicated that ^{239}Pu is present as airborne dust east of the plant site in concentrations ranging up to approximately 10^{-9} pCi/ml.

The reason this incident is worth reporting is that considerable public concern was developed (Shapley, 1971) with a general failure of confidence in the credibility of the information released by the plant management and the AEC. The source of the first public knowledge that there was off-site plutonium contamination was a citizens' group with access to highly trained radiochemists and highly sophisticated radiochemical equipment. The basic failure was in the lack of adequate reporting of the off-site levels of plutonium contamination by the government and its contractor. It is no longer possible to conceal information of this kind from the public, intentionally or otherwise.

Chapter 17

Methods of Environmental Surveillance

A wide assortment of instruments and methods are available with which to (a) monitor the presence and ecological behavior of radionuclides that are present in the environment; (b) identify the pathways by which human exposure results; and (c) estimate the dose to humans.

Broadly speaking, one can differentiate two basic types of monitoring programs: those that are regional or global in nature and those that provide information about the immediate vicinity of a particular facility from which radioactive substances may be emitted. Ideally, the individual plant-site monitoring programs should blend into surveillance programs operated by the local or state government to provide information on a county- or statewide basis, and these programs, in turn, should blend into national and international surveillance programs operated by national governments or even supernational organizations such as the Pan-American Health Organization, the World Health Organization (WHO), or the International Atomic Energy Agency (IAEA). Such international agencies have in fact assembled environmental monitoring data, the most outstanding example being that of the United Nations Scientific Committee on the Effects of Atomic Radiation (UNSCEAR), which periodically publishes compendia of surveillance information from all parts of the world. However, the usefulness of the assembled information has been handicapped in the past by a lack of uniformity in the methods of gathering and reporting data. This has been true not only at the international level, but at the regional and local level as well.

The types of environmental monitoring programs may be further differentiated into the following: (1) preoperational surveillance designed to provide baseline radiological information against which any changes caused by a particular nuclear activity can be ascertained; (2) postoperational surveillance; and (3) a surveillance program that can be held in abeyance pending an emergency in which extraordinary information is required.

In the postoperational surveillance program, a primary objective is to determine if relevant governmental regulations are being observed. It should be emphasized constantly that surveillance programs at best provide only an *estimate* of population exposure, with the limits of uncertainty of the estimate dependent on the kind and number of observations made, the type of instrumentation employed, and the methods used in analyzing the data. In view of this, an environmental surveillance program should be flexible so that efforts will not be wasted in providing dose estimates that have more precision than is meaningful at very low levels of exposure while maintaining the capability of intensifying the surveillance program to provide more accurate estimates at higher levels.

The basic principles of environmental monitoring have been laid down by ICRP (International Commission on Radiological Protection, 1965) and EPA (Environmental Protection Agency, 1972), and there is an extensive literature in which health physicists have described their experiences with a variety of surveillance programs (Reinig, 1970; International Atomic Energy Agency, 1966). Fundamentally, the programs should be designed to provide assurance that the dose to man is below some prescribed level, usually specified in a governmental regulation or guideline. In practice, one usually finds that there are one or more critical pathways between the point of discharge of nuclides to the environment and some vector of human exposure, such as a particular food, a bathing beach, or perhaps airborne dust. Within the critical pathway there will also be one or two critical nuclides which, in conjunction with a critical population, serve as the primary focus of the environmental surveillance program. Sometimes there may be secondary pathways involving nuclides of lesser significance which nevertheless require investigation. A not uncommon situation near nuclear power plants is that the critical nuclide is radioiodine and the critical pathway is deposition of this nuclide on pastures that nourish cows whose milk is furnished to children, the critical population. A secondary pathway might be the use of river water containing ^{65}Zn that passes to humans by way of crop irrigation. Although the iodine–milk pathway may be critical in that the highest dose is delivered to humans by this route, it may also be desirable to monitor the secondary pathway which, though it results in lower dose to humans, might nevertheless be of potential importance.

This chapter will focus primarily on the requirements for environmental surveillance programs in the vicinity of nuclear facilities, but the basic principles can be applied to the design of regional, national, or supernational monitoring networks.

The subject can be discussed under several headings as follows.

1. The preoperational phase, in which the environment is studied to identify the critical pathways in relation to: (a) the physical and chemical identity of the radioactive material that will be discharged to the environment; (b) the living habits of the potentially exposed population; and (c) the ecological mechanisms for dilution and concentration.

2. The postoperational stage, the design of which is based on information obtained in the preoperational study, modified as necessary by the performance of the facility and continuing observation of critical pathway kinetics.

3. A surveillance plan to be used in emergencies.

The design of every surveillance program is based on these fundamental questions: (a) what should be measured; (b) where should the measurements be made; (c) with what frequency; and (d) how are the data thus obtained converted to dose estimates.

Some General Observations

PREOPERATIONAL PHASE

Enough information must be obtained in the preoperational survey to identify the pathways by which humans will be exposed to the effluents from the proposed facility. In most cases the critical pathway will be obvious, but one must always be on guard for some subtlety that may influence the radiation exposure of human populations. An example of the latter would be the use of uranium mill tailings as a building construction material (Chapter 8) or consumption of laverbread made from a seaweed that tends to concentrate ^{106}Ru (Chapter 12). Figures 17-1 and 17-2 identify the basic pathways by which man may be exposed to radioactive environmental contaminants.

The design of the preoperational monitoring program will depend on the kind of facility being planned and whether sources other than natural ones are having an effect on the radiation background. The principal man-made source that has affected the natural background up to the present has been testing of nuclear weapons. In only rare situations will one facility be built so close to another as to be affected by its radioactive releases.

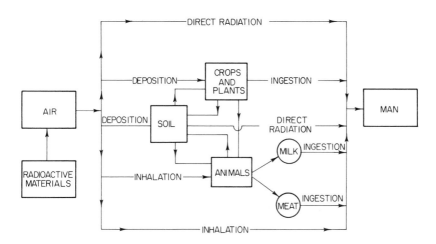

Fig. 17-1. Pathways between radioactive materials released to the atmosphere and man (Environmental Protection Agency, 1972).

Because weapons testing is spasmodic, there is no reason to undertake long series of measurements to document the preoperational variations due to differences in fallout from such tests. When weapons testing is in progress, there is apt to be considerable variability in the day-to-day rate of fallout, although much less variability in the cumulative deposits of long-lived nuclides. Thus, in principle, the preoperational surveillance requirement is for a system of measurements that will (a) determine if there are any natural radioactive anomalies in the region; (b) define the ambient γ-radiation levels, apportioned between naturally occurring and man-made nuclides; and (c) obtain baseline data for some of the more important nuclides, both natural and man-made. This data can be accumulated in the year prior to start-up of the facility by techniques that will be described later in this chapter.

The preoperational study should also be used to gather the kind of environmental information that will be needed for critical-pathway dose calculations. A basic objective of the preoperational survey should be to develop predictive techniques by means of which the dose to the critical organ can be estimated per unit release of the critical nuclide, whether it be in the air or water. For example, what will be the dose to an individual consuming fish from a stream into which a given quantity of ^{137}Cs is annually discharged? or what is the dose to a child's thyroid from a given release of ^{131}I discharged from a stack of given height? In order to enable one to make these calculations, preoperational meteorological and hydro-

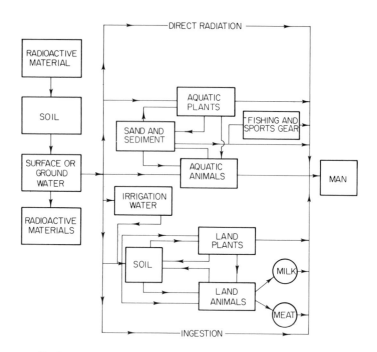

Fig. 17-2. Pathways between radioactive materials released to ground and surface waters (including oceans) and man (Environmental Protection Agency, 1972).

logical observations are needed for a period of 1 year or longer. The climatological data should include a record of wind speed, wind direction, and class of stability. Additionally, the mixing characteristics of the receiving waters should be ascertained by dye studies, models, or other means.

It should be noted that during the preoperational phase one can often obtain important information about critical pathways by tracing the behavior of nuclides introduced by weapons testing. When these nuclides can be detected in water, suspended solids, sediments, benthic organisms, or fish, the information will frequently add significantly to an understanding of the critical-pathway processes. Such measurements will prove particularly useful in locations where estimates are available on the rates of fallout and the cumulative deposits.

POSTOPERATIONAL PHASE

The ICRP (International Commission on Radiological Protection, 1965) states succinctly that the information to be gathered in the postoperational monitoring program should be guided by the following three principles. (1) The information is needed to assess the actual or potential exposure of the critical groups averaged over extended periods, e.g., 1 year. (2) Only the exposure by the critical pathways needs routine examination. (3) In addition to estimating exposures, it may sometimes be necessary to follow trends.

These are rational principles which, when adopted, greatly simplify the design of monitoring programs conducted in the vicinity of nuclear facilities. Regrettably, there has developed over the past 2 decades the feeling that it is desirable to undertake expensive measurements, not necessarily dictated by the above principles, in order to satisfy some vague public relations need. All too frequently, in the place of well-designed logical programs of monitoring, determined by the need for critical-pathway analyses, a wasteful approach is taken in which relatively massive sampling programs are accompanied by analytical programs based on an apparent faith in the persuasive power of sheer masses of numerical data that in fact provide far less information than would be the case if fewer data were collected on a more selective basis.

The operator of a nuclear facility is normally responsible for monitoring the environs of the plant within the area influenced by its radioactive emissions. Governmental responsibility should involve sufficient monitoring close to the plant to validate the reports of the facility operator and should also include a monitoring program that covers the total geographical area for which the agency is responsible. This type of monitoring program can often be coordinated with air, water, and food surveillance programs conducted for more general reasons, such as analyses for trace metals, pesticides, or other chemical substances.

One of the most important factors that influences the design of a surveillance program is the required sensitivity of the procedures for dose estimation. The contemporary trend toward limiting the maximum exposure to individuals in the general population to somewhere between 1 and 10% of the ICRP recommendations poses technical difficulties. To meet contemporary requirements, surveillance programs should be capable of estimating doses on the order of 1 mrem/year to the maximum exposed individual, as well as integral population doses of the order of a few hundred man-rem per year, averaged over a population of 1 million people. The concept of "lowest practicable dose" has been distorted through misunderstanding on the part of many scientists, governmental officials, and members of the public at large.

It is immediately evident that direct measurement will rarely be possible if quantitative data is required at levels of a few millirem per year and that for the most part one must rely on calculations based on critical-pathway measurements and a quantitative understanding of critical-pathway behavior. The main portion of this chapter will be concerned with the manner in which such measurements can be made.

Methods of Environmental Surveillance

RATES OF RADIOACTIVE EMISSIONS

Knowledge of the rates of radioactive emissions will frequently provide the principal assurances that population exposure criteria are not being exceeded, and when the liquid and gaseous releases are below some pre-determined limit, the requirements for environmental surveillance can be minimal, with sufficient environmental sampling to confirm that the actual exposures do not exceed those predicted from knowledge of the release rates.

Estimates of the amounts of the individual nuclides being discharged in gaseous- or liquid-waste streams can be made by measuring the total β or γ radioactivity of the wastes, supplemented with periodic analyses by γ spectrometry to provide information about isotopic composition. For reactors such as the BWR that discharge gaseous wastes continuously, the frequency of isotopic analysis of the gases should depend on the condition of the fuel cladding and the rate at which the composition is known to be varying. Where the concentration of radioactivity is reasonably constant in relation to the power level, monthly analyses may be sufficient. On the other hand, if the rate of release is increasing due to progressive fuel-cladding failure, more frequent analyses may be desirable and, in particular, a close watch should be maintained for the appearance of the radioiodines. When radioactive gases are released in batches, samples from each batch should be analyzed by γ spectrometry.

Similarly, the frequency of isotopic analysis of liquid wastes should depend on the extent to which the radionuclide composition of the coolant is known to be varying. Following start-up of a new core, the presence of "tramp" fissionable material in the form of surface contamination can result in the coolant becoming contaminated with traces of fission products. As the core ages, minor imperfections in the cladding can increase in number and, in the PWR's, leaks in the heat exchangers may develop that allow traces of primary coolant to pass to the secondary system. Factors such as these can result in abrupt changes in the radionuclide spectrum of the wastes. In addition to periodic analyses by γ spectrometry, radio-chemical analyses of the coolant should be made for α radioactivity and for β emitters such as ^{90}Sr, ^{89}Sr, tritium, and electron-capture nuclides such as ^{55}Fe.

DIRECT γ RADIATION

The ambient γ-radiation levels in the vicinity of a new facility should be documented preoperationally. The presence of radiation anomalies can be ascertained very quickly, usually by driving or flying over the surrounding area with a scintillometer having a rapid response time. Such anomalies will in most cases be due to outcroppings of granitic rock, but the possibility of other anomalies cannot be ruled out until an area-wide survey has been made.

Estimates of the annual dose rate at a number of benchmark locations should then be made. One method of making such estimates utilizes thermoluminescent dosimeters (TLD's) which can be exposed at these sites for periods of several weeks or months (Attix *et al.*, 1968; Cameron *et al.*, 1968). The sensitivity of thermoluminescent techniques can be expected to improve but contemporary techniques do not permit measurement of a

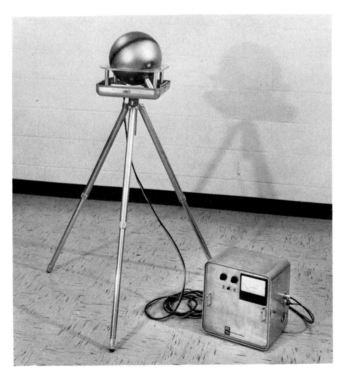

Fig. 17-3. Ionization chamber and electrometer. The chamber is pressurized with 40 atm of argon and is capable of measuring γ-radiation differences of about 1 mR/hr (Reuter-Stokes, Cleveland, Ohio).

change of less than about 50 mR/year, unless unusual precautions are taken in the handling, transportation, and readout of the dosimeters.

The pressurized ionization chamber (Fig. 17-3) is a more sensitive method of measuring ambient γ radiation and, properly used, can detect changes of less than 5 mR/year. Beck et al. (1971) use a stainless-steel-walled 10 in. in diameter sphere filled with ultrapure argon to a pressure of about 40 atm. With a good electrometer it should be possible to measure hourly changes of about 0.1 μR/hr, equivalent to about 1 mR/year. The normal fluctuations in background, owing to changes in soil moisture and other factors, can be documented by operating the pressurized ionization chamber continuously in conjunction with a strip-chart recorder. Alternatively, one might make monthly or semimonthly measurements over a period of 1 or 2 years.

Fallout from weapons testing produces major perturbations in the ambient γ radiation. In the preoperational phase of surveillance it is possible to document the contribution of this source of contamination by analyzing γ spectra obtained with a suitable detector and pulse-height analyzer. The techniques for separating the total γ-radiation dose rate into the various contributing components have been described by Beck et al. (1971), who have used both thallium-activated sodium iodide scintillators and Ge(Li) detectors. Figure 17-4 illustrates the remarkable resolution of which

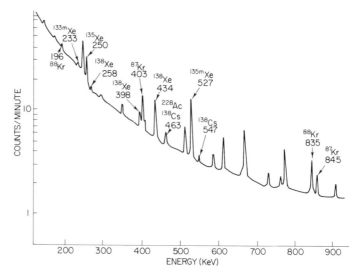

Fig. 17-4. In situ Ge(Li) spectrum taken near stack of BWR power plant with relatively short noble gas holdup (<1 hr). Peaks not identified are due to natural and fallout emitters (Harold Beck, U.S. Atomic Energy Commission, Health and Safety Laboratory).

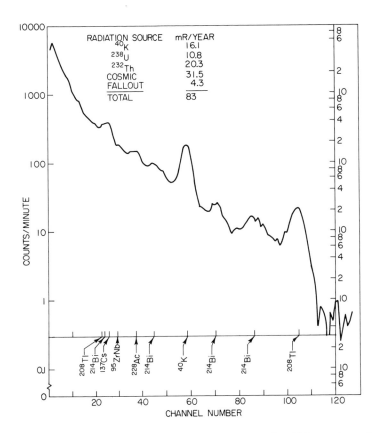

Fig. 17-5. Typical preoperational γ spectrum at reactor site. This spectrum, in conjunction with a pressurized ion chamber measurement of the total γ dose rate, can be reduced to provide the portion of the total dose contributed by the five sources of background radiation (Environmental Analysts Inc.)

the Ge(Li) system is capable for *in situ* field spectrometry. Figure 17-5 shows a spectrum obtained with a NaI crystal and indicates the contribution to the ambient γ background from the principal sources of radiation.

SURFACE DEPOSITION

Various simple devices have been used for collecting samples of radioactive particulates settling to the surface of the earth as dry dust or in precipitation. If one is concerned with the possibility of food-chain contamination, fallout samples are a convenient method of documenting the amount of contamination per unit area of ground surface. These techniques are more suitable for regional monitoring and for documenting deposition of long-lived fallout constituents. A simple method is shown in Fig. 17-6, in which a 1 ft² acetate film covered with a sticky substance is mounted horizontally on a frame about 3 ft above the ground. Coatings are available commercially which retain their adhesive properties when wet, and dust particles may be entrapped efficiently even though contained in raindrops. This method of collection, because of its simplicity, was used for many years throughout the world in collecting hundreds of daily samples of the fallout from weapons testing (Eisenbud and Harley, 1953). A major disadvantage of the technique is that the more soluble components can be washed off the film by rain. The average collection efficiency of the gummed films was estimated to be 70% under a wide range of conditions.

Other types of collectors such as the funnel shown in Fig. 17-7 have been designed to overcome the defects in the gummed film collector. The fallout can be washed from the funnels directly into flasks and then transported

Fig. 17-6. Gummed-film fallout monitor. The cellulose acetate film is covered with an adhesive that retains its sticky properties when wet and catches the dust particles that deposit either in dry form or in raindrops (U.S. Atomic Energy Commission, Health and Safety Laboratory).

Fig. 17-7. Fallout collection devices. On the right is a simple stainless-steel pot. The funnel-type collector on the left is designed to permit the fallout collector to be washed through the ion-exchange column immediately below the funnel (John H. Harley).

to the laboratory for radiochemical analysis or the rainwater or wash water can be passed through an ion-exchange column which concentrates the fission products (Welford and Collins, 1960). In order to collect large samples of fallout, some investigators have covered entire roofs with plastic film and have allowed the rainwater to run off the roof into collection barrels. The method of choice will be dictated by the purpose of the sampling program and the desired sensitivity.

One method of obtaining a cumulative sample of fallout is to collect surface soil for radiochemical analysis. Soil makes an excellent sampling medium, but it is, unfortunately, a difficult matrix from which to extract the radioactive substances for radiochemical assay.

Alexander (1959; Alexander *et al.*, 1960) has considered the problems of sampling soil in connection with his investigation of the worldwide distribution of ^{90}Sr from weapons testing. He recommends the selection of sites that have a good vegetative cover and are nearly level. Sites subject to overwash from higher ground or flooding should be avoided. Soils that pack when dry should also be avoided, as should sites that have a high population of worms that might affect the vertical distribution of isotopes. These criteria are particularly important where it is desired to obtain, by means of soil sampling, an estimate of the total amount of fallout in a given area. The soil should be sampled to a known depth, which can be accomplished with the use of augers by which precisely dimensioned borings can be obtained. Multiple borings covering an area of from 1 to 2 ft² to a depth of 6 in. were found to provide representative samples.

Because radiochemical analysis of soils for radionuclides is a complicated procedure, it is desirable to reduce the bulk of the original sample to the extent practical. For many purposes it will suffice to sample to a depth of 2 to 4 in.*

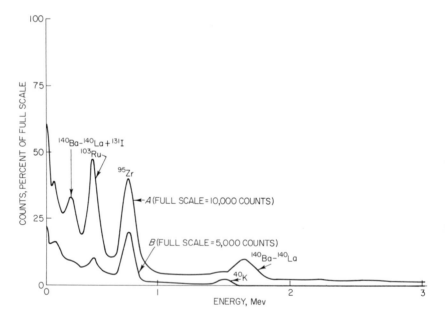

Fig. 17-8. γ-ray spectra obtained from a 500 g sample of grass cuttings when the Soviet Union weapons testing was under way in October, 1961. (*a*) The spectrum obtained during the period of actual fallout; (*b*) the spectrum as it appeared approximately 6 months later. Note that in the second spectrum the ^{95}Zr (65-day half-life) has diminished but is still present as is the ^{103}Ru (40-day half-life). ^{140}Ba–^{140}La and ^{131}I have all but disappeared. The ^{40}K that was masked by the ^{140}Ba–^{140}La in the first spectrum is more conspicuous in the second (Gerard Laurer).

*For more detailed descriptions of sampling and radiochemical procedures, see Harley (1967).

Whether or not one removes vegetation and organic debris from the surface soil sampled depends on the purpose of the sampling. Where one wishes to estimate the amount of fallout deposited on a given area, the vegetation and debris should be analyzed along with the soil. However, there should be no reason for retaining pebbles or twigs.

Grass is an excellent medium for trapping surface fallout and can be used as a sensitive indicator of fresh fallout. Figure 17-8 shows a γ-ray spectrum obtained from a 500 g sample of grass cuttings when Soviet weapons testing was in progress in October, 1961. Such a sample can be obtained and its spectrum determined in a few minutes by means of γ-ray spectrometry.

ATMOSPHERIC SAMPLING

The general procedure used in sampling the atmosphere (International Atomic Energy Agency, 1967a) for radioactive particulate activity is to draw air through a filter at a known rate for a known period of time. The radioactivity of the filters may then be counted and the activity per unit volumes of air ascertained. The measurement of both α and β activity is complicated by the fact that the atmosphere normally contains short-lived radon and thoron daughters in concentrations that may be higher than the permissible concentrations of some of the long-lived nuclides for which one may be monitoring. For example, if one draws air through a filter paper at a rate of 1 cfm (see Fig. 7-1) and if the air contains 5×10^{-5} pCi/cm^3 of radon in equilibrium with radium A (^{218}Po) through radium C' (^{214}Po), the α activity of the filter paper will increase to about 100 pCi at the end of 2 hr. This might be interpreted erroneously as a concentration of 3×10^{-5} pCi/cm^3, about 500 times the maximum permissible concentration (MPC) for exposure of the general population to ^{239}Pu. Thus, the natural α activity of the atmosphere tends to mask the presence of long-lived α emitters, but, fortunately, the natural α-emitting radionuclides in the atmosphere are short-lived, and if one permits them to decay overnight, the short-lived nuclides will no longer present a problem.

Monitoring for atmospheric α activity in the vicinity of power reactors is unnecessary. Although it is theoretically possible for plutonium and other toxic transuranic elements to be emitted to the environment, such emissions would be associated with overwhelming amounts of β- and γ-emitting nuclides. Except in the event of a massive accidental discharge, α measurements need not be undertaken in surveillance programs in the vicinity of reactors.

Several types of filter media are used for air sampling (American National Standards Institute, 1969). Those most commonly employed for collection

TABLE 17-1

COLLECTION EFFICIENCY AND FLOW RESISTANCE OF SELECTED AIR-SAMPLING FILTER MEDIA[a]

Filter class	Filter designation	Supplier	Collection efficiency % for retaining 0.3 μm, DOP Velocity (cm/sec)				Flow resistance (mmHg) Velocity (cm/sec)			
			10.7	26.7	53	106	35	53	71	106
Cellulose	Whatman 41	W & R Balston, Ltd. England	64	72	84	98	24	36	48	72
	SS 589/1	Scheicher and Schuell Germany	46	56	66	80	18	27	37	56
	TFA-41	The Staplex Co. United States	62	74	86	98	23	40	48	81

Cellulose-asbestos	S-P Rose	Establissements Schneider Poelman France	99.18	99.28	99.52	99.75	38	57	75	112
	HV-70	Hollingsworth and Vose United States	96.6	98.2	99.2	99.8	44	64	87	127
Glass-fiber	MSA 1106B	Mine Safety Appliance Co. United States	99.968	99.932	99.952	99.978	20	30	40	61
	Gelman E	Gelman Instrument Co. United States	99.974	99.964	99.970	99.986	19	28	38	57
Membrane	Millipore AA (Pore size 0.8 μm)	Millipore Filter Corp. United States	99.992	99.985	99.980	—	98	142	195	285
	Gelman AM-1 (Pore size 5 μm)	Gelman Instrument Co. United States	88	88	92	95	56	84	117	190

[a] Lockhart et al. (1964).

of radioactive dusts and fumes are either cellulose, glass, or mineral fibers or millipore membranes which are made of plastic and have a very high collection efficiency for particles in the submicron range. A disadvantage of the millipore filter is its fragility and relatively high resistance to air flow. The cellulose filters are advantageous in that they can be dissolved in reagents or ashed if it is desired to undertake radiochemical analysis. The plastic, glass, and mineral fibers tend to have higher collection efficiencies than cellulose. Table 17-1 summarizes the collection efficiency and flow resistance characteristics of selected air-sampling filter media (Lockhart *et al.*, 1964). The uncertainties in estimating the human hazard from inhaling radioactive dust are so great that small differences of the order of 10 to 20% owing to imperfection in filter performance are relatively unimportant and would not affect one's evaluation of a given set of data. All the commercial filter media, when used properly, have efficiencies that are more than adequate to serve the purpose. Radioiodine released from freshly irradiated fuel is apt to exist in vapor form and will not be retained by filter media such as are suitable for dusts. Activated charcoal has been shown to be effective for radioiodine sampling and may be used in series with dust filters (Cowser, 1964).

Air samplers may be fixed or portable and may be devices that simply collect the dust on filter papers for counting in a laboratory or they may be equipped with automatic counting and recording devices.

Where power is available and where relatively larger samples are desired, the "high-volume" air sampler shown in Fig. 17-9 has proved to be very popular. When used without the carbon filter, the sampling rate varies from 20 to 60 cfm, depending on the filter media used. The unit may be used at

Fig. 17-9. "High-volume" air sampler equipped with activated charcoal backing to a fibrous filter. This type of sampler can collect up to 60 cfm through various types of filter media (Los Alamos Scientific Laboratory).

Fig. 17-10. A continuous air monitor that filters dust on a fixed filter mounted adjacent to a Geiger–Müller counter. The cumulative radioactivity of the filter paper is measured on the strip chart (Nuclear Measurements Co. and the Consolidated Edison Company).

the lower flow rate with Wattman No. 41 filter paper with which the pump may be operated continuously for approximately 1 hr without significant overheating.

Continuous atmospheric monitors are available which draw samples through moving tapes and automatically count the sample after permitting time for decay of the natural radioactivity. A continuous air sampler that is much simpler in construction draws air through a fixed filter adjacent to which is mounted a suitable radioactivity counter (Fig. 17-10). The radioactivity of the filter is continuously recorded, and provision is made to activate an alarm when the radioactivity level is more or less than predetermined values. The latter condition might indicate a loss of flow through the filter, causing its equilibrium radioactivity to drop below the expected background because of natural radioactivity.

Aquatic Monitoring

If liquid wastes are being discharged into a body of water, the critical pathway for human exposure should be thoroughly understood so that the sampling program can be designed with specific reference to the require- ments for estimating human exposure (Foster *et al.*, 1971). Samples may be required of the water, sediments, benthic organisms, vegetation, and fin fish or shellfish. Generally speaking, samples of water need be analyzed only if the water is being taken for human consumption or is being used for irrigation. In most cases, samples of shellfish or fin fish will serve to document the principal route of human exposure. Samples of sediments or vegetation may be taken as an indicator of potential exposure, since in many situations the vegetation and sediments will have higher concentrations of the critical nuclides than the food consumed by man. In practice, the design of the sampling program will probably be based on the pre- and postoperational ecological studies and will result from consultation with regulatory authorities.

If the wastes are being discharged into a flowing stream that serves as a source of potable water, a continuous sampler at the point of intake to the water supply may be desirable, but if so it would seem logical that this be maintained by the health authorities rather than the plant operator. The continuous sampler is desirable for a flowing stream because grab samples may miss slugs of waste being discharged intermittently. On the other hand, the sediments and biota may serve to provide an integrated measure of the variable exposure during any period of time, and in most instances it will be a suitable substitute for water sampling.

Water samples can be passed through ion-exchange columns to remove the cesium and transition elements, which can then be measured by γ spectrometry. ^{89}Sr or ^{90}Sr require radiochemical separations and β counting. Analysis for tritium will usually require electrolytic concentration followed by β counting in a liquid scintillation system.

Samples of sediment can be collected with dredges designed for that purpose or by retrieving cores obtained by driving cylinders into the sediments.

Aquatic vegetation and fish should be dried and ashed at 400°C for a 24 hr preparatory to γ scintillation counting or radiochemical analysis. Temperatures above 400°C should be avoided to prevent volatilization of the ^{137}Cs. If ^{131}I is thought to be present, γ spectra should be obtained prior to ashing. The size of the sample must be determined based on the expected concentration of radioactivity.

Ground-Water Sampling

We have seen in an earlier chapter that transport of radioactive substances through soils proceeds at a slow rate. There is little basis for concern over possible ground-water contamination except within very short distances (thousands of feet) of sites of ground storage of large quantities of radioactive wastes. However, as we have seen in Chapter 7, deep wells have been known to be the source of natural radioactivity, principally from radium and radon.

FOOD SAMPLING

It is sometimes necessary to sample foods to evaluate the extent to which humans are exposed. The scale of such a sampling program will vary, depending on the circumstances. If it is established from knowledge of the kinds and amounts of radionuclides discharged that significant food-chain contamination is not possible, no food-sampling program should be necessary. This will, in fact, be the case for most places in which only minimal if any environmental contamination is occurring. In other instances the food-sampling program may be limited to the produce of relatively few farms. However, if the opportunity for widespread contamination exists, the samples may be collected from very large areas, namely, hundreds of square miles in the case of a major spent-fuel-reprocessing plant or a whole country in the case of fallout from the explosion of nuclear weapons.

The danger of contamination of foods with radioactive substances released from industrial or institutional applications of atomic energy is normally eliminated by restricting the discharge of liquid and solid radioactive effluents, and a program of sampling foods grown in the environs will not ordinarily be necessary, although possibly desirable from the point of view of relationships with the public. However, in rare instances, food may, in fact, provide the most meaningful of all possible environmental samples. For example, it is desirable to sample shellfish harvested from an estuary into which radioactive corrosion products such as ^{60}Co or ^{65}Zn are being discharged. Such radionuclides are known to concentrate in shellfish, samples of which will yield more meaningful information than other biota. In such situations it may be possible to plant shellfish in or directly in front of the waste discharge canal on the theory that shellfish so exposed would have higher concentrations of radioactivity than shellfish harvested elsewhere (Salo and Leet, 1969). This should be done only if it is certain that shellfish metabolism is normal under these artificial conditions.

If ^{131}I is being released into the atmosphere in proximity to cow pastures, the ^{131}I concentration in the milk produced will provide far more meaningful information than air samples, deposition samples, or samples of the forage. Significant contamination of pastures by ^{131}I will result in detectable contamination of milk within 24 hr.

Under any given circumstances, the need for food sampling and the methods by which such sampling should be conducted should be based on a thorough understanding of agricultural practices in the area, and the program should be guided by competent specialists.

The needs for radiological analyses of food are apt to be highly variable, depending on the numbers and types of nuclear facilities in the area and whether testing of nuclear weapons is resulting in significant environmental contamination. Thus, the regional or national monitoring program should be flexible so that its extent can be varied as the needs require. Because radiochemical analysis involves relatively expensive procedures, careful attention should be given to the design of the sampling program when one is required to sample a large geographical area such as a state or country. Apart from economic considerations, there is the need to avoid overloading radiochemical laboratory facilities, which are always of limited capacity and may be unable to cope with the analytical load imposed by badly designed sampling systems.

Basically, there are four questions which must be answered before starting a food-sampling program. (1) What foods should be sampled? (2) Where should they be collected? (3) How many samples are required? (4) What kinds of analyses should be performed, and with what sensitivity, accuracy, and precision?

The foods one selects as part of a large-scale sampling program should depend on the characteristic behavior of the significant radionuclides in food chains and on the dietary habits of the population. Although it is of scientific interest to follow as many of these isotopes as possible through the various steps in the food chain, not all foods or all radionuclides need be studied for the purpose of surveillance. Some of the radionuclides present will be of little public health significance because they are present in such small amounts or because they are chemically inactive and do not enter metabolically into food chains. Since the basic purpose of a regional monitoring program will be to estimate the absorbed dose among the general population, only those foods should be sampled and only those nuclides analyzed that contribute significantly to population exposure. We have seen from previous chapters that in most cases the important nuclides are ^{90}Sr, ^{137}Cs, and ^{131}I. In many countries, fresh milk is the food of choice for obtaining samples of a population's total intake of ^{90}Sr, ^{137}Cs, and ^{131}I. This greatly simplifies the sampling problem, because the very nature of the milk distribution system in most large cities results in considerable blending from various sources. The extent of the network and the frequency of sampling depends entirely on the circumstances and the objectives of the program (Eisenbud et al., 1963b; Stein et al., 1971).

The presence of most of the nuclides of interest in the vicinity of a nuclear generating station can be ascertained by γ spectrometry. One notable exception is tritium, which is a weak β emitter and usually measured in a liquid scintillation counter. ^{89}Sr and ^{90}Sr are also pure β emitters, but these nuclides are present in light water reactor wastes in such small amounts that environmental measurements should not be necessary. ^{137}Cs, the rate of production and half-life of which is comparable with ^{90}Sr, is far more labile and can be expected to be present in reactor coolants in concentrations several orders of magnitude greater than ^{90}Sr. Thus, ^{90}Sr need not be measured in waste solutions or environmental samples unless very high concentrations of ^{137}Cs are encountered.

DIRECT HUMAN MEASUREMENTS

The purpose of food analysis is to enable one to estimate the body burden to the exposed population, using standard-man parameters, modified as necessary by local demographic data. The absorbed dose can be calculated from the body burden. Radiochemical analysis of human tissues that become available at autopsy or in the course of surgical procedures is one more direct means of obtaining such correlations. Another method of estimating body burdens in living individuals is by γ-ray spectrometry, usually using thallium-activated sodium iodide crystals in well-shielded rooms that have come to be called "whole-body counters".

Whole-body counting is particularly useful for measuring body burdens of nuclides having penetrating γ radiations (Meneely and Linde, 1965), but it has also been applied (Laurer and Eisenbud, 1968; Ramsden, 1969) to measurement of nuclides such as ^{241}Am, ^{329}Pu, and ^{210}Pb, all which have in common that the electromagnetic radiation is relatively soft (13 to ~60 keV) and require special techniques specifically applicable to this portion of the spectrum. Whole-body γ-ray spectrometers are sophisticated instruments of which relatively few exist in the world (International Atomic Energy Agency, 1970b). Such instruments have been used routinely to monitor the ^{131}I burden of human thyroids (Laurer and Eisenbud, 1963) and to estimate the skeletal burdens of ^{241}Am (Wrenn et al., 1972) and ^{210}Pb (Eisenbud et al., 1969). A number of investigators have also used whole-body counting to monitor various populations for their ^{137}Cs burden, as was described in Chapter 14. Whole-body counters are useful for research, but they have little place in a program of routine surveillance of the general population except under unusual conditions.

Environmental Surveillance under Emergency Conditions

Radiological emergencies potentially include a broad spectrum of circumstances. At one end is the innocuous incident in which there is no opportunity for significant public exposure, but in which excitement is engendered by the knowledge that an accident has occurred in which radioactive materials are involved. At the other end of the spectrum might be a serious accident involving a nuclear weapon or major nuclear facility in which there is substantial potential for human injury and where quick action is required to obtain the basic environmental information on which decisions will be made regarding the countermeasures to be taken. Intermediate between these examples would be an accident at a nuclear facility, where the amounts of material released to the environment are substantially in excess of those approved for routine circumstances and where a quick appraisal is required of the extent of public risk. Experience has shown that regardless of the severity of the accident these incidents have in common the need for an emergency plan designed to deal with the full spectrum of accidents that might occur at a given facility. The design of the emergency plan should take into consideration the characteristics of the site with respect to geography, population density, governmental organization, and the information media. From the negative point of view, all radiological incidents that involve the general public are capable of producing misunderstandings, excitement, and frequently a degree of press attention that may be all out of proportion to the true importance of the incident. It is essential that the emergency plan be designed to provide prompt, accurate information which can be dispensed candidly and efficiently to the public agencies and the media.

Since it is not possible in a work such as this to cover all of the possible contingencies that may arise, this section will deal with certain general principles which should be observed, and foremost among these is that the emergency plan should be reduced to writing and have the concurrence of all key individuals both within the organization operating the facility and among the various governmental organizations that might be involved. In many states, responsibility for coordinating the off-site surveillance activities, digesting the data, preparing public information releases, and making decisions about required countermeasures is that of the Health Department. The state police, the local police, and numerous other governmental agencies will also be involved. The interrelationships of these various organizations should be clearly understood by everyone concerned.

The full course of the emergency can be separated into two phases. Initially, the organization having operational responsibility for the facility involved in the accident must rely on its own resources, pending arrival of outside assistance. This phase may last a few minutes or several hours, depending on the geographical situation and the degree of advance preparation.

In the second phase, the somewhat limited facilities of the local organization will be augmented by outside assistance, and some governmental organization, such as the state Health Department, will usually assume responsibility for managing the off-site aspects of the emergency.

PREPARATION FOR EMERGENCIES

An essential first step is to analyze the kinds of accidents that can take place and their radiological consequences. This has been done for many years as part of the AEC licensing procedure, and for all nuclear reactors, fuel-processing plants, reprocessing plants, or other facilities licensed by the AEC, the accident analysis is an essential part of the Safety Analysis Report submitted by the applicant. The accident analysis provides the basic descriptions of the contingencies with which the plan must deal.

An early step in development of the plan should be consultation with the various governmental organizations that would become involved in the event of an accident. This will vary from place to place, but most states have standby emergency procedures to which the radiological emergency plan should be adapted. Thus, in many localities the state police, an organization whose staff normally provides continuous emergency readiness, can be relied on to receive information about emergencies and to place into effect various standard operating procedures and contingency plans. The police organization might also have responsibility for controlling traffic in the vicinity of the accident, arranging for detours, and arranging emergency transportation for certain key people.

It is likely that the Health Department will have prime responsibility for evaluating the significance of off-site radiological data and deciding what actions should be taken. The response to any emergency should be graded according to its severity, and the actions that are taken to reduce the dose to the off-site population should be determined in advance and be included in the emergency plan that is developed jointly by the facility operator and the governmental organization that has responsibility for dealing with the emergency.

For large facilities, an important part of the plan is liaison with supporting radiological assistance teams. Accidents of the severity of the SL-1 or the Windscale production reactor, as described in Chapter 16, require relatively large numbers of skilled personnel to gather the necessary information. In the United States the AEC and other governmental organizations maintain a network of emergency teams located at all major facilities. These teams are prepared to offer assistance to others.

The written plan should describe the kinds of radiological monitoring equipment that is available locally for emergency use. The equipment should be stored in a place that would be accessible regardless of the kind of accident and where opportunities for radioactive contamination are minimized. It requires great discipline to maintain such standby equipment in proper operating conditions, and the emergency plan should include the necessary maintenance and inspection schedules.

The plan should identify the principal pathways of human exposure so that the radiological monitoring efforts can be properly directed. By the time the facility goes into operation, enough radioecological information should be available so that tables or charts can be developed that make it possible for the plant operator and health physicist to estimate the potential dose commitment to the surrounding population from any type of release. The tables of dose commitment to the surrounding population can be prepared so that for unit quantities of whatever mixtures of radionuclides are released, either by air or water, the dose to the population can be calculated. This will require identifying the locations of water intakes, dairy farms, and other vectors of exposure which should be listed in the original emergency plan.

Chapter 18

Environmental Radioactivity in Retrospect and Prospect

This concluding chapter provides an opportunity to summarize the radiation dose now received by mankind from both natural and artificial sources and to offer some predictions as to the manner in which human exposure may be increased by the future growth of the nuclear industry.

Human Radiation Exposure from Various Sources

The extent to which humans are exposed to ionizing radiation will be summarized according to the various sources of exposure, including natural radioactivity, occupational exposure, nuclear power production, the governmental centers of research and production, miscellaneous uses of radioactive substances, and medical applications.

Natural Radioactivity

Our knowledge of the natural radioactive background serves as a logical starting point from which to examine the significance of the additional sources of ionizing radiations that result from man's exploitation of the atom.

Until the beginning of this century, man's only exposure to ionizing radiation originated from the natural sources which, in most habitats near sea level, result in a total body dose of about 100 mrem/year. In cities like Denver that are at relatively high altitudes, the annual dose can be somewhat higher than this, and in a few places known for the abnormally high radioactivity of soils and rocks, the whole-body dose can be ten or more times higher than the average sea-level figure. It is also known that in certain areas water obtained from deep wells contains sufficient radium to contribute additionally to the skeletal dose received by inhabitants. The sources of natural radioactivity were discussed in Chapter 7.

Natural radioactivity would at first seem to provide ready-made opportunities for studies of the effects of low levels of radiation exposure. Since it is known that the dose from natural radioactivity varies from place to place, the question of how man is affected by small doses of ionizing radiation could conceivably be answered by epidemiological studies in which the incidence of diseases such as leukemia, bone cancer, and certain effects that are known to be genetic in origin could be studied in relation to the natural level of ionizing radiation to which the indigenous population is exposed. Fortunately for mankind, though unfortunately for the scientific investigator, we saw in Chapter 2 that these effects occur so infrequently (if at all) from doses of radiation in the range of natural exposure that it is not feasible to obtain data on the effects of such low levels by epidemiological or experimental means.

Until comparatively recently, the only changes in human radiation exposure resulted from migration, changes in eating habits, and changes in the materials with which man built his homes and workshops. Thus, as the early North American settlers pushed their way west from the eastern seaboard to the high plains of Wyoming and Colorado, the whole-body dose they received from nature increased from less than 100 mrem/year to almost twice that value in some places.

The natural radiation exposure can be significantly modified by one's method of house construction. If a house is built of limestone, a material low in radium and thorium, the dose received while indoors is reduced to some extent because the limestone provides shielding from external radiation. However, some building materials, such as granite and certain types of concrete, have high radium and thorium contents that add γ-radiation exposure to an extent that more than offsets absorption of radiation from without.

The wide variability in exposure to naturally occurring ionizing radiations was discussed in Chapter 7. For the purpose of providing a yardstick with which to measure the significance of man-made sources of radiation, a dose rate of 130 mrem/year will be assumed to represent the "average" dose from natural sources of ionizing radiation.

OCCUPATIONAL EXPOSURE

This text has been concerned with the environment external to nuclear facilities and until now has dealt only in passing with the subject of occupational health and safety. However, some discussion of this subject is appropriate at this stage because it relates in a number of ways to the more general subject of environmental radioactivity. First, the population of radiation workers now numbers more than 750,000 in the United States and thus represents a sizable sample of the adult population. Because this portion of the population is allowed a much higher dose than the general population, the health records of radiation workers may be relevant to the question of whether effects would be expected to occur at lower dose rates among members of the general population. Second, one would expect that accidents in the nuclear industry would be more likely to result in exposure of the employees than the public. An analysis of the record of occupational safety can thus serve as an indicator of the prudence with which the industry discharges its responsibilities for protection of health and safety generally.

During the first 40 years of this century, the applications of ionizing radiations were relatively few, and the available radioactive materials were limited to elements, such as radium, uranium, and thorium, that could be obtained from mineral sources. These substances were used in amounts that were insufficient to result in significant levels of environmental contamination, and the risk from their use was limited to the individuals who were in direct contact with the materials, e.g., radium dial painters, radiologists, X-ray technicians, and patients. Although only about 1400 Ci of ^{226}Ra were taken from the earth's crust during that period (see Chapter 2 and 10), many people lost their lives because of the hazards attending the misuse of this material. The total number of deaths from the use of radium in the first 40 years of this century will probably never be known, but it certainly exceeded 100 persons. The number of physicians and their assistants who were victims of uncontrolled X-ray exposure during this period was probably much larger.

In contrast to that tragic record, a few statistics serve to illustrate the degree of control that has been achieved over the dangers of radiation exposure during the 3 decades in which the modern atomic energy program has existed and has artificially produced the radioactive equivalent of many billions of curies of radium.

It is estimated that in the United States there are 772,000 persons exposed to ionization radiations in the course of their employment (Environmental Protection Agency, 1972) of which about 100,000 are employed in facilities operated by the AEC. The employees in the AEC program are involved with duties that range from clerical activities to research, manufacturing, and heavy construction. An additional 50,000 "radiation workers" are employed in the civilian nuclear energy industry (U.S. Atomic Energy Commission, 1970a). Thus, about 600,000 persons are exposed to radiations in the course of their employment outside of the atomic energy industry, and of this group the great majority work with X rays in the medical profession.

Since World War II, the AEC has reported regularly on the radiation exposure of its employees in statistical reports that also contain data on industrial accidents of all kinds, those that involve radiation exposure as well as those in which radiation was not a factor (U.S. Atomic Energy Commission, 1971d).

The overall accident experience of the AEC and contractor employees is shown in Fig. 18-1, in which the frequency of lost-time accidents of *all kinds* is compared with the national average accident frequency for all industry. The AEC accident record has been consistently between one-half

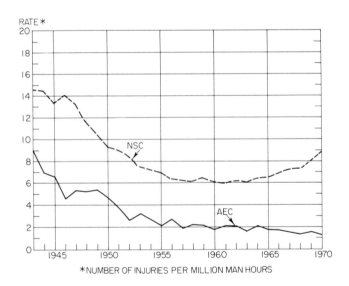

Fig. 18-1. Atomic Energy Commission and National Safety Council national industrial average injury frequency rates, 1943–1970 (U.S. Atomic Energy Commission, 1971d).

and one-fourth of the national average, which is all the more outstanding an accomplishment when one considers that the AEC program involves much heavy construction, chemical processing, and other types of industrial work that is potentially more hazardous than average industry. In recent years, of the 42 industries for which the National Safety Council publishes accident statistics, none have had a frequency of lost-time accidents lower than the AEC plants and laboratories (U.S. Atomic Energy Commission, 1971d; National Safety Council, 1971).

Despite the relative excellence of this 28-year-old program, there have been a total of 295 work-related fatal accidents, a large number of which (61%) occurred in construction activities because of falls or falling objects motor vehicle mishaps, and electric shock. Of these 295 deaths, six were due to radiation (U.S. Atomic Energy Commission, 1971d). A seventh radiation-caused death occurred in a privately operated industrial plant in 1964 (Karas and Stanbury, 1965). Of the six radiation-connected deaths in the AEC program, three resulted from the SL-1 accident, which was described in Chapter 16, and two were associated with relatively hazardous laboratory work in a weapons laboratory early in the program.

Thus, although radiation is potentially the most serious of all occupational hazards in the atomic energy program, it is significant that during the 28 years from 1943–1970 the fatalities owing to radiation exposure accounted for about 2% of all deaths due to occupational accidents. It is also significant that the fatal accidents were associated primarily with experimental programs, and that, as of this writing, the last of the fatal radiation accidents in the AEC program occurred more than 11 years ago at the SL-1 in 1961.

In addition to the six fatalities, accidents have resulted in 12 cases of clinically observable radiation injuries to employees during the period 1943–1970. These cases included β burns of the skin, excision of tissue in which plutonium had lodged mechanically, and the acute radiation syndrome.

So much for the results of accidental massive exposure to radiation; there remains the important question of the frequency with which injuries or deaths can be expected to occur because of the delayed effects of repeated small doses. Except for the mining industry, there have been no known injuries from the delayed effects of radiation in the atomic energy industry. Ironically, uranium mining is the one part of the atomic energy program that had reason to take meticulous precautions, based on experience in the European mines, as discussed in Chapters 2 and 8. Had the uranium mine atmospheres been controlled so as to meet the standard established in 1941 to control the radon hazard in another industry (National Committee on

Radiation Protection, 1941), the tragic epidemic of lung cancer among the uranium miners of Southwestern United States would have been avoided. Regrettably, regulation of the mines was not preempted by the AEC but was left to the states, which lacked either the means or the will to deal with the problem in an effective manner.

The absence of known cases of delayed injury from ionizing radiation exposure in the atomic energy industry is somewhat assuring. The industry is approaching 30 years of age, which should be long enough for such cases to become manifest if they are to occur in significant numbers from the practices adopted early in the program. It is, of course, possible that isolated injuries or deaths because of chronic radiation exposure have occurred, or will occur, and a few cases here and there might not be recognized as such against the background of normal morbidity and mortality. However, one can find comfort in the records of personnel radiation exposure, as summarized in Table 18-1. During the 24-year period 1947–1970, less than

TABLE 18-1

RADIATION EXPOSURE OF AEC AND AEC CONTRACTOR PERSONNEL TO WHOLE-BODY PENETRATING RADIATION[a]

| Year | \multicolumn{5}{c}{Rem} |
	0–1	1–5	5–10	10–15	>15
1947–1954	130,128	5,311	284	32	6
1955	56,708	3,157	285	41	1
1956	38,225	2,312	100	4	3
1957	45,510	2,424	83	5	1
1958	59,455	6,271	159	10	12
1959	71,600	3,912	66	2	1
1960	77,522	4,629	41	2	1
1961	90,651	5,174	40	3	8
1962	122,437	5,707	113	0	8
1963	107,786	5,472	80	0	1
1964	122,711	6,157	86	11	0
1965	128,360	6,671	175	8	0
1966	130,552	7,218	167	0	2
1967	101,764	6,513	108	1	0
1968	103,206	4,776	4	0	0
1969	98,625	4,288	4	1	0
1970	92,185	4,464	12	0	0
	1,577,425	84,456	1,807	120	44

[a] U.S. Atomic Energy Commission (1971d).

0.2% of the employees received an annual dose of more than 5 rem, and almost 95% received an annual dose of 1 rem or less. There have been 44 annual exposures greater than 15 rem, most of which were associated with the accidents that caused the deaths and injuries noted earlier.

Using the data of Table 18-1, one can estimate the per capita and integral doses to the population of atomic energy workers by assuming that the mean annual dose is at the midpoint for each class of exposure for all individuals, except those exposed to greater than 15 rem. The latter group will be excluded because the 44 cases are well documented and many of the individuals were known to have received much more than 15 rem. Thirteen of the cases, including the three deaths, were known to have received more than 100 rem. Thus, it would not be proper to include these individuals in the category of "low-level exposure." Excluding these 44 individuals and based on the stated assumption, a total of 1,663,808 man-years of exposure during the 24-year period 1947–1970 resulted in a total of 1,013,000 man-rems of exposure or a per capita annual exposure of 0.6 rem for those occupationally exposed. This is undoubtedly a conservative assumption because most of the integrated exposure is in the 0–1 rem class, where the mean exposure is probably lower than 0.5 rem/year. This inference is supported by the work of an interagency Special Studies Group appointed by the Federal Radiation Council* (Environmental Protection Agency, 1972), which has analyzed the AEC records of employee radiation exposure for 1969 in somewhat greater detail than is possible from the data of Table 18-1. It was concluded that the per capita whole-body dose for that year was 197 mrem, which is about one-third of the estimate obtained by the method used above.

The same group has examined *all* sources of occupational radiation exposure in the United States and concluded that the per capita dose to 772,000 radiation workers was 210 mrem/year. The integral dose was 163,883 man-rem, and the contribution to the annual per capita dose for the entire United States population was 0.79 mrem.

Since leukemia is perhaps the most grave of the somatic effects of whole-body irradiation, it is revealing to consider the implications of radiation exposure in the AEC program insofar as this disease is concerned. We have seen in Chapter 2 that if one assumes the dose-response curve for leukemia production to be linear, without a threshold, and independent of dose rate, one can estimate that 1 million man-rem (rounded off from the estimated 1,013,000 man-rem) could result in approximately 20 cases of leukemia.

---

* As noted in Chapter 3, the Federal Radiation Council has recently been absorbed by the Environmental Protection Agency.

Assuming these cases developed in the 30-year period 1942–1972, the average annual rate would be about two-thirds of a case per year in a population of about 100,000 workers.* The incidence of leukemia in the United States population is about 6 cases/100,000 persons/year, so that in the 1,663,000 man-years at risk, approximately 96 cases would be expected to occur. As an upper limit, one might, thus, expect that the leukemia incidence in this population would increase by about 20%; a difference too small to be identified with any degree of confidence (Chapter 2). The actual increase could be zero if a threshold exists or if the dose-response curve is markedly dose rate dependent. If we assume, as was suggested in Chapter 2, that five miscellaneous radiation-induced cancers will occur for each case of leukemia, a total of about 100 additional neoplasms might be expected to result from the 30-year exposure period. It would seem worthwhile to require that the radiation exposure records and medical histories of radiation workers be maintained meticulously and be studied every decade or so to determine if there is evidence that radiation-associated disease is occurring with increased frequency. However, the fact remains that the relatively high incidence of neoplastic disease in the general population is likely to mask any effects at these levels of exposure. The death rate from all cancers (unadjusted for the age distribution of atomic energy workers) is about 160/100,000 persons/year (U.S. Bureau of Census, 1971) or 4800 cases in a 30-year period.

One might also look at the accumulated mortality statistics of atomic energy workers, but such data are apt to be affected by a number of selective factors that tend to link mortality statistics and overall socioeconomic status. Thus, Larson *et al.* (1966) examined the health records of atomic energy workers at three Oak Ridge, Tennessee installations for the period 1950 through 1965. Based on age-adjusted mortality rates, one would have predicted 992 deaths from all causes during the 16-year period included in the study, whereas 692 deaths actually occurred. This significantly lower death rate is no doubt attributable to many factors, including possibly higher mean educational level, a higher percentage of skilled people earning higher salaries, and other socioeconomic factors including the higher standards of medical care that one would expect in a modern planned community such as Oak Ridge. All one can conclude is that the additional occupational risk owing to ionizing radiation exposure in the nuclear energy industry is more than offset by other factors characteristic of the industry. Some of

---

* This is obviously an oversimplification, since some cases from exposure during the last decade or so might not as yet have developed. Thus, one might wish to use a rate of 1 case per year, which does not significantly alter the argument developed for purposes of illustration.

these factors, such as the highly selective personnel practices of the industry, may, in fact, be related to the dangers of radiation exposure. The bias in selection that reduced the incidence of alcoholism or accident proneness are examples of factors that are likely to be associated with reduced mortality.

EXPOSURE OF THE GENERAL POPULATION TO RADIATIONS FROM THE NUCLEAR POWER INDUSTRY

Although various national and international regulatory and standards-setting organizations have proposed guidelines that would limit the per capita dose received by the general population to 170 mrem/year, it has become evident that the nuclear power industry can be managed in such a way that exposure of the general population can be controlled to much less than 1% of this limit. As was discussed in Chapter 3, the ICRP, NCRP, FRC, and AEC all stress the concept that any radiation exposure should be considered potentially harmful and that exposures should be reduced to the lowest practicable extent. The nuclear industry has followed this philosophy in the design and operation of its facilities.

In discussing exposure from generation of nuclear power, one must consider not only the power reactors but also the uranium mines and mills, the fuel fabrication plants, fuel reprocessing, and waste management. It is not sufficient to examine the potential of the reactors to produce environmental contamination; the entire fuel cycle must be examined (see Chapter 8, 9, 11, and 12).

Mining, Milling, and Fuel Fabrication

Uranium mining and milling does not, in general, result in a significant increase in environmental radioactivity outside the immediate area of the mines themselves, and the EPA (Environmental Protection Agency, 1972) has concluded that radiation levels in the mining communities are the same as in other communities in the same geographical area. The external radiation and radon levels in the vicinity of the mills were considered to be so low that, considering the sparse settlements, the population dose is indistinguishable from background.

However, as noted in Chapter 8, mill tailings have been used in building construction in some of the mining communities (Joint Committee on Atomic Energy, 1971), resulting in higher than normal radon and γ radiation exposures to inhabitants of these buildings. Although the numbers of persons exposed in relation to the level of exposure is not sufficient to affect the national statistics of population exposure, the fact remains that the mining industry has resulted in some elevation of the radiation exposure of some members of the general population. The exact numbers of individuals so involved and their levels of exposure have not as yet been summarized; but the evidence thus far is that while the practice of using mill tailings for construction purposes has elevated the radiation dose received by many individuals who live near the mills, the exposures are within the range of those received elsewhere from natural sources.

Power Reactor Operation

Estimates of the population dose from operation of nuclear power reactors can be made using the known operating experience of 13 civilian power reactors for which data are available. These data were used by the EPA Special Studies Group referred to earlier in conjunction with estimates of future nuclear generating capacity in the United States and abroad.

An important factor that controls exposure of the general population to radiations from nuclear reactors was the decision by the AEC and FRC to limit the dose to the maximum exposed individual to 500 mrem/year. It was assumed that if the maximum dose is less than 500 mrem/year, the average will be less than 170 mrem/year, an assumption that introduced a very large safety factor. To illustrate, consider the case of a boiling-water reactor (BWR) from which the principal source of exposure to the general population is a cloud of passing radioactive gases discharged from the plant stack. If the BWR stack is surrounded by a circular fence 500 m distant, at which the annual dose is 500 mrem, an individual sitting continuously on the fence would be the "critical population." From the known behavior of gaseous effluents diffusing from point sources, it can be assumed that the dose rate beyond the fence will, on the average, diminish inversely with the 1.8 power of distance from the stack. In addition, some reduction in dose rate will result from radioactive decay of short-lived nuclides. The per capita doses to nearby populations have been calculated for three extreme cases in which 10^5, 10^6, and 10^7 people are uniformly distributed around the fence at a density of 1000 persons/km². If we assume that the wind blows uniformly from all directions, the per capita dose rates for the three populations will be 25, 4.1, and 0.62 mrem/year, respectively. This, in fact, overestimates the per capita dose rates, because the 500 mrem/year exposure rate would actually be permissible only in the direction of maximum wind direction, which will be perhaps one-eighth of the plant's circumference. For seven-eights of the circumference, the dose will be less than 500 mrem/year at the fence line, and the per capita dose at greater distances will be proportionately reduced. Thus, the per capita dose to nearby residents will be very much less than one-third of the dose to the maximum exposed individual, as assumed by AEC.

Other examples could be given to illustrate the substantial safety factors that are inherent in the method of administering the radiation protection standards. Thus, in the case of liquid radioactive emissions, the maximum exposed individual is frequently assumed to be a person who consumes fish caught at the outlet of a reactor-cooling water-discharge canal. It is at such a location that one would expect to find fish that have absorbed more radioactivity than at other places in the receiving body of water. The critical-pathway dose calculations are made on the conventional assumption that the individual each day consumes 50 g of flesh from the fish caught at this location. This is usually a highly hypothetical set of assumptions that would hardly apply to the general population, which probably consumes no fish caught near the outfall. Thus, the dose to the general population from liquid radioactive effluents would be very much less than one-third of the dose received by the assumed maximum exposed individual, and it might, in fact, be zero.

Until the late 1960's, there was a widespread unwritten assumption that if the dose to the population living in the immediate vicinity of a nuclear facility could be demonstrated to be less than about 10% of the maximum permissible dose, i.e., 10 to 20 mrem/year, this fulfilled the requirement that the dose be maintained at the lowest practicable value. In all but very few cases, this condition was more than satisfied. However, as a result of increased public concern about the general subject of environmental pollution, a good deal of pressure developed in 1969–1970 for a stricter interpretation of the requirement that all radiation exposures be reduced literally to the lowest practicable level. This, in turn, required that better data be available on the extent to which people were exposed to radiations from nuclear facilities. In earlier times, one would be satisfied to demonstrate that the exposure to the individuals living near the plant was less than 10% or so of the permissible dose, and no effort would be made to obtain quantitative dose estimates below this level. However, in 1971 the AEC issued proposed guidelines that defined the lowest practicable dose in the vicinity of light-water reactors and, in effect, limited offsite exposure to 5 to 10 mrem/year. As a result, increased efforts have been made to define critical pathways in the vicinity of nuclear facilities and to estimate the doses to the surrounding population at much lower levels than had been thought previously to be necessary. Considerable effort has been expended recently to obtain dose estimates on the order of 0.1 mrem/year or less, as will be seen later in this chapter.

It is predicted that the world nuclear generating capacity will increase from about 20,000 MWe (megawatt equivalent) in 1970 to 2,000,000 MWe in the year 2000. Population dose forecasts have been made by the Special Studies Group (Environmental Protection Agency, 1972) on the basis of the projected growth of nuclear generating capacity. Although reactor technology should continue to improve and future effluents may be somewhat reduced over those at present, this was not taken into consideration.

The study estimated that the whole-body γ dose to the population of the United States from operation of nuclear power reactors would increase in future decades as shown in Table 18-2, in which the estimated per capita dose by the year 2000 will reach 0.20 mrem/year. Approximately 10% of the dose in the year 2000 will be achieved by the year 1980, at which time the techniques for making these dose estimates should be so considerably improved as to make it possible to refine the doses projected for subsequent decades.

Based on a population of 321 million people in the year 2000, the rate of integrated exposure of the United States population will be 64,000 man-rem/year. If we assume again that 1 million man-rem result in a maximum of 20 cases of leukemia, 64,000 man-rem would annually commit the United States population to 1.2 cases of leukemia. However, the 1.2 cases would be spread over a 20-year period. Assuming further, for the purpose of illustration, that the dose rate in year 2000 is delivered to the population of the United States during a 70-year lifetime, during which cases would develop over a period of 60 years, the total number of cases would be $60 \times 1.2 = 72$ in a population which we will assume remains constant at 300 million inhabitants and in which the sponteneous leukemia incidence

TABLE 18-2

ESTIMATED EXTERNAL γ WHOLE-BODY DOSES TO POPULATIONS IN THE VICINITY OF POWER REACTORS[a]

Year	Percent of United States population at risk	Annual man-rem	Annual average dose to United States population (mrem)
1960	1.5	16.4	0.0001
1970	22.3	430	0.002
1980	100	6,080	0.027
1990	100	22,780	0.089
2000	100	56,000	0.20

[a] From Environmental Protection Agency (1972).

remains unchanged over a 70-year period. The 72 cases of radiation-induced leukemia would occur against a background of (60 cases/1 million persons/ year) × (300 million persons) × (70 years) = 1.26 million cases.

Fuel Reprocessing

Exposure of the general public to radiations from a fuel-reprocessing plant can occur as a result of external γ exposure from gaseous releases, from ground deposition, and by the usual pathways for internal exposure. The great quantities of waste materials that will become available in the course of fuel reprocessing have been discussed in Chapters 11 and 12.

Until 1972, only one privately operated fuel-reprocessing plant had been operated in the United States. However, the number of nuclear power reactors is expected to increase to about 400 plants at 200 different sites by the end of the century. The amount of fuel that becomes available for reprocessing is expected to increase from about 200 metric tons/year in 1970 to 20,000 metric tons/year by 2000 A.D. During this 30-year period the type of fuel will gradually shift from the present uranium oxide fuels used in light-water reactors to the plutonium fuels of the fast breeder. By the end of the century, it is estimated that of the 20,000 tons/year of fuel to be reprocessed, 17,000 tons will be in the form of plutonium and 3000 tons in the form of uranium (Oak Ridge National Laboratory, 1970).

The only operating data for a commercial fuel-reprocessing plant comes from the privately owned facility in upstate New York. A survey conducted jointly by the New York State Department of Health, the United States Public Health Service, and the AEC (Shleien, 1970b) was described in Chapter 11 and concluded that the dose to the "typical individual" around the plant in 1968 did not differ significantly from that for the average adult population of New York State. However, the dose commitment to local hunters eating venison and fish collected near the site was slightly higher than the dose to the typical individual.

Using the best available information, the EPA Special Study Group estimated that the annual dose to the United States population from fuel reprocessing will be as shown in Table 18-3. The doses have been calculated for skin, lung, respiratory lymph nodes, bone, and thyroid. The estimated integral whole-body dose to the United States population by the year 2000 is 65,000 man-rem, which is about the same as the forecast exposure from power reactor operation. Although the fuel-reprocessing plants discharge more waste than reactors, there will be far fewer reprocessing plants.

A dozen or so fuel-reprocessing plants distributed throughout the United States will become point sources for liquid and atmospheric wastes that affect the population as an inverse function of distance from the plant.

TABLE 18-3

ESTIMATED ANNUAL DOSE TO THE UNITED STATES POPULATION FROM FUEL REPROCESSING[a]

Year	Whole-body dose (mrem/person)	Man-rem to United States population	Other body organ doses (mrem/person)				
			Skin	Lung	Respiratory[b] lymph nodes	Bone[b]	Thyroid
1970	0.0008	170	0.008	0.001	0.05	0.001	0.001
1980	0.02	5,000	0.1	0.03	0.8	0.02	0.1
1990	0.09	25,000	0.4	0.1	1.4	0.09	0.5
2000	0.2	65,000	0.8	0.2	2.3	0.2	1.1

[a] From Environmental Protection Agency (1972).
[b] Dose received over 50 years following exposure.

At a distance of 100 km or so, the effect of the plant will be imperceptible. However, each plant can interact with chemical-reprocessing plants in other places throughout the world by causing a general elevation in the global inventories of ^{85}Kr and ^{3}H. It is estimated that the dose from the worldwide inventory of tritium will have stabilized at a rate of about 0.04 mrem/year for an integrated dose of about 14,000 man-rem for the United States population, and that the per capita dose from ^{85}Kr by the end of the century will be 0.04 mrem/year to the whole body and 1.6 mrem/year to the skin (Environmental Protection Agency, 1972). The integrated exposure from ^{85}Kr will then be 12,000 man-rem to the whole body and 48,000 man-rem to the skin. There are, of course, enormous uncertainties in these kinds of estimates since the experience to date, as noted earlier, has been with only one reprocessing plant. However, considering that the technology is more likely to improve than not and because of the general conservativeness with which the dose estimates were made, any errors should give values that err on the high side. These estimates can be reevaluated from time to time in the light of actual developments both in technology and radiobiology. If necessary, the emissions of ^{85}Kr and tritium can be reduced, as was discussed in Chapter 12.

MISCELLANEOUS ATOMIC ENERGY COMMISSION FACILITIES

The AEC operates about 25 major nuclear facilities, including (1) the National Laboratories at Brookhaven, Oak Ridge, and Argonne; (2) production plants such as the Feed Materials Production Center; (3) military nuclear reactor research and development centers; and (4) nuclear weapons production facilities. In addition, the AEC operates the Nevada test site, conducts Plowshare demonstrations, and has conducted nuclear rocket tests, also in Nevada. The EPA Special Studies Group has examined the reports of environmental radioactivity in the vicinity of these facilities and has estimated the contribution to the national per capita whole-body dose to be about 0.01 mrem from 1960 through the year 2000. The national annual exposure integral would thus increase to about 3000 man-rem by the year 2000. In addition, it was estimated that in 1960 and 1970 the annual dose to the lungs of about 5 million people living near these facilities was about 0.6 mrem, and that the bone dose received by between 100,000 and 260,000 people was an average of 4.5 mrem.

MEDICAL USES OF IONIZING RADIATION

The relative importance of exposure from uses of ionizing radiations in the healing arts was clearly expressed by Charles E. Edwards, Food and

Drug Commissioner of the United States (Edwards, 1971) when he said: "More than 90% of all human exposure to manmade radiation comes from the diagnostic use of X-ray in contrast to about 1% from radioactive discharges from nuclear power plants, about which there has been so much concern." Diagnostic radiology is the major source of ionizing radiation exposure in all economically developed countries, but the genetically significant dose is evidently higher in the United States than in other countries where the dose in 1964 ranged from 12 mrad/year (New Zealand) to 38 mrad/year (Sweden), in contrast to 55 mrad/year in the United States (U.S. Public Health Service, 1969b). The past and projected genetically significant doses from medical diagnostic radiology have been assumed by the EPA to remain constant to the year 2000 at a per capita dose of 72 mrem. With the expanding population, the integrated dose will increase from 14.8 million man-rem in 1970 to 32.1 million man-rem by the end of the century.

In a nation that has been genuinely concerned about the potentially harmful effects of low-level radiation, it is hard to understand why the dose from diagnostic radiology shows no signs of abating. The reasons are manyfold and are related to poor technique on the part of the physician responsible for the X-ray unit, as well as to the increased use of X rays and radioactive pharmaceuticals. Heller *et al.* (1964) called attention to the fact that in a survey of techniques used in diagnostic radiology in the City of New York the field size in chest examinations was so inadequately controlled that the gonads were unnecessarily exposed in about half of the films taken. Other faulty techniques have been emphasized by Blatz (1962a,b). When a fluoroscopic screen ages, it loses brightness, which may require that the X-ray beam intensity be increased. The beam intensity must also be increased unnecessarily if the examining physician does not properly acclimatize his eyes to the required darkness. The United States Public Health Service (1969b) estimated that the genetically significant dose in the United States could be reduced from 55 to 19 mrem if the X-ray beam would be reduced to an area no larger than necessary for the selected film size.

Based on a 1964 survey conducted by the United States Public Health Service, Shleien estimated that the mean dose to the bone marrow from diagnostic medical radiology was 165 mrem/year to an exposed population of 67.3 million persons. The integrated bone marrow exposure to this population was, thus, 11 million man-rem.

The rapidly increasing use of radiopharmaceuticals for diagnostic purposes was discussed in Chapter 10. It is difficult to estimate the extent to which these radioactive drugs will be used in the future, but based on continuation of the present rate of increased usage, Shleien arrived at an integral dose from this source of 3.3 million man-rem by about 1980.

Population exposure to ionizing radiation administered for therapeutic purposes will not be considered here. The individual doses are high, but the patients are usually advanced in age, and in many cases the radiation therapy is required to mediate a grave prognosis.

FALLOUT FROM NUCLEAR WEAPONS TESTING

In Chapter 14 we discussed worldwide fallout at length and reviewed the extent to which the general population has been exposed to radiations from this source. The dose commitment from tests conducted thus far have been made, but projections based on future tests are not possible since they depend on the question of whether or not open-air nuclear weapons testing will be resumed. If one assumes that the test-ban agreement of 1962 will continue indefinitely and that nonparticipant powers will not add significantly to contamination that now exists, the annual whole-body dose will diminish from a peak of about 13 mrem/year in 1963 to 4.9 mrem/year by the end of the century. The integrated population exposure will decrease from 2.4 million man-rem/year in 1963 to 1.6 million man-rem/year by the year 2000.

MISCELLANEOUS SOURCES OF EXPOSURE

Additional radiation exposure is received from color television receivers (Neill *et al.*, 1971), radioactive wristwatches, and air travel. The total annual whole-body doses from these and other miscellaneous sources of

TABLE 18-4

TOTAL ANNUAL WHOLE-BODY DOSES FROM MISCELLANEOUS RADIATION[a]

Source	Annual doses for years				
	1960 (mrem)	1970 (mrem)	1980 (mrem)	1990 (mrem)	2000 (mrem)
Television	0	0.1	0.1	0.1	0.1
Consumer products	1.6	1.9	1.0	0.01	0.01
Air transport	0.05	0.05	0.05	0.05	0.05
	1.6	2.0	1.2	0.2	0.2

[a] From Environmental Protection Agency (1972).

TABLE 18-5

PER CAPITA DOSE RATES AND INTEGRAL DOSES IN THE UNITED STATES FOR YEARS 1960, 1970, AND 2000[a]

Source of exposure	1960 183 × 10^6 [b]		1970 205 × 10^6 [b]		2000 321 × 10^6 [b]	
	Per capita dose rate (mrem/year)	Integral dose (million man-rem/year)	Per capita dose rate (mrem/year)	Integral dose (million man-rem/year)	Per capita dose rate (mrem/year)	Integral dose (million man-rem/year)
Natural	130	24	130	27	130	42
Occupational	0.75	14×10^{-2}	0.8	16×10^{-2}	0.9	27×10^{-2}
Nuclear power	0.0001	18×10^{-6}	0.002	41×10^{-5}	0.2	64×10^{-3}
Fuel reprocessing	—	—	0.0008	17×10^{-5}	0.2	64×10^{-3}
AEC activities other than open-air weapons testing	0.01	18×10^{-4}	0.01	20×10^{-4}	0.01	32×10^{-4}
Open-air weapons testing	13.0[c]	2.4	4.0	0.82	4.9	1.6
TV, consumer products, air travel	1.6	29×10^{-3}	2.6	40×10^{-3}	1.1	6×10^{-2}
Diagnostic radiology	72	13.3	72	14.8	72	23.1
	217	40	209	43	208	66

[a] From Environmental Protection Agency (1972).
[b] Population of United States.
[c] 1963.

TABLE 18-6

SOURCES OF PER CAPITA DOSE IN THE UNITED STATES[a]

	1960	1970	2000
	Per capita annual dose rate		
	182 mrem/year	194 mrem/year	224 mrem/year
Fraction due to:			
Natural sources	0.60	0.62	0.62
Medical sources	0.33	0.34	0.35
Power generation and fuel processing	0.0000005	0.00002	0.002
Fallout from weapons tests	0.06	0.02	0.02

[a] Compiled from Environmental Protection Agency (1972).

radiation exposure, as estimated by the EPA Special Studies Group, is given in Table 18-4, where it is seen that the total is expected to diminish from about 1.6 mrem/year in 1960 to 0.2 mrem/year at the end of the century. The difference is due mainly to the assumed reduction in exposure to the use of radium in consumer products.

SUMMARY OF POPULATION EXPOSURE

Estimates of the per capita and population integral doses in the United States for the period 1960 through 2000, as given in Tables 18-5 and 18-6, are based on the work of the Special Studies Group referred to previously. The outstanding feature of Table 18-6 is the high fraction contributed by medical sources of radiation exposure and the relatively insignificant fraction due to nuclear energy.

Speculation about Future Sources of Exposure

Technological innovation is likely to occur to such an extent in the years ahead that it would be naive to attempt to forecast what new sources of ionizing radiations may come into being. Thirty years ago, very few people would have forecast the many technical developments we have experienced, such as the birth and growth of the nuclear energy program, lasers, and microwaves. Many other developments undoubtedly lie ahead, some of which may influence the atomic energy industry.

Generation of electrical power by controlled fusion processes may become a reality, although probably not before the end of the century. Although many people look to this method of producing power as the ultimate in "clean" energy, the process will involve production of radioactive materials on a scale that, as yet, has not been adequately evaluated. In particular, the dense neutron fluxes will undoubtedly produce activation products in copious amounts.

The breeder reactor, which should be available commercially by 1980, will result in production of enormous quantities of plutonium, which will involve new environmental problems for which adequate controls must be developed. However, experience at the large plutonium production centers, such as Hanford, which have been highly successful in safeguarding both employees and the public, will be directly applicable to the breeder program.

No broad prospective view of the subject would be complete without some mention of the possibility of a catastrophic nuclear accident. As was discussed in Chapter 9, the most severe type of accident to the present generation of reactors is considered to be the result of sudden loss of coolant, in which case the risk to the public would depend on the extent to which fission products would be released from the reactor core and escape to the environment. If present concepts are valid, the severity of such an accident would be due to the effects of ^{131}I, a nuclide that could produce high exposures to human thyroids. If this possibility is sufficiently credible, it should be possible to stockpile tablets of potassium iodide or other iodine compounds that could be administered to a potentially exposed population within an hour or two after an accident that threatens to release iodine in copious quantities (Blum and Eisenbud, 1967). It has been shown that doses in the range of 50 to 200 m of potassium iodide can effectively prevent the thyroid from absorbing a significant dose of radioactive iodine.

While, on the one hand, it must be recognized that the possibility of a catastrophic reactor accident exists, it is also clear that the probability of such an accident is very low. Attempts to quantify the probabilities of accidents of various severities have been made, and Starr (1969, 1971) has related these estimates to the more common risks of everyday living. Starr notes that for many types of risks man demonstrates an intuitive response which causes him to take actions to stabilize the risk when, in his collective intuitive judgment, the risk outweighs the benefits. Starr differentiates between involuntary risks, such as those from the great natural disasters or war, and voluntary risks, such as those from the use of tobacco or general aviation. He concludes, from extensive analyses of accident records, that somehow the upper bound of risk that society will voluntarily accept is about the same order as that of death from disease, 10^{-6} fatalities per person per hour. This is the upper bound of acceptance, and it is about three orders of magnitude higher than many other risks that are voluntarily accepted. Starr argues that the risk of death from a reactor accident compares favorably with the risks accepted by society, an argument that few can quarrel with, based on experience thus far. However, the fact remains that we are dealing with a combination of high severity and low probability, the magnitudes of which cannot be ascertained by the usual actuarial methods because of the very low probability of occurrence and the consequent inability to obtain sufficient actuarial data over a reasonable period of time.

If the overriding risk in the event of a severe reactor accident is, in fact, due to radioiodine, it should be emphasized that this nuclide might increase the probability of thyroid cancer, which is not ordinarily a fatal disease. The risk to individuals living near a nuclear power plant has been analyzed by Beattie (1967), who concluded that the maximum risk would be to a child living at about 1000 yards from the reactor. He assumed that a major reactor accident would occur with a frequency of $1/10^4$ reactor years. This seemed to be a high estimate of probability, based on other estimates that appear in the literature. Beattie assumed a release of 500 Ci of ^{131}I and that exposure took place by inhalation (which is also a reasonable assumption, since cows' milk produced in the area affected by a nuclear emergency would be removed from the market). It was calculated that the probability of a child developing thyroid cancer because it is living close to the reactor is on the order of 10^{-7}/year. This would represent about a 10% increase in the normal risk of developing thyroid cancer in individuals under the age of 20. In this age group, deaths from cancer of the thyroid are very rare, the incidence being less than 0.01/100,000/year. The risk of ^{131}I exposure following a reactor accident could be reduced by at least another two orders of magnitude if potassium iodide or equivalent stable iodine compounds could be administered prophylactically within 2 hr after the accident.

Fortunately, the more volatile radionuclides produced in a reactor are short-lived and, in the case of the noble gases, relatively inert. Long-lived bone-seeking nuclides such as ^{90}Sr and the transuranic elements are orders of magnitude less volatile and far less likely to be dispersed to the environment.

Thus, when one considers both low-level exposure from normal operation of the nuclear industry and the danger of abnormal releases, one concludes that at least for the foreseeable future the levels of environmental radioactive contamination from the civilian uses of nuclear energy can be managed by techniques which have been shown to be effective during the past 3 decades. When one considers the benefits that will accrue to mankind from civilian utilization of atomic energy on an extensive scale, the risks due to environmental radioactivity, though finite, are miniscule compared to the benefits that will be realized. However, when one contemplates the effects of radioactivity from the military uses of nuclear energy and, more particularly, the interaction of the effects of radioactivity with the effects due to blast and fire, the cost to humanity of military exploitation of the atom in war would be so great as to make war unacceptable in relation to any benefits that could possibly accrue to any one country or to the world.

More than 25 years ago, on June 14, 1946, United States Representative Bernard M. Baruch presented a program for international control of atomic energy to the infant United Nations (U.S. Department of State, 1946). It was then only 10 months since Hiroshima and only 5 months since the preceding January, when the first General Assembly of the United Nations convened in London. Mr Baruch's address was delivered in the twilight of a long and productive life intimately involved with the history of his times. His introductory words, among the first ever spoken at the United Nations on the need for nuclear sanity, provide an appropriate closing for this volume, in which we have dealt with the environmental atom in both war and peace.

"My Fellow Citizens of the World:

We are here to make a choice between the quick and the dead . . ."
"Behind the black portent of the new atomic age lies a hope which, seized upon with faith, can work our salvation. If we fail, then we have damned every man to be the slave of Fear. Let us not deceive ourselves: We must elect World Peace or World Destruction . . ."

That choice remains to be made.

Appendix

The Properties of Certain Radionuclides

The following pages summarize some of the important properties of the nuclides that are of principal interest to the environmentalist. The reader should bear in mind that the biological and physical constants are subject to revision and should be prepared to consult the primary sources of such information.

Explanations of the subheadings used, and the principal citations, are as follows.

Half-Life
Physical (Lederer *et al.*, 1968).
Biological (ICRP, 1960; ICRP, 1968).
Specific Activity (elemental): For naturally occurring radionuclides, the elemental specific activity is given in units of picocuries per gram of element (Rankama, 1963).
Principal Human Metabolic Parameters (as defined in ICRP, 1960):

f_1: transfer coefficient from the gastrointestinal tract to the blood.
f_2: fraction in organ of reference of that in total body.
f_2': transfer coefficient from blood to organ of reference.
f_w: fraction of that taken into the body by ingestion that is retained in the critical organ.
f_a: fraction of that taken into the body by inhalation.

Maximum Permissible Body Burden: (ICRP (1960); ICRP (1968).

Miscellaneous Numerical Values: For documentation, unless given, see elsewhere in this text or the more recent reports of the United Nations Scientific Committee on the Effects of Atomic Radiation (1966, 1968).

Dosimetric Considerations: Given as dose equivalent rate to the critical organ from nuclides naturally present in the body or per microcurie inhaled or ingested.

Maximum Permissible Concentrations in Air (MPC_a) and Water (MPC_w): From the Code of Federal Regulations (10CFR20) as promulgated by the U.S. Atomic Energy Commission for exposure in unrestricted areas.

Tritium (^{3}H)

Half-Life

    Physical: 12.36 years.

    Biological: 9.5 ± 4.1 days, total body for HTO.

Specific Activity in Nature: 5–10 pCi/liter of H_2O in lakes and streams prior to advent of bomb testing.

Sources: Cosmic-ray interactions with N and O; ternary fission; spallation from cosmic rays, ^{6}Li (n, α) ^{3}H. World inventory of naturally produced tritium is approximately 80 MCi (Harley and Lowder, 1971).

Principal Modes and Energies of Decay (MeV): β^- 0.018.

Special Chemical and Biological Characteristics: Combines readily with electronegative substances. Not selectively concentrated in any organ. Metabolized as H_2O; skin absorption is 80% of lung absorption rate.

Principal Human Metabolic Parameters: $f_1 = 1.0$, $f_2 = 1.0$, $f_2' = 1.0$, $f_w = 1.0$, $f_a = 1.0$.

Principal Organ: Total body.

Amount of Element in Body: 7×10^3 g.

Dosimetric Considerations: 0.167 mrem/μCi ingested (total body).

MPC_a: 2×10^{-7} μCi/ml.

MPC_w: 3×10^{-3} μCi/ml.

Other: Production rates, 6.6 MCi/megaton of thermonuclear bombs (Jacobs, 1968); in light water reactors, produces primarily by ternary fission and secondarily by neutron interactions with light elements. 5–10 Ci/MWe year for PWR; 0.3–0.9 Ci/MWe year for BWR.

Carbon-14

Half-Life

    Physical: 5730 years.

Biological: Total body, 10 days; fat, 40 days.
Specific Activity of Living Carbon and Atmospheric CO_2 prior to Era of Nuclear Energy: 7.5 ± 2.7 pCi/g. Lower values in urban areas owing to fossil fuel combustion.
Sources: Cosmic-ray neutron activation, ^{14}N (n, p) ^{14}C; production rate at earth's surface, 1.6 atoms/cm^2 sec (UNSCEAR, 1964), nuclear weapons testing 3.4 × 10^4 Ci/megaton (Klement, 1959).
Principal Modes and Energies of Decay (MeV): β^- 0.155.
Special Characteristics: Metabolism varies considerably with the chemical compound of which it is a part, thereby determining the critical organ.
Principal Human Metabolic Parameters: $f_1 = 1.0$, $f_2 = 0.1$, $f_2' = 0.025$, $f_w = 0.025$, $f_a = 0.02$.
Maximum Permissible Body Burden: 400 μCi.
Normal Daily Intake of Stable Element: 400 g.
Average Natural Tissue and/or Body Burden: ~0.1 μCi.
Amount of Stable Element in Body: 1.3 × 10^4 g.
Dosimetric Considerations: From natural sources: gonads, 0.7 mrem/year; whole body, 1.0 mrem/year; endosteal cells, 1.6 mrem/year; bone marrow, 1.6 mrem/year (UNSCEAR, 1966). As $^{14}CO_2$, 0.576 mrem to body fat per microcurie ingested.
Special Ecological Aspects: Exists naturally as CO_2 with no latitude gradient.
MPC_a: 1 × 10^{-7} μCi/ml (soluble).
MPC_w: 8 × 10^{-4} μCi/ml (soluble).

Potassium-40

Half-Life
  Physical: 1.26 × 10^9 years.
  Biological: Whole body, 58 days.
Specific Activity: 853 pCi/g K
Sources: Naturally occurring primordial nuclide.
Principal Modes and Energies of Decay (MeV): β^- 1.31 (89%); EC (11%); γ 1.46 (11%).
Special Chemical and Biological Characteristics: The element is distributed throughout the body, mainly in muscle and bone.
Principal Human Metabolic Parameters: $f_1 = 1.0$, $f_2 = 0.65$, $f_2' = 0.65$, $f_w = 0.65$, $f_a = 0.49$.
Mean Daily Intake = 1.4–6.5 g.
Typical Normal Body Burden: 0.12 μCi (70 kg man).
Critical Organs: Whole body, muscle.
Amount of Element in Body: 2 × 10^{-3} g/g wet tissue or 140 g total body.

Dosimetric Considerations: (From natural body burdens)—20 mrem/year whole body; 8 mrem/year endosteal cells; 15 mrem/year bone marrow.

Special Aspects: Body burden decreases with age and in muscle-wasting diseases.

Miscellaneous Information: Stable $K = 2.59\%$ of earth's crust and 380 ppm in sea water; ^{40}K concentration in soils 1–30 pCi/g. Potassium-40 is the predominant radioactive component in normal foods and human tissue.

Manganese-54

Half-Life

  Physical: 303 days.

  Biological: Whole body, 17 days; liver, 25 days; kidneys, 7 days.

Sources: Nuclear weapons testing; activation product–light water reactors ^{54}Cr (p, n) ^{54}Mn, ^{54}Fe (n, p) ^{54}Mn.

Principal Modes and Energies of Decay (MeV): EC; γ Cr X rays, 0.835 (100%).

Special Characteristics: Stable manganese is an essential trace element for both plants and animals.

Effective Energy (MeV): 0.13 GI tract.

Principal Human Metabolic Parameters: $f_1 = 0.1$, $f_2 = 0.35$, $f_2' = 0.24$, $f_w = 0.02$, $f_a = 0.07$ (liver).

Maximum Permissible Body Burden: 20 μCi (liver).

Normal Daily Intake of Stable Element: 3.1×10^{-3} g/day.

Critical Organs: GI tract (LLI), liver.

Amount of Element in Body: 0.026 g.

Special Ecological Aspects: Shellfish are principal pathway for human exposure. Mn-54 entering water is precipitated as oxide but can be released from sediments with increasing salinity.

MPC_a: 1×10^{-9} μCi/ml.

MPC_w: 1×10^{-4} μCi/ml.

Iron-55

Half-Life

  Physical: 2.6 years.

  Biological: Spleen, 600 days; whole body, 800 days.

Sources: Nuclear weapons testing, 1.7×10^7 Ci/megaton produced in light water reactors by ^{54}Fe (n, γ) ^{55}Fe and ^{56}Fe $(n, 2n)$ ^{55}Fe.

Principal Modes and Energies of Decay: EC 6.0 keV, Auger electron 5.0 keV.
Principal Human Metabolic Parameters: $f_1 = 0.1$, $f_2 = 0.02$, $f_2' = 0.02$, $f_w = 2.0 \times 10^{-3}$, $f_a = 6.0 \times 10^{-3}$ (spleen).
Maximum Permissible Body Burden: 1×10^3 μCi.
Normal Daily Intake of Stable Element: 0.027 g/day.
Critical Organs: Spleen, GI (LLI).
Amount of Stable Element in Body: 4.9 g.
Dosimetric Considerations: Dose to red blood cell precursors increased by deposition of energy of Auger electrons (Wrenn, 1968).
Special Ecological Aspects: Direct foliar deposition is the most important pathway for terrestrial plant contamination. Readily taken up into marine food chains due to the low stable iron content of ocean waters. Concentration in the blood of Lapps and Eskimos high owing to air–lichen–reindeer (caribou)–man food chain. Also higher in consumers of high fish diets.
MPC_a: 3×10^{-8} μCi/ml (soluble and insoluble).
MPC_w: 8×10^{-4} μCi/ml (soluble), 2×10^{-3} μCi/ml (insoluble).

Iron-59

Half-Life
  Physical: 45.6 days.
  Biological: Spleen, 600 days; whole body, 800 days.
Sources: Nuclear weapons testing, 2.2×10^6 Ci/megaton; activation product ^{58}Fe (n, γ) ^{59}Fe.
Principal Modes and Energies of Decay (MeV): β^- 0.13 (1%), 0.27 (46%), 0.46 (53%), 1.56 (0.3%); γ 0.14 (0.9%), 0.19 (2.4%), 0.34 (0.3%), 1.10 (56%), 1.29 (44%).
Effective Energy (MeV): 0.34 (spleen), 0.0065 (total body).
Principal Human Metabolic Parameters: See ^{55}Fe.
Maximum Permissible Body Burden: 20 μCi.
Normal Daily Intake: See ^{55}Fe.
Critical Organs: spleen, GI (LLI).
Amount of Stable Element in Body: 4.9 g.
Dosimetric Considerations: 14 mrem to spleen per microcurie ingested; 42 mrem to spleen per microcurie inhaled.
Special Ecological Aspects: See ^{55}Fe.
MPC_a: 5×10^{-9} μCi/ml (soluble), 2×10^{-9} μCi/ml (insoluble).
MPC_w: 6×10^{-5} μCi/ml (soluble), 5×10^{-5} μCi/ml (insoluble).

Cobalt-60

Half-Life
    Physical: 5.26 years.
    Biological: 9.5 days plus possible long-term component.
Sources: Nuclear weapons testing; activation product ^{59}Co $(n,\ \gamma)$ ^{60}Co; ^{60}Ni $(n,\ p)$ ^{60}Co.
Principal Modes and Energies of Decay (MeV): β^- 0.312 (99+%); γ 1.173 (100%); 1.332 (100%).
Critical Human Metabolic Parameters: $f_1 = 0.3$, $f_2 = 1.0$, $f_2' = 1.0$, $f_w = 0.3$, $f_a = 0.4$ (total body).
Maximum Permissible Body Burden: 10 μCi.
Normal Intake of Element: 7×10^{-6} g/day.
Amount of Element in Body (grams): 1.8×10^{-3}.
Critical Organs: GI (LLI), total body.
Dosimetric Considerations: 4×10^{-3} rem to total body/μCi inhaled; 3×10^{-3} rem to total body/μCi ingested.
Special Ecological Aspects: Released from nuclear reactors in cooling waters— uptake in marine food web; used clinically to measure the absorption and retention of labeled vitamin B-12 in man.
MPC_a: 1×10^{-8} μCi/ml (soluble), 3×10^{-10} μCi/ml (insoluble).
MPC_w: 5×10^{-5} μCi/ml (soluble), 3×10^{-5} μCi/ml (insoluble).

Zinc-65

Half-Life
    Physical: 245 days.
    Biological: Whole body, 933 days.
Sources: Nuclear weapons testing, activation product ^{64}Zn $(n,\ \gamma)$ ^{65}Zn, ^{63}Cu $(^2$H, $\gamma)$ ^{65}Zn.
Principal Modes and Energies of Decay (MeV): EC (98.5%); β^+ 0.325 (1.5%); γ 0.51 from β^+ (3%), 1.11 (45%).
Critical Human Metabolic Parameters: $f_1 = 0.1$, $f_2 = 1.0$, $f_2' = 1.0$, $f_w = 0.1$, $f_a = 0.3$.
Maximum Permissible Body Burden: 60 μCi.
Normal Daily Intake of Element: 0.017 g/day.
Critical Organ: Total body.
Amount of Stable Isotopes in Body: 2.4 g.
Dosimetric Considerations: 0.012 rem to total body/μCi inhaled; 0.004 rem to total body/μCi ingested.
Special Ecological Aspects: Concentrates in the visceral portions of shellfish.

MPC_a: 4×10^{-9} μCi/ml (soluble), 2×10^{-9} μCi/ml (insoluble).
MPC_w: 1×10^{-4} μCi/ml (soluble), 2×10^{-4} μCi/ml (insoluble).

Krypton-85

Half-Life
  Physical: 10.73 years.
Sources: Primarily from fission. A minor product of cosmic-ray interactions.
Principal Modes and Energies of Decay (MeV): β^- 0.672; γ 0.514 (0.43%).
Special Chemical Characteristics and Biological Characteristics: Inert noble
  gas. Diffusion is sole mechanism of absorption into body; partition
  coefficients blood/gas 0.05, tissue/blood 1.0, fat/blood 10.0.
Effective Energy (MeV): 0.24.
Principal Human Metabolic Parameters: f_1 = 1.0, f_2 = 1.0, f_2' = 1.0, f_w =
  1.0, f_a = 0.75.
Critical Organ: Total body (submersion skin dose).
Dosimetric Considerations: Doses from immersion in ^{85}Kr cloud of 0.3
  pCi/ml for 1 year would be: 7 mrem—total body, 500 mrem—body
  surface, 4×10^{-4} mrem—gonads, and 4×10^{-4} mrem to the lens of
  eye (Hendrickson, 1971).
Special Ecological Aspects: Present atmospheric inventory is about 60 MCi
  (Kirk, 1972).
MPC_a: 3×10^{-7} μCi/ml (submersion).
Other: Air concentration in U.S. in 1970 $\simeq 15 \times 10^{-6}$ pCi/ml, world in-
  ventory from cosmic production <2 MCi, nuclear explosives ~3
  MCi, and reactors produce approximately 300 Ci/MW(e)/year.

Strontium-89

Half-Life
  Physical: 50.5 days.
  Biological: Bone, 1.8×10^4 days; whole body, 1.3×10^4 days.
Source: Fission product.
Principal Modes and Energies of Decay (MeV): β^- 1.46 (100%).
Special Chemical Characteristics and Biological Characteristics: Alkaline
  earth element, with tendency to concentrate in bone.
Effective Energy (MeV): 2.8 bone.
Principal Human Metabolic Parameters: f_1 = 0.3, f_2 = 0.99, f_2' = 0.7,
  f_w = 0.21, f_a = 0.28 (bone).
Maximum Permissible Body Burden: 4 μCi (bone).

Normal Daily Intake: 1×10^{-3} g/day.
Critical Organ: Bone.
Amount of Stable Isotopes in Body: 0.34 g.
Dosimetric Considerations: 0.15 rem to bone/μCi inhaled; 0.11 rems to bone/μCi ingested.
MPC$_a$: 3×10^{-10} μCi/ml (soluble), 1×10^{-9} μCi/ml (insoluble).
MPC$_w$: 3×10^{-6} μCi/ml (soluble), 3×10^{-5} μCi/ml (insoluble).

Strontium-90

Half-Life
   Physical: 28.1 years.
   Biological: See [89]Sr.
Principal Modes and Energies of Decay (MeV): β^- 0.54 (100%).
Special Chemical and Biological Characteristics: See [89]Sr.
Effective Energy (MeV): 5.5.
Principal Human Metabolic Parameters: $f_1 = 0.3$, $f_2 = 0.99$, $f_2' = 0.3$, $f_w = 0.09, f_a = 0.12$ (bone).
Maximum Permissible Body Burden: 2 μCi.
Normal Annual Intake: See [89]Sr.
Critical Organ: Bone.
Amount of Stable Isotopes in Body: See [89]Sr.
Dosimetric Considerations: 15.6 rem to bone/μCi inhaled; 11.7 rem to bone/μCi ingested.
Special Ecological Aspects: Incorporates into calcium pool of the biosphere.
Discrimination Factors: Cows' milk/grass = 0.13: human bone/cows' milk = 0.25.
MPC$_a$: 3×10^{-11} μCi/ml (soluble), 2×10^{-10} μCi/ml (insoluble).
MPC$_w$: 3×10^{-7} μCi/ml (soluble), 4×10^{-5} μCi/ml (insoluble).

Iodine-131

Half-Life
   Physical: 8.05 days.
   Biological: 138 days.
Sources: A fission product produced by nuclear weapons at a rate of 125 MCi/megaton. Attains equilibrium of 26 kCi/MWt of steady-state reactor operation.
Principal Modes and Energies of Decay (MeV): γ 0.36 (80%), 0.64 (9%), 0.72 (3%).

Biological Characteristics: Soluble form readily absorbed through the skin, lung, and alimentary tract.

Principal Human Metabolic Parameters: $f_1 = 1.0$, $f_2 = 0.2$, $f_2' = 0.3$, $f_w = 0.3$, $f_a = 0.23$.

Maximim Permissible Body Burden: 0.7 μCi.

Normal Intake of Stable Element: 2×10^{-4} g/day.

Critical Organ: Thyroid.

Amount of Stable Isotopes in Body: 4.6×10^{-3} g in 16 g thyroid.

Dosimetric Considerations: rem/μCi ingested = 15.5 to thyroid of 6-month-old child; 1.9 to thyroid of adult male; 3.5×10^{-3} to total body of adult male; 1.3 rem to adult thyroid/μCi inhaled. Dose commitments approximately 10 times higher for a 1-year-old child.

Special Ecological Aspects: Principal route of human absorption is via fresh milk; grass–cow–milk–man chain with rapid distribution of milk a primary consideration. Milk content of radioiodine reaches a peak 3 days after deposition on forage (Soldat, 1965). Effective half-time of removal from grass is about 5 days. 9.1×10^{-2} μCi ^{131}I/liter milk can be expected per μCi ^{131}I deposited on 1 m^2 of grass. The ratio pCi/kg grass: pCi/m^3 air = 4200. The ratio pCi/liter milk: pCi/kg grass = 0.15 (Soldat, 1963).

MPC_a: 1×10^{-10} μCi/ml (soluble), 1×10^{-8} μCi/ml (insoluble).

MPC_w: 3×10^{-7} μCi/ml (soluble), 6×10^{-5} μCi/ml (insoluble).

Cesium-134

Half-Life

    Physical: 2.5 years.

    Biological: Total body adults, 50–150 days; total body children, 44 days.

Sources: An activation product produced in nuclear reactors at a rate of 3 Ci/MWd, ^{134}Cs/^{137}Cs = 0.4–0.6.

Principal Modes and Energies of Decay (MeV): β^- 0.662 (61%); γ 0.6 (98%), 0.8 (98%).

Special Chemical Characteristics: Alkali metal with properties similar to K and Rb; most salts are soluble.

Biological Characteristics: Metabolism resembles that of potassium—distributed throughout body mainly in muscle ($\sim$60% of total in muscle mass). Convenient to express concentration in biological material as pCi ^{134}Cs/g K.

Effective Energy (MeV): 1.1 (total body).

Principal Human Metabolic Parameters: $f_1 = 1.0$, $f_2 = 1.0$, $f_2' = 1.0$, $f_w = 1.0$, $f_a = 0.75$ total body.
Maximum Permissible Body Burden: 20 μCi.
Critical Organ: Total body.
Amount of Stable Element in Body: 1.5 $\times$ 10^{-3} g.
MPC_a: 1 $\times$ 10^{-9} μCi/ml (soluble), 4 $\times$ 10^{-10} μCi/ml (insoluble).
MPC_w: 9 $\times$ 10^{-6} μCi/ml (soluble), 4 $\times$ 10^{-5} μCi/ml (insoluble).

Cesium-137

Half-Life
  Physical: 30 years.
  Biological: Total body adults, 50–150 days; total body children, 44 days.
Sources: Nuclear weapons testing 0.17 MCi ^{137}Cs/megaton fission. Produced in thermal reactors at rate of about 1.2 pCi/MWt/year (Also see Cs-134).
Principal Modes and Energies of Decay (MeV): β^- 1.17 (7%), 0.51 (92%); γ 0.66 (82%).
Special Chemical Characteristics and Biological Characteristics: See ^{134}Cs.
Principal Human Metabolic Parameters: See ^{134}Cs.
Maximum Permissible Body Burden: 30 μCi.
Critical Organ: Total body.
Amount of Stable Isotopes in Body: 1.5 $\times$ 10^{-3} g.
Dosimetric Considerations: 1 pCi/^{137}Cs/g K = 0.02 mrad/year; 0.046 rem to total body/μCi inhaled; 0.061 to total body/μCi ingested.
Special Ecological Aspects: Higher body burdens in Lapps and Eskimos due to air–lichen–reindeer (caribou)–man food chain; plant uptake from soil small; foliar absorption is most significant route of entry into food web.
MPC_a: 2 $\times$ 10^{-9} μCi/ml (soluble), 5 $\times$ 10^{-10} μCi/ml (insoluble).
MPC_w: 2 $\times$ 10^{-5} μCi/ml (soluble), 4 $\times$ 10^{-5} μCi/ml (insoluble).

Lead-210 (Ra D)

Half-Life
  Physical: 22.2 years.
  Biological: From bone, power function with apparent ultimate value of 10–12 years.
Sources: Naturally occurring from U decay chain.

Principal Modes and Energies of Decay (MeV): β^- 0.015 (81%), 0.061 (19%); γ 0.0465 (3.86%); Bi X-rays—0.013 av. (23.35%).

Special Chemical Characteristics: Behaves similarly to alkaline earth elements.

Biological Characteristics: Replaces Ca in bone matrix.

Effective Energy (MeV): (Including daughters) 5.2 (total body).

Principal Human Metabolic Parameters: $f_1 = 0.08$, $f_2 = 0.06$, $f_2' = 0.14$, $f_w = 0.01$, $f_a = 0.04$ (kidney).

Maximum Permissible Body Burden: 0.4 μCi.

Normal Ingested Intake of Stable Element: 3×10^{-4} g/day.

Average Natural Tissue and/or Body Burden: 100–200 pCi in skeleton; $(1.5 \times 10^{-2}$ pCi/g fresh bone).

Critical Organ: Kidney.

Amount of Stable Element in Body: 0.12 g.

Dosimetric Considerations: Dose rate from naturally present ^{210}Pb $= 0.3$ mrem/year to the gonads and 3.6 mrem/year to the bone.

Special Ecological Aspects: Present in the skeletons of uranium miners as a result of direct inhalation and short-lived radon daughter decay; concentrations higher in skeletons of smokers, and consumers of lichen-eating reindeer.

MPC_a: 4×10^{-12} μCi/ml (soluble), 8×10^{-12} μCi/ml (insoluble).

MPC_w: 1×10^{-7} μCi/ml (soluble), 2×10^{-4} μCi/ml (insoluble).

Polonium-210 (Ra F)

Half-Life
  Physical: 138 days.
  Biological: Spleen, 60 days; whole body, 30 days.

Sources: Naturally occurring from uranium decay chain and ^{209}Bi (n,γ) ^{210}Bi β^- ^{210}Po.

Principal Modes and Energies of Decay (MeV): α 5.3 (100%), γ 0.8 (0.0011%).

Special Chemical Characteristics: Tendency for radiocolloid formation.

Principal Human Metabolic Parameters
  For spleen: $f_1 = 0.06$, $f_2 = 0.07$, $f_2' = 0.04$, $f_w = 2.0 \times 10^{-3}$, $f_a = 0.01$.
  For kidney: $f_1 = 0.06$, $f_2 = 0.13$, $f_2' = 0.07$, $f_w = 4.0 \times 10^{-3}$, $f_a = 0.02$.

Range of Normal Intake: 1–10 pCi/day.

Average Natural Tissue and/or Body Burden: 735 pCi.

Critical Organs: Spleen, kidney.

Dosimetric Considerations: (From natural body burden) 3.0 mrem/year whole body; 21.0 mrem/year endosteal cells; 3.0 mrem/year bone marrow. 12.0 rem to kidney/μCi inhaled; 2.4 rem to kidney/μCi ingested. Builds in from [210]Pb.

Special Ecological Aspects: Air–lichen–reindeer–man pathway results in abnormally high body burdens; direct fallout from atmospheric [222]Rn decay causes relatively high concentrations in broad leaf vegetation; present in tobacco smoke.

Radon-222

Half–Life
    Physical: 3.82 days.

Sources: Naturally occurring from uranium decay chain.

Principal Modes and Energies of Decay (MeV): α 5.49 (100%).

Special Chemical Characteristics: Inert noble gas, somewhat soluble in body fat. Under most conditions, the principal dose is from α emissions by short-lived daughter products inhaled as attachments to inert dust normally present in the atmosphere.

Normal Range of Atmospheric Randon Content: 0.1–0.5 pCi/liter.

Critical Organ: Lung.

Dosimetric Considerations: A series of short-lived daughter products approach equilibrium with [222]Rn in a few hours. (see Chapter 7 and Fig. 7-1). A special "unit," used in United States uranium mines, is the Working Level, defined as any combination of radon daughters in 1 liter of air that will produce 1.3 $\times$ 10[5] MeV of α energy. This would be equivalent to a [222]Rn concentration of 100 pCi/liter, in equilibrium with its short-lived α-emitting progeny (see Table 7-2). According to Altshuler *et al.* (1964) and Jacobi (1964), the critical lung tissue is the basal cells of the bronchial epithelium, to which 1 working level will deliver about 20 rem in a year of 2000 exposure hours. At a continuous background level of 0.1 pCi/liter, the dose to the bronchial epithelium would thus be about 85 mrem.

MPC_a: 3 $\times$ 10[-9] μCi/ml.

Other: When [226]Ra is deposited in the human skeleton, approximately 67% of the radon produced diffuses to the blood and is eliminated in the expired breath. One microcurie of [226]Ra in the skeleton results in a concentration of 13 pCi/liter of [222]Rn in exhaled air.

Radium-226

Half–Life
    Physical: 1620 years.

Biological: Whole body, 900 days; bone, power function with apparent ultimate value of 10–12 years.

Sources: Naturally occurring from ^{238}U decay chain.

Principal Modes and Energies of Decay (MeV): α 4.78 (94.3%), 4.59 (5.7%); γ 0.188 (4%), 0.26 (0.007%).

Biological Characteristics: Deposits in bone with nonuniform distribution. Following decay of ^{226}Ra in bone, approximately 67% of ^{222}Rn diffuses to the blood and is exhaled.

Effective Energy (MeV): 110 (bone) (Ra + daughters).

Principal Human Metabolic Parameters: $f_1 = 0.3$, $f_2 = 0.99$, $f_2' = 0.1$, $f_w = 0.04$, $f_a = 0.03$ (bone).

Maximum Permissible Body Burden: 0.1 μCi.

Normal Intake: 590 pCi/year.

Average Natural Tissue and/or Body Burden: $\sim$36 pCi.

Critical Organ: Bone.

Dosimetric Considerations: ^{226}Ra + daughters (at 35% equilibrium), osteocytes, 10 mrem/year; Haversian canals, 5.4 mrem/year; trabecular marrow, 0.6 mrem/year; gonads, 0.5 mrem/year; 30 rem to bone/ μCi inhaled; 40 rem to bone/μCi ingested.

MPC$_a$: 3 × 10^{-12} μCi/ml (soluble), 2 × 10^{-12} μCi/ml (insoluble).

MPC$_w$: 3 × 10^{-8} μCi/ml (soluble), 3 × 10^{-5} μCi/ml (insoluble).

Thorium-232

Half-Life
   Physical: 1.41 × 10^{10} years.
   Biological: Bone, 7.3 × 10^4 days; whole body, 5.7 × 10^4 days.

Specific Activity: 1.11 × 10^5 pCi/g Th.

Sources: Naturally occurring.

Principal Modes and Energies of Decay (MeV): α 4.01 (77%), 3.95 (23%), γ Ra LX-rays.

Special Chemical Characteristics: Hydroxides and oxides are insoluble; nitrates, sulfates, chlorides, and perchloride salts are readily soluble.

Biological Characteristics: Tendency to concentrate on bone surfaces.

Effective Energy (MeV): 270 (bone) (^{232}Th + daughters).

Principal Human Metabolic Parameters: $f_1 = <10^{-4}$, $f_2 = 0.9$, $f_2' = 0.7$, $f_w = 20 × 10^{-5}$, $f_a = 0.18$ (bone).

Maximum Permissible Body Burden: 0.04 μCi.

Normal Annual Intake: $\cong$19 μg.

Average Natural Tissue and/or Body Burden: $\simeq$215 μg (total body) or 24 pCi.

Critical Organs: Bone (endosteum), liver.

Dosimetric Considerations: 6500 rem to bone/μCi of soluble Th inhaled; 2500 rem to bone/μCi of soluble Th ingested.

Special Ecological Aspects: Thorium-bearing minerals result in anomolously high natural radiation levels in Brazil and India.

MPC_a: 1×10^{-12} μCi/ml (soluble and insoluble).

MPC_w: 2×10^{-6} μCi/ml (soluble), 4×10^{-5} μCi/ml (insoluble).

Other: 1 μCi body burden results in concentration in expired air of 5.6×10^{-3} μCi/liter. One curie natural thorium $= 1$ Ci ^{232}Th $+ 1$ Ci ^{228}Th by ICRP definition. Depending on the type of rock, the concentration of ^{232}Th in the earth's crust is 0.1–15 ppm.

Uranium-235

Half-Life
  Physical: 7.1×10^8 years.
  Biological: Kidney, 15 days; whole body, 100 days.

Specific Activity (Elemental) pCi/g: 1.54×10^4 pCi/g ^{nat}U.

Sources: Normal constituent (0.72%) of uranium in earth's crust. Enriched with respect to ^{238}U in fissionable materials for reactors and weapons.

Principal Modes and Energies of Decay (MeV): α 4.58 (8% doublet), 4.40 (57%), 4.37 (18%); γ 0.143 (11%), 0.185 (54%), 0.204 (5%).

Effective Energy (MeV): 46 (kidney).

Principal Human Metabolic Parameters: $f_1 = <10^{-2}, f_2 = 0.065, f_2' = 0.11, f_w = 1.1 \times 10^{-3}, f_a = 0.028$.

Maximum Permissible Body Burden: 0.03 μCi.

Normal Annual Elemental Intake $= 470$ μg.

Critical Organ: Kidney.

Amount of Element in Body: 54 μg ^{nat}U.

Dosimetric Considerations: Since ^{nat}U has a low specific activity, chemical damage to the kidney is likely to be more important than radiation damage. However, radiation injury to the lung or kidney must be considered if exposure is to enriched U.

MPC_a: 2×10^{-11} μCi/ml (soluble), 4×10^{-12} μCi/ml (insoluble).

MPC_w: 3×10^{-5} μCi/ml (soluble and insoluble).

Uranium-238

Half-Life
  Physical: 4.5×10^9 years.

Biological: Kidney, 15 days; whole body, 100 days.
Specific Activity (Elemental) (pCi/g): 3.0×10^5 pCi/g ^{nat}U.
Sources: Naturally occurring in earth's crust. ^{238}U is present to extent of 99.28% by weight in natural uranium.
Principal Modes and Energies of Decay (MeV): α 4.20 (75%), 4.15 (25%).
Effective Energy (MeV): 43 (kidney).
Principal Human Metabolic Parameters: $f_1 - <10^{-2}, f_2 = 0.065, f_2' = 0.11, f_w = 1.1 \times 10^{-3}, f_a = 0.028.$
Maximum Permissible Body Burden: 5×10^{-3} μCi.
Normal Annual Intake $= 470$ μg ^{nat}U.
Average Natural Tissue and/or Body Burden: 54 μg ^{nat}U, primarily in the skeleton.
Critical Organ: Kidney.
MPC_a: 3×10^{-12} μCi/ml (soluble), 5×10^{-12} μCi/ml (insoluble).
MPC_w: 4×10^{-5} μCi/ml (soluble and insoluble).
Other: See ^{235}U.

Plutonium-238

Half-Life
  Physical: 86.4 years.
  Biological: Bone, 7.3×10^4 days.
Sources: Minor constituent of nuclear weapons fallout. Produced for power sources (SNAP) by ^{237}Np (n,β) ^{238}Pu.
Principal Modes and Energies of Decay (MeV): α 5.50 (72%), 5.46 (28%).
Special Chemical Characteristics: Member of the actinide series of rare-earth elements. Forms insoluble fluorides, hydroxides, and oxides. Soluble complexes with citrate.
Biological Characteristics: Partitioned in body equally between bone and liver.
Effective Energy (MeV): 280 (bone).
Principal Human Metabolic Parameters: $f_1 = 3 \times 10^{-5}, f_2 = 0.9, f_2' = 0.8, f_w = 2.4 \times 10^{-5}, f_a = 0.2$ (to bone).
Maximum Permissible Body Burden: 0.04 μCi (bone).
Critical Organs: Bone and lung.
Special Ecological Aspects: Occurs in environment as oxide; firmly adsorbed on soils.
MPC_a: 7×10^{-14} μCi/ml (soluble), 1×10^{-12} μCi/ml (insoluble).
MPC_w: 5×10^{-6} μCi/ml (soluble), 3×10^{-5} μCi/ml (insoluble).

Plutonium-239

Half-Life
  Physical: 2.44 × 10⁴ years.
  Biological: See ²³⁸Pu.
Sources: Produced in thermal reactors by neutron irradiation of ²³⁸U. Used in nuclear weapons, and as fuel for fast reactors.
Principal Modes and Energies of Decay (MeV): α 5.06 (11%), 5.13 (17%), 5.15 (73%).
Special Chemical Characteristics and Biological Characteristics: See ²³⁸Pu.
Effective Energy (MeV): 270 (bone).
Principal Human Metabolic Parameters (for bone deposition): $f_1 = 3 \times 10^{-5}$, $f_2 = 0.9, f_2' = 0.8, f_w = 2.4 \times 10^{-5}, f_a = 0.2$.
Maximum Permissible Body Burden: 0.04 μCi.
Critical Organ: Bone (soluble) and lung (insoluble).
Dosimetric Considerations: 0.86 rem to bone/μCi ingested; 7100 rems to bone/μCi inhaled. Lung dose calculations of questionable meaning in view of particulate characteristics of insoluble Pu and the short range of the α emissions.
MPC_a: 6 × 10⁻¹⁴ μCi/ml (soluble), 1 × 10⁻¹² μCi/ml (insoluble).
MPC_w: 5 × 10⁻⁶ μCi/ml (soluble), 3 × 10⁻⁵ μCi/ml (insoluble).

Bibliography

Aarkrog, A. (1971). Prediction models for ^{90}Sr and ^{137}Cs levels in the human food chain. *Health Phys.* **20,** 297–312.

Adams, J. A. S., and Lowder, W. M. (1964). "The Natural Radiation Environment." Univ. of Chicago Press, Chicago, Illinois.

Agnedal, P. O. (1967). Calcium and strontium in Swedish waters and fish and accumulation of ^{90}Sr. *Radioecol. Concentr. Processes, Proc. Int. Symp.*, 1966, p. 879.

Agnedal, P. O., and Bergström, S. O. W. (1966). Recipient capacity of Tvären, a Baltic Bay. *Proc. Disposal of Radioactive Wastes into Seas, Oceans, and Surface Waters, 1966* p. 753. IAEA, Vienna.

Albenesius, E. L. (1959). Tritium as a product of fission. *Phys. Rev. Lett.* **3,** 274.

Albert, R. E. (1966). "Thorium—Its Industrial Hygiene Aspects." Academic Press, New York.

Albert, R. E., and Arnett, L. C. (1955). Clearance of radioactive dust from the human lung. *AMA Arch. Ind. Health* **12,** 99.

Alexander, L. T. (1959). Strontium-90 distribution as determined by the analysis of soils. *In* "Fallout from Nuclear Weapons Tests," p. 278. Hearings before Joint Committee on Atomic Energy.

Alexander, L. T. (1967). "Depth of Penetration of the Radioisotopes Strontium-90 and Cesium-137," Fallout Program Quarterly Summary, Rep. HASL-183. USAEC, Washington, D.C.

Alexander, L. T., Hardy, E. P., and Hollister, H. L. (1960). Radioisotopes in soils: Particularly with reference to Strontium-90. *In* "Radioisotopes in the Biosphere" (R. S. Caldicott and L. A. Snyder, eds.), p. 3. Univ. of Minnesota Press, Minneapolis.

Allardyce, C., and Trapnell, E. R. (no date). "The First Pile." Div. Tech. Inform., USAEC, Washington, D.C.

Allen, R. E. (1971). "Summary Information on Accidental Releases of Radioactivity to the Atmosphere from Underground Nuclear Detonations Designed for Containment." August 5, 1963-June 30, 1971, WASH-1183. USAEC, Washington, D.C.

Alpher, R. A., and Herman, R. C. (1953). Origin and abundance distribution of elements. *Annu. Rev. Nucl. Sci.* **2,** 1.

Altshuler, B., Nelson, N., and Kuschner, M. (1964). Estimation of the lung tissue dose from the inhalation of radon and daughters. *Health Phys.* **10,** 1137–1162.

American National Standards Institute. (1969). "Guide to Sampling Airborne Radioactive Materials in Nuclear Facilities," ANSI N13.1. Amer. Nat. Stand. Inst., New York.

American Nuclear Society. (1969). "Nucl. Appl. Technol." 7, No. 3. Amer. Nucl. Soc.

American Nuclear Society. (1971). Symposium for reactor containment spray system technology. Nucl. Technol. 10, 400.

Anderson, E. C. (1953). The production and distribution of natural radiocarbon. Annu. Rev. Nucl. Sci. 2, 63.

Anderson, R. E. (1971). Symposium on the delayed consequences of ionizing radiation (pathology studies at the Atomic Bomb Casualty Commission, Hiroshima and Nagasaki, 1945–1970). Hum. Pathol. 2, 469.

Anderson, W., and Turner, R. C. (1956). Radon content of the atmosphere. Nature (London) 178, 203.

Anderson, W., Mayncord, W. V., and Turner, R. C. (1954). The radon content of the atmosphere. Nature (London) 174, 424.

Angelovic, J. W., White, J. C., and Davis, E. M. (1969). Interaction of ionizing radiation, salinity, and temperature on the estuarine fish Fundulus heteroclitus. Radioecol., Proc. Nat. Symp., 2nd, 1969 U.S. AEC CONF-670503.

Anonymous. (1969). Radioisotopes: Production and development of large scale uses. Part II. Applications. Isotop. Radiat. Technol. 6, 238.

Anonymous. (1969–1970). Radioisotope-powered cardiac pacemaker. Isotop. Radiat. Technol. 7, 192–193.

Archer, V. E., and Lundin, F. E., Jr. (1967). Radiogenic lung cancer in man: Exposure-effect relationship. Environ. Res. 1, 370.

Atomic Industrial Forum, Detroit (1961). Memo, p. 27.

Atomic Power Development Associates, New York (1967). "October 5, 1966 Fuel Damage Incident at the Enrico Fermi Atomic Power Plant," NP-16750. At. Power Develop. Ass.

Auerbach, S. I., Nelson, D. J., Kaye, S. V., Reichle, D. E., and Coutant, C. C. (1971). Ecological considerations in reactor power plant siting. Proc. Environm. Aspects Nucl. Power Sta., 1971. IAEA, Vienna.

Auxier, J. A. (1965). Nuclear accident at Wood River Junction. Nucl. Safety 6, 298–315.

Baker, L., Jr., Rose, D., and Miller, C. E., Jr. (1970). The liquid-metal fast breeder reactor safety program. Nucl. Safety 11, No. 1, 1–11.

Baron, T., Gerhard, E. R., and Johnstone, H. F. (1949). Dissemination of aerosol particles dispersed from stacks. Ind. Eng. Chem. 41, 2403.

Barreira, F. (1961). Concentration of atmospheric radon and wind directions. Nature (London) 190, 1092.

Bartlett, B. O., and Russell, R. Scott (1966). Prediction of future levels of long-lived fission products in milk. Nature (London) 209, 1062.

Barton, C. J., Jacobs, D. G., Kelly, M. J., and Struxness, E. G. (1971). Radiological considerations in the use of natural gas from nuclearly stimulated wells. Nucl. Technol. 11, 335–344.

Beasley, T. M., and Palmer, H. E. (1966). Lead-210 and polonium-210 in biological samples from Alaska. Science 152, 1062.

Beasley, T. M., Jokela, T. A., and Eagle, R. J. (1971). Radionuclides and selected trace elements in marine protein concentrates. Health Phys. 21, 815.

Beattie, J. R. (1967). Risks to the population and the individual from iodine releases. Nucl. Safety 8, 573–576.

Beaver, R. J. (1961). Contamination of fuel-element surfaces. Nucl. Safety 3, 61.

Beck, H. L. (1966). Environmental gamma radiation from deposited fission products, 1960–1964. Health Phys. 12, 313–322.

Beck, H. L., and dePlanque, G. (1968). "The Radiation Field in Air Due to Distributed Gamma-Ray Sources in the Ground," Rep. HASL-195. USAEC, Washington, D.C.

Beck, H. L., Lowder, W. M., Bennett, B. G., and Condon, W. J. (1966). "Further Studies of External Environmental Radiation," Rep. HASL-170. USAEC, Washington, D.C.

Beck, H. L., Lowder, W. M., and McLaughlin, J. E. (1971). "In Situ External Environmental Gamma Ray Measurements Utilizing Ge (Li) and NaI (Tl) Spectrometry and Pressurized Ionization Chambers," IAEA SM/148-2. IAEA, Vienna.

Becquerel, H., and Curie, P. (1901). Action physiologique des rayons du radium. *C. Re. Acad. Sci.* **132,** 1289.

Behounek, F. (1970). History of exposure of miners to radon. *Health Phys.* **19,** 56.

Behrens, C. F. (1959). "Atomic Medicine." Williams & Wilkins, Baltimore, Maryland.

Belter, W. G. (1965). U.S. operational experience in radioactive waste management, 1958–1963. *Proc. 3rd Int. Conf. Peaceful Uses At. Energy, 1964.* Vol. 14, p. 31. United Nations, New York.

Belyaev, V. E., Kolesnikov, A. G., and Nelepo, B. A. (1965). Estimate of the intensity of radioactive pollutants in the oceans on the basis of new data on decay processes. *Proc. 3rd Int. Conf. Peaceful Uses At. Energy, 1964.* Vol. 14, pp. 83–86. United Nations, New York.

Beninson, D., Vander Elst, E., and Cancio, D. (1966). Biological aspects in the disposal of fission products into surface waters. *Proc. Disposal of Radioactive Wastes into Seas, Oceans, and Surface Waters,* p. 337. IAEA, Vienna.

Bennett, B. G., (1971). "Strontium-90 in the Diet," Rep. HASL-242. USAEC, Washington, D.C.

Bethel, A. L. *et al.* (1959). Shippingport power station (PWR) waste disposal facilities. *In* "Industrial Radioactive Waste Disposal." Hearings before Joint Committee on Atomic Energy.

Bieri, R. H., Koide, M., and Goldberg, E. D. (1966). The noble gas contents of pacific seawaters. *J. Geophys. Res.,* **71,** No. 22.

Björnerstedt, R., and Engstrom, A. (1960). Radioisotopes in the skeleton: Dosage implications based on microscopic distribution. *In* "Radioisotopes in the Biosphere," Chapter 27. Univ. of Minnesota Press, Minneapolis.

Blanchard, R. L., and Holaday, D. A. (1960). Evaluation of radiation hazards created by thoron and thoron daughters. *Amer. Ind. Hyg. Ass., Quart.* **21,** 201.

Blanchard, R. L., and Moore, J. B. (1970). ^{210}Pb and ^{210}Po in tissues of some Alaskan residents as related to consumption of caribou or reindeer meat. *Health Phys.* **18,** 127.

Blasewitz, A. G., McElroy, J. L., and Schneider, K. J. (1971). "Solidification of High-Level Liquid Radioactive Wastes," Hanford Eng. Develop. Lab., Rep. HEDL-SA-236. USAEC, Washington, D.C.

Blatz, H. (1962a). Reduction of dose in medical and industrial radiography. *Amer. J. Pub. Health* **52,** 1385–1390.

Blatz, H. (1962b). Radiation hazards from the standpoint of the practicing physician. *N.Y. State J. Med.* pp. 3893–3898.

Blatz, H. (1964). "Introduction to Radiological Health." McGraw-Hill, New York.

Blifford, I. H., Lockhart, L. B., *et al.* (1952). On the natural radioactivity of the air. *Nav. Res. Lab. Rep.* **4036.**

Blomeke, J. O., and Harrington, F. E. (1968). Waste management at nuclear power stations. *Nucl. Safety* **9,** 239–248.

Blum, M., and Eisenbud, M. (1967). Reduction of thyroid irradiation from ^{131}I by potassium iodide. *J. Amer. Med. Ass.* **200**, 1036.

Bogorov, V. G., and Kreps, E. M. (1958). Concerning the possibility of disposing of radioactive waster in ocean trenches. *Proc. 2nd U.N. Int. Conf. Peaceful Uses At. Energy, 1958* p. 2058. United Nations, New York.

Bradshaw, R. L., Empson, F. M., McClain, W. C., and Houser, B. L. (1970). Results of a demonstration and other studies of the disposal of high level solidified radioactive wastes in a salt mine. *Health Phys.* **18**, 63–67.

Branch, I. L., and Connor, J. A. (1961). Nuclear safety in space. *Nucleonics* **19**, 64.

Brecker, R., and Brecker, E. (1969). "The Rays: A History of Radiology in the United States and Canada." Williams & Wilkins, Baltimore, Maryland.

Breslin, A. J., and Glauberman, H. (1970). Investigation of radioactive dust dispersed from uranium tailings piles. *In* "Environmental Surveillance in the Vicinity of Nuclear Facilities" (W. C. Reinig, ed.). Thomas, Springfield, Illinois.

Brewer, A. W. (1949). Evidence for a world circulation provided by the measurements of helium and water vapour distribution in the stratosphere. *Quart. J. Roy. Meteorol. Soc.* **75**, 351–363.

Briggs, G. A. (1969). "Plume Rise." USAEC, Washington, D.C.

Briggs, G. A., Van der Hoven, I., Engelman, R. J., and Halitsky, J. (1968). Processes other than natural turbulence affecting effluent concentrations. *In* "Meteorology and Atomic Energy" (D. H. Slade, ed.). USAEC, Washington, D.C.

Bright, G. O. (1971). Light-water-reactor safety. *Nucl. Safety* **12**, 433–438.

Brittan, R. O., and Heap, J. C. (1958). Reactor containment. *Proc. 2nd U.N. Int. Conf. Peaceful Uses At. Energy, 1958* p. 437. United Nations, New York.

Broseus, R. W. (1970). Cesium-137/strontium-90 ratios in milk and grass from Jamaica. M. S. Thesis, New York University.

Brown, J. M., Thompson, J. F., and Andrews, H. L. (1962). Survival of waste containers at ocean depths. *Health Phys.* **7**, 227.

Bruce, F. R. (1960). Origin and nature of radioactive wastes in the United States atomic energy programme. *Proc. Disposal of Radioactive Wastes, 1959.* IAEA, Vienna.

Brues, A. (1958). Critique of the linear theory of carcinogenesis. *Science* **128**, 693.

Brues, A. (1959). *In* "Low Level Irradiation," Publ. No. 59. Amer. Ass. Advan. Sci., Washington, D.C.

Bryan, G. W., Preston, A., Templeton, W. L. (1966). Accumulation of radionuclides by aquatic organisms of economic importance in the United Kingdom. *Proc. Disposal of Radioactive Wastes into Seas, Oceans, and Surface Waters, 1966.* IAEA, Vienna.

Bryant, P. M. (1964). Derivation of working limits for continuous release rates of iodine-131 to atmosphere in a milk producing area. *Health Phys.* **10**, 249–258.

Bryant, P. M. (1966). Derivation of working limits for continuous release rates of ^{90}Sr and ^{137}Cs to atmosphere in a milk producing area. *Health Phys.* **12**, 1393–1405.

Bryant, P. M. (1969). Data for assessments concerning controlled and accidental releases of ^{131}I and ^{137}Cs to atmosphere. *Health Phys.* **17**, 51.

Bryant, P. M. (1970a). The derivation and application of limits and reference levels for environmental radioactivity in the United Kingdom. *Proc. Health Phys. Aspects Nucl. Facility Siting, 1970* p. 634.

Bryant, P. M. (1970b). Derivation of working limits for continuous release rates of [129]I to atmosphere. *Health Phys.* **19**, 611.

Buchanan, J. R. (1963). SL-1 final report. *Nucl. Safety* **4**, 83–86.

Buck, C. (1959). Population size required for investigating threshold dose in radiation induced leukemia. *Science* **129,** 1357.

Burch, P. R. J. (1969). Ionizing radiation and life shortening. *Nucl. Safety* **10,** 161–170.

Burnett, T. J. (1970). A derivation of the "Factor of 700" for [131]I. *Health Phys.* **18,** 73.

Burns, F. J., Albert, R. E., and Heimbach, R. D. (1968). The RBE for skin tumors and hair follicle damage in the rat following irradiation with alpha particles and electrons. *Radiat. Res.* **36,** 225–241.

Bustard, T. S., Princiotta, F. T., and Barr, H. N. (1970). Reentry protection for radio-isotope heat sources. *Nucl. Appl. Technol.* **9,** 572–583.

Caldwell, R. D., Crosby, R. F., and Lockhard, M. P. (1970). Radioactivity in coal mine drainage. *In* "Environmental Surveillance in the Vicinity of Nuclear Facilities" (W. C. Reinig, ed.). Thomas, Springfield, Illinois.

Cameron, J. R., Suntharalingam, N., and Kenney, G. N. (1968). "Thermoluminescent Dosimetry." Univ. of Wisconsin Press, Madison.

Cantril, S. T., and Parker, H. M. (1945). "The Tolerance Dose," MDDC-1100 USAEC, Washington, D.C.

Casarett, A. P. (1968). "Radiation Biology." Prentice-Hall, Englewood Cliffs, New Jersey.

Casarett, G. W. (1965). Experimental radiation carcinogenesis. *Tumor Res.* **7,** 49–82.

Chadwick, R. C., and Chamberlain, A. C. (1970). Field loss of radionuclides from grass. *Atmos. Environ.* **4,** 51–56.

Chamberlain, A. C. (1955). Aspects of travel and deposition of aerosol and vapor clouds. *U.K. At. Energy Auth., Res. Comp. Rep.* **HP/R1261.**

Chamberlain, A. C. (1960). Aspects of the deposition of radioactive and other gases and particles. *Int. J. Air Pollut.* **3,** 63.

Chamberlain, A. C. (1970). Interception and retention of radioactive aerosols by vegetation. *Atmos. Environ.* **4,** 57–58.

Chamberlain, A. C., and Chadwick, R. C. (1966). Transport of iodine from atmosphere to ground. *Tellus* **18,** 226–237.

Chamberlain, A. C., and Dunster, H. J. (1958). Deposition of radioactivity in north-west England from the accident at Windscale. *Nature (London)* **182,** 629–630.

Chastain, J. W. (1958). "U.S. Research Reactors Operations and Use." Addison-Wesley, Reading, Massachusetts.

Chester, C. V., and Chester, R. O. (1970). Civil defense implications of a pressurized water reactor in a thermonuclear target area. *Nucl. Appl. Technol.* **9,** 786–795.

Chipman, W. A. (1960). Biological aspects of disposal of radioactive wastes into marine environments. *Proc. Disposal of Radioactive Wastes, 1959.* IAEA, Vienna.

Clapp, C. A. (1934). "Cataract." Lea & Febiger, Philadelphia, Pennsylvania.

Clark, H. M. (1954). The occurrence of an unusually high-level radioactive rainout in the area of Troy, N.Y. *Science* **119,** 619.

Clayton, G. D., Arnold, J. R., and Patty, F. A. (1955). Determination of sources of particulate atmospheric carbon. *Science* **122,** 751.

Clegg, J. W., and Foley, D. D. (1958). "Uranium Ore Processing." Addison-Wesley, Reading, Massachusetts.

Codman, E. A. (1902). A study of the cases of accidental x-ray burns hitherto recorded. *Philadelphia Med. J.*, pp. 438–442.

Cole, D. W., Mott, W. E., and Sagan, L. A. (1970). Factors relating to the application of radioisotopes to circulatory-support systems. *Isotop. Radiat. Technol.* **7,** 138–145.

Coleman, J. R., and Liberace, R. (1966). Nuclear power production and estimated krypton-85 levels. *Radiol. Health Data Rep.* **7,** 615–621.

Collins, J. C. (1960). "Radioactive Wastes, Their Treatment and Disposal." Wiley, New York.

Colorado Committee for Environmental Information. (1970). "Report on the Dow Rocky Flats Fire: Implications of Plutonium Releases to the Public Health and Safety. Subcommittee on Rocky Flats, Boulder, Colo., Jan. 13, 1970," Rep. HASL-235 (Ref. 2, p. 38). USAEC, Washington, D.C.

Comar, C. L., and Wasserman, R. H. (1960). Radioisotope absorption and methods of elimination: differential behavior of substances in metabolic pathways. *In* Radioisotopes in the Biosphere," p. 526. Univ. of Minnesota Press, Minneapolis.

Comar, C. L., Wasserman, R. H., and Nold, M. M. (1956). Strontium calcium discrimination factors in the rat. *Proc. Soc. Exp. Biol. Med.* **92,** 859.

Comar, C. L., Trum, B. F., Kuhn, U.S.G., III, Wasserman, R. H., Nold, M. M., and Schooley, J. C. (1957). Thyroid radioactivity after nuclear weapons tests. *Science* **126,** 16.

Conard, R. A. *et al.* (1960). Medical survey of Rongelap people five and six years after exposure to fallout. *Brookhaven Nat. Lab. Rep.* **BNL-609.**

Conard, R. A., Dobyns, B. M., and Sutow, W. W. (1970a). Thyroid neoplasia as late effect of exposure to radioactive iodine in fallout. *J. Amer. Med. Ass.* **214,** 316–324.

Conard, R. A. *et al.* (1970b). Medical survey of the people of Rongelap and Utirik Islands thirteen, fourteen, and fifteen years after exposure to fallout radiation (March 1967, March, 1968 and March, 1969). *Brookhaven Nat. Lab. Rep.* **BNL 50220 (T-562).**

Conlon, F. B., and Pettigrew, G. L. (1971). Summary of federal regulations for packaging and transportation of radioactive materials. *U.S., Pub. Health Serv., Rep.* **BRH/DMRE 71-1.**

Corbett, B. L. (1971). Fuel meltdown at St. Laurent. I. *Nucl. Safety* **12,** No. 1, 35–39.

Cornish, A. C., and Simens, H. (1971). A guide to the U.S. transportation regulations. *Proc. Int. Symp. Packaging & Transport. Radioactive Mater., 3rd, 1971.* USAEC CONF-710801, Vol. 1.

Cottrell, W. B., Browning, W. E., Jr., Parker, G. W., Castleman, A. W., Jr., and Junkins, R. L. (1965). U.S. experience on the release and transport of fission products within containment systems under simulated reactor accident conditions. *Proc. 3rd Int. Conf. Peaceful Uses At. Energy, 1964.* United Nations, New York.

Court Brown, W. M., Doll, R., and Hill, A. B. (1960). Incidence of leukaemia after exposure to diagnostic radiation in utero. *Brit. Med. J.,* **2,** 1539.

Cowser, K. E. (1964). "Current Practices in the Release and Monitoring of ^{131}I at NRTS, Hanford, Sanannah River, and ORNL," ORNL-NSIC3. Oak Ridge Nat. Lab., Oak Ridge, Tennessee.

Cowser, K. E., Kaye, S. V., Rohwer, P. S., Snyder, W. S., and Struxness, E. G. (1967). "Dose-Estimation Studies Related to Proposed Construction an Atlantic-Pacific Interoceanic Canal with Nuclear Explosives: Phase I," Contr. No. W-7405-eng-26. Health Phys. Div., Oak Ridge Nat. Lab., Oak Ridge, Tennessee.

Cox, W. M., Blanchard, R. L., and Kahn, B. (1970). Relation of radon concentration in the atmosphere to total moisture detention in soil and atmospheric thermal stability. *Advan. Chem. Ser.* **93,** 436–446.

Crabtree, J. (1959). The travel and diffusion of the radioactive material emitted during the Windscale accident. *Quart. J. Roy. Meteorol. Soc.* **85,** 362.

Crocker, G. R., O'Connor, J. D., and Freiling, E. C. (1966). Physical and radiochemical properties of fallout particles. *Health Phys.* **12**, 1099.

Cronkite, E. P. (1961). Evidence for radiation and chemical as leukemogenic agents. *Environ. Health* **3**, 297.

Cronkite, E. P. *et al.* (1956). "Some Effects of Ionizing Radiation on Human Beings." USAEC, Washington, D.C.

Cuthbert, F. L. (1958). "Thorium Production Technology." Addison-Wesley, Reading, Massachusetts.

Davis, J. J., Perkins, R. W., Palmer, R. F., Hanson, W. C., and Cline, J. F. (1958). Radioactive materials in aquatic and terrestrial organisms exposed to reactor effluent water. *Proc. 2nd U.N. Int. Conf. Peaceful Uses At. Energy, 1958.* United Nations, New York.

de Bartoli, M. C., and Gaglone, P. (1969). Snap plutonium-238 fallout at Ispra, Italy. *Health Phys.* **16**, 197–204.

de Laguna, W. (1970). Radioactive waste disposal by hydraulic fracturing. *Nucl. Safety* **11**, 391.

de Laguna, W., Binford, F. T., Weeren, H. O., Witkowski, E. J., and Struxness, E. G. (1971). "Safety Analysis of Waste Disposasl by Hydraulic Fracturing at Oak Ridge," ONRL-4665. Oak Ridge Nat. Lab., Oak Ridge, Tennessee.

de Villiers, A. J., and Windish, J. P. (1964). Lung cancer in fluorspar mining community. *Brit. J. Ind. Med.* **21**, 94.

Dietrich, J. R., and Zinn, W. H. (1958). "Solid Fuel Reactors." Addison-Wesley, Reading, Massachusetts.

DiNunno, J. J., Anderson, F. D., Baker, R. E., and Waterfield, R. L. (1962). "Calculation of Distance Factors for Power and Test Reactor Sites," Rep. TID-14844. USAEC, Washington, D.C.

Dobson, G. M. B. (1956). Origin and distribution of the polyatomic molecules in the atmosphere. *Proc. Roy. Soc., Ser. A* **236**, 187–193.

Dolphin, G. W. (1968). The risk of thyroid cancer following irradiation. *Health Phys.* **15**, 219.

Dolphin, G. W. (1971). Dietary intakes of iodine and thyroid dosimetry. *Health Phys.* **21**, 711.

Dolphin, G. W., and Marley, W. G. (1969). Risk evaluation in relation to the protection of the public in the event of accidents at nuclear installations. *Proc. Environ. Contam. by Radioactive Mater., 1969.* IAEA, Vienna.

Dominick, D. D. (1971). Statement before Joint Committee on Atomic Energy, United States Congress, Subcommittee on Raw Materials.

Dorn, D. W. (1964). Heavy isotope production by nuclear explosive devices. *In* "Engineering with Nuclear Explosives," Rep. TID-7695. USAEC, Washington, D.C.

Drew, R. T., and Eisenbud, M. (1966). The normal radiation dose to indigenous rodents on the Morro do Ferro, Brazil. *Health Phys.* **12**, 1267–1274.

Drinker, P., and Hatch, T. (1954). "Industrial Dust," 2nd ed. McGraw-Hill, New York.

Drobinski, J. C., Jr. (1966). Radiocarbon in the environment. *Radiol. Health Data Rep.* **7**, 10–12.

Dudley, R. A. (1959). "Natural and Artificial Radioactivity Background of Man," *Proc. AAAS Symp.*

Dunning, G. M. (1962). "Fallout from USSR 1961 Nuclear Tests," Rep. TID-14377. USAEC, Washington, D.C.

Dunster, H. J. (1958). The disposal of radioactive liquid wastes into coastal waters.

Proc. 2nd U.N. Int. Conf. Peaceful Uses At. Energy, 1958. United Nations, New York.

Dunster, H. J. (1969). United Kingdom studies on radioactive releases in the marine environment. *In* "Biological Implications of the Nuclear Age." USAEC, Washington, D.C.

Dunster, H. J., Howells, H., and Templeton, W. L. (1958). District surveys following the Windscale Incident, October, 1957. *Proc. 2nd U.N. Int. Conf. Peaceful Uses At. Energy, 1958.* United Nations, New York.

Duursma, E. K., and Cross, M. C. (1971). Marine sediments and radioactivity. *In* "Radioactivity in the Marine Environment," Nat. Acad. Sci., Washington, D.C.

Eckert, J. A., Coogan, J. S., Mikkelsen, R. L., and Lem, P. N. (1970). Cesium-137 concentrations in Eskimos, Spring, 1968. *Radiol. Health Data Rep.* **11,** 219–225.

Edwards, C. C. (1971). Public statement made on the occasion of publication by the Food and Drug Administration in the Federal Register of the proposed X-Ray Equipment Performance Standards, Oct. 7.

Edwards, M. A. (1966). "Tabulation of Data on Announced Nuclear Detonations by all Nations through 1965," UCRL-14786. Univ. of California, Livermore.

Eisenbud, M. (1959). Deposition of Strontium-90 through October, 1958. *Science* **130,** 76.

Eisenbud, M. (1964). Radioactive fallout problems in food, water and clothing. *Arch. Environ. Health* **8,** 606.

Eisenbud, M., and Harley J. H. (1953). Radioactive dust from nuclear detonations. *Science* **117,** 141.

Eisenbud, M., and Harley, J. H. (1956). Radioactive fallout through September, 1955. *Science* **124,** 251.

Eisenbud, M., and Petrow, H. G. (1964). Radioactivity in the atmospheric effluents of power plants that use fossil fuels. *Science* **144,** 288.

Eisenbud, M., Mochizuki, Y., Goldin, A. S., and Laurer, G. R. (1962). Iodine-131 dose from Soviet nuclear tests. *Science* **136,** 370.

Eisenbud, M., Mochizuki, Y., and Laurer, G. R. (1963a). I-131 dose to human thyroids in New York City from nuclear tests in 1962. *Health Phys.* **9,** No. 12, 1291.

Eisenbud, M., Pasternack, B., Laurer, G. R., Mochizuki, Y., Wrenn, M. E., Block, L., and Mowafy, R. (1963b). Estimation of the distribution of thyroid doses in a population exposed to I-131 from weapon tests. *Health Phys.* **9,** 1281–1290.

Eisenbud, M., Pasternack, B., Laurer, G. R., and Block, L. (1963c). Variability of the I-131 concentrations in the milk distribution system of a large city. *Health Phys.* **9,** No. 12, 1303.

Eisenbud, M., Petrow, H., Drew, R. T., Roser, F. X., Kegel, G., and Cullen, T. L. (1964). Naturally occurring radionuclides in foods and waters from the Brazilian areas of high radioactivity. *In* "The Natural Radiation Environment" (J. A. S. Adams and W. M. Lowder, eds.), p. 837. Univ. of Chicago Press, Chicago, Illinois.

Eisenbud, M., and Wrenn, M. E. (1963). Biological disposition of radioiodine: A review. *Health Phys.* **9,** 1133.

Eisenbud, M., and Quigley, J. (1956). Industrial hygiene of uranium processing. *AMA Arch. Ind. Health* **14,** 12.

El Paso Natural Gas Co. (1965). "Project Gasbuggy." U.S. AEC Bureau of Mines, Lawrence Radiation Laboratory of the University of California.

El-Wakil, M. M. (1962). "Nuclear Power Engineering." McGraw-Hill, New York.

Engelmann, E. (1961). Estimate of the dietary intake of radium-226 for New York City infants. *Radiol. Health Data* **2**, 391.

Environmental Protection Agency. (1972). "Estimates of Ionizing Radiation Doses in the United States 1960–2000," Report of Special Studies Group, Division of Criteria & Standards, Office of Radiation Programs.

Environmental Protection Agency. (1972). "Environmental Radioactivity Surveillance Guide." Report of Surveillance & Inspection Division, Office of Radiation Programs.

Essig, T. H. (1971). "Radiological Impact of Hanford Waste Disposal on Groundwater Quality," Rep. BNWL-SA-3744. Battelle Pacific Northwest Lab., Richland, Washington.

Etherington, H., ed. (1958). "Nuclear Engineering Handbook." McGraw-Hill, New York.

Evans, R. D. (1943). Protection of radium dial workers and radiologists from injury by radium. *J. Ind. Hyg. Toxicol.* **25**, 253.

Evans, R. D. (1966). The effect of skeletally deposited alpha-ray emitters in man. *Brit. J. Radiol.* **39**, 881.

Evans, R. D. (1967). The radium standard for boneseekers–evaluation of the data on radium patients and dial painters. *Health Phys.* **13**, 267–278.

Evans, R. D., and Raitt, R. W. (1935). The radioactivity of the earth's crust and its influence on cosmic ray electroscope observations made near ground level. *Phys. Rev.* **48**, 171.

Evans, R. D., Keane, A. T., Kolenkow, R. J., Neal, W. R., and Shanaham, M. M. (1969). Radiogenic tumors in the radium and mesothorium cases studied at M.I.T. *In* "Delayed Effects of Bone-Seeking Radionuclides" (C. W. Mays *et al.*, eds.). Univ. of Utah Press, Salt Lake City.

Failla, G. (1960). Discussion. *In* "Selected Materials on Radiation Protection Criteria and Standards: Their Basis and Use." Hearings before Joint Committee on Atomic Energy.

Failla, G., and McClement, P. (1957). The shortening of life by chronic whole body irradiation. *Amer. J. Roentgenol., Radium Ther. Nucl. Med.* [N.S.] **78**, 946.

Faul, H. (1954). "Nuclear Geology." Wiley, New York.

Federal Radiation Council. (1960). "Background Material for the Development of Radiation Protection Standards," Rep. No. 1. US Govt. Printing Office, Washington, D.C.

Federal Radiation Council. (1961). "Background Material for the Development of Radiation Standards," Rep. No. 2. US Govt. Printing Office, Washington, D.C.

Federal Radiation Council. (1962a). "Health Implications of Fallout from Nuclear Weapons Testing through 1961," Rep. No. 3. US Govt. Printing Office, Washington, D.C.

Federal Radiation Council. (1962b). "Pathological Effects of Thyroid Irradiation." US Govt. Printing Office, Washington, D.C.

Federal Radiation Council. (1963). "Estimates and Evaluation of Fallout in the United States from Nuclear Weapons Testing Conducted Through 1962," Rep. No. 4. US Govt. Printing Office, Washington, D.C.

Federal Radiation Council. (1964a). "Background Material for the Development of Radiation Protection Standards," Rep. No. 5. US Govt. Printing Office, Washington, D.C.

Federal Radiation Council. (1964b). "Revised Fallout Estimates for 1964–1965 and

Verification of the 1963 Predictions," Rep. No. 6. US Govt. Printing Office, Washington, D.C.

Federal Radiation Council. (1965). "Background Material for the Development of Radiation Protection Standards—Protective Action Guides for Strontium-89, Strontium-90, and Cesium-137," Rep. No. 7. US Govt. Printing Office, Washington, D.C.

Federal Radiation Council. (1966). "Pathological Effects of Thyroid Irradiation," Revised Report. US Govt. Printing Office, Washington, D.C.

Federal Radiation Council. (1967). "Guidance for the Control of Radiation Hazards in Uranium Mining," Rep. No. 8. US Govt. Printing Office, Washington, D.C.

Feeley, H. W. (1960). Strontium-90 content of the stratosphere. *Science* **131**, 645.

Feldt, W. (1971). Research on the Maximum Radioactive Burden of Some German Rivers. *Proc. Symp. Environ. Aspects Nucl. Power Sta., 1971.* IAEA. Vienna.

Fenimore, J. W. (1964). Land burial of solid radioactive waste during a 10-year period. *Health Phys.* **10**, 229–236.

Ferri, E. S., and Baratta, E. J. (1966). Polonium 210 in tobacco, cigarette smoke, and selected human organs. *Publ. Health Rep.* **81**, 121–127.

Finkel, A. J., Miller, C. E., and Hasterlik, R. J. (1969). Radium-induced malignant tumors in man. *In* "Delayed Effects of Bone-Seeking Radionuclides" (C. W. Mays *et al.*, eds.). Univ. of Utah Press, Salt Lake City.

Fisenne, I. M., and Keller, H. W. (1970). "Radium-226 in the Diet of Two U.S. Cities," Rep. HASL-224. USAEC, Washington, D.C.

Folsom, T. R., and Vine, A. C. (1957). Tagged water masses for studying the oceans. *Nat. Acad. Sci.—Nat. Res. Council., Rep.* **551**.

Food and Agriculture Organization. (1960). "Radioactive Materials in Food and Agriculture." FAO, United Nations.

Foster, R. F. (1959). Distribution of reactor effluent in the Columbia River. *In* "Industrial Radioactive Waste Disposal." Hearings before Joint Committee on Atomic Energy.

Foster, R. F., Ophel, I. L., and Preston, A. (1971). Evaluation of human radiation exposure. *In* "Radioactivity in the Marine Environment." Nat. Acad. Sci., Washington, D.C.

Fox, C. H. (1969). "Radioactive Wastes. Understanding the Atom Series," rev. ed. USAEC, Washington, D.C.

Fredrikson, L., *et al.* (1958). Studies of soil-plant-animal interrelationship with respect to fission products. *Proc. 2nd U.N. Int. Conf. Peaceful Uses At. Energy, 1958,* p. 177. United Nations, New York.

Freiling, E. C., and Kay, M. A. (1965). "Radionuclide Fractionation in Air Burst Debris," Rep. USNRDL-TR-933. U.S. Nav. Radiol. Defense Lab.

Fresco, J., Jetter, E., and Harley, J. (1952). Radiometric properties of the thorium series. *Nucleonics* **10**, 60.

Freudenthal, P. C. (1970a). "Aerosol Scavenging by Ocean Spray," HASL-232. USAEC, Washington, D.C.

Freudenthal, P. C. (1970b). Strontium 90 concentrations in surface air: North America versus Atlantic Ocean from 1966 to 1969. *J. Geophys. Res.* **75**, 4089–4096.

Gahr, W. N. (1959). Uranium mill wastes. *In* "Industrial Radioactive Waste Disposal." Hearings before Joint Congressional Committee on Atomic Energy.

Gallaghar, R. G., and Saenger, E. L. (1957). Radium capsules and their associated hazards. *Amer. J. Roentgenol., Radium Ther. Nucl. Med.* [N.S.] **77**, 511–523.

Gallegos, A., Whicker, F., and Hakonson, T. (1970). Accumulation of radiocesium in rainbow trout via a non-food chain pathway. *Proc. 5th Annu. Midyear Topical Symp., 1970* Vol. II, p. 477.

Garner, R. J. (1960). An assessment of the quantities of fission products likely to be found in milk in the event of aerial contamination of agricultural land. *Nature (London)* **186,** 1063.

Gifford, F. A. (1961). Use of routine meteorological observations for estimating atmospheric dispersion. *Nucl. Safety* **2,** No. 4.

Glasstone, S. (1955). "Principles of Nuclear Reactor Engineering." Van Nostrand-Reinhold, Princeton, New Jersey.

Glasstone, S. (1962). "The Effects of Nuclear Weapons." USAEC, Washington, D.C.

Glauberman, H., and Loysen, P. (1964). The use of commercial incinerators for the volume reduction of radioactively contaminated wastes. *Health Phys.* **10,** 237.

Glueckauf, E. (1961). "Atomic Energy Waste." Wiley (Interscience), New York.

Gold, S., Barkhau, H. W., Shleien, B., and Kahn, B. (1964). Measurement of naturally occurring radionuclides in air. *In* "The Natural Radiation Environment" (J. A. S. Adams and W. M. Lowder, eds.) p. 369. Univ. of Chicago Press, Chicago, Illinois.

Goodjohn, A. J., and Fortescue, P. (1971). "Environmental Aspects of High Temperature Gas-Cooled Reactors," *Proc. Amer. Power Conf.* Illinois Inst. of Technol., Chicago, Illinois.

Gosline, C. A., Falk, L. L., and Helmer, E. N. (1956). *In* "Air Pollution Hand-book" (P. A. Magill, F. R. Holden, and C. Ackley, eds.), Chapter 5. McGraw-Hill, New York.

Grella, A. W. (1971). Today's role of the DOT in regulating the transport of radioactive material. *Proc., Int. Symp. Packaging & Transport. Radioactive Mater., 3rd, 1971.* USAEC CONF-710801, Vol. 1.

Groves, L. R. (1962). "Now It Can Be Told: The Story of the Manhattan Project." Harper, New York.

Grubbé, E. H. (1933). Priority in the therapeutic use of x-rays. *Radiology* **21,** 156.

Gustafson, P. F. (1969). Cesium-137 in freshwater fish during 1954–1965. *Proc. Nat. Symp. Radioecol., 2nd, 1969.* USAEC CONF-670503, p. 249.

Gustafson, P. F., and Miller, J. E. (1969). The significance of ^{137}Cs in man and his diet. *Health Phys.* **16,** 167–184.

Gustafson, P. F., Nelson, D. M., Brar, S. S., and Muniak, S. E., (1970). "Recent Trends in Radioactive Fallout," ANL-7760, Part III, p. 246. Argonne Nat. Lab.

Gwaltney, R. C. (1969). Missile generation and protection in light-water-cooled reactors. *Nucl. Safety* **10,** 300.

Haas, P. (1971). The Rulison project in retrospect. *Nucl. News (Hinsdale, Ill.)* pp. 55–59.

Hagis, W., Dobry, T., and Dix, G. (1961). Nuclear safety analysis of SNAP III for space missions. *J. Amer. Rocket Soc.* **31,** 1744.

Halitsky, J. (1968). Gas diffusion near buildings (Sect. 5.5). *In* "Meteorology and Atomic Energy" (D. H. Slade, ed.). USAEC, Washington, D.C.

Hallden, N. A., Fisenne, I. M., Ong, L. D. Y., and Harley, J. H. (1961). "Radioactive Decay of Weapons Debris," Rep. HASL-117 USAEC, Washington, D.C.

Hamilton, E. I. (1971). The relative radioactivity of building materials. *Amer. Ind. Hyg. Ass.,* **32,** 398.

Hamilton, E. I. (1972). The concentration of uranium in man and his diet. *Health Phys.* **22,** 149.

Hansen, W. G. *et al.* (1964). Farming practices and concentrations of fission products in milk. *U.S., Pub. Health Serv., Publ.* **999-R-6.**

Hardy, E., and Alexander, L. T. (1962). Rainfall and deposition of strontium-90 in Challam County, Washington. *Science* **136,** 881.

Harley, J. H. (1952). A study of the airborne daughter products of radon and thoron. Unpublished Doctoral Thesis, Rensselaer Polytechnic Institute, Troy, New York.

Harley, J. H. (1956). "Operation Troll," NYO-4656. USAEC, Washington, D.C.

Harley, J. H., ed. (1967). "Manual of Standard Procedures," 2nd ed., NYO 4700. Health & Safety Lab., USAEC, Washington, D.C.

Harley, J. H. (1969). Radionuclides in food. *AEC Symp. Ser.* **16.**

Harley, J. H. (1971). "Worldwide Plutonium Fallout from Weapons Tests," Proc. Environ. Plutonium Symp. Los Alamos Sci. Lab., Los Alamos, New Mexico.

Harley, J. H., and Lowder, W. M. (1971). "Natural Radioactivity and Radiation," HASL-242. USAEC, Washington, D.C.

Harper, W. R. (1961). "Basic Principles of Fission Reactors." Wiley (Interscience), New York.

Harrison, F. (1969). Accumulation and distribution of ^{54}Mn and ^{65}Zn in freshwater clams. *Radioecol., Proc. Nat. Symp., 2nd, 1969.* U.S. AEC CONF-670503, p. 198.

Hartung, F. H., and Hesse, W. (1879). Die Lungenkrebs, die Bergkrankenheit, in den Schneeberger Gruben. *Vierteljahresschr. Gerichtl. Med. Oeff. Gesundheitwessen* [N. S.] **30,** 296.

Harvey, D. G., and Morse, J. G. (1961). Radionuclide power for space missions. *Nucleonics* **19,** 69 (1961).

Harvey, R. S. (1964). Uptake of radionuclides by freshwater algae and fish. *Health Phys.* **10,** 243–247.

Harvey, R. S. (1970). Temperature effects on the sorption of radionuclides by freshwater algae. *Health Phys.* **19,** 293–297.

Hashizume, T., Maruyama, T., Kumamoto, Y., Kato, Y., and Kawamura, S. (1969). Estimation of gamma-ray dose from nuetron-induced radioactivity in Hiroshima and Nagasaki. *Health Phys.* **17,** 761–771.

Hawkins, M. (1961). "Procedures for the Assessment and Control of the Shorter Term Hazards of Nuclear Warfare Fallout in Water Supply Systems." University of California Research Institute of Engineering, Berkeley.

Hawley, C. A., Jr., Sill, C. W., *et al.* (1964). "Controlled Environmental Radiodine Tests National Reactor Testing Station," IDO-12035. USAEC, Washington, D.C.

Healy, J. W., and Fuquay, J. J. (1958). Wind pickup of radioactive particles from the ground. *Proc. 2nd U.N. Int. Conf. Peaceful Uses At. Energy, 1958,* p. 391. United Nations, New York.

Heller, M. B., Blatz, H., Pasternack, B., and Eisenbud, M. (1964). Radiation dose from diagnostic medical x-ray procedures to the population of New York City. *Amer. J. Pub. Health Nat. Health* **54,** 1551–1559.

Hemplelmann, L. H. (1968). Risk of thyroid neoplasms after irradiation in childhood. *Science* **160,** 159–165.

Hendrickson, M. M. (1971). The dose from ^{85}Kr released to the earth's atmosphere. *Proc. Environ. Aspects Nucl. Power Sta., 1971.* IAEA, Vienna.

Hendrickson, M. M., and Strenge, D. L. (1970). Reasons for differences in calculated estimates of the cloud dose. *Proc. Health Phys. Aspects Nucl. Facility Siting, 1970* pp. 566–576.

Hewlett, R. C., and Anderson, O. E. (1962). "The New World" (Vol. I of a History of the Atomic Energy Commission). Penn. State Univ. Press, University Park, Pennsylvania.

Hill, C. R. (1966). Polonium-210 content of human tissues in relation to dietary habit. *Science* 152, 1261.

Hillel, D. (1971). "Soil and Water: Physical Principles and Processes." Academic Press, New York.

Hilsmeier, W. F., and Gifford, F. A., Jr. (1962). "Graphs for Estimating Atmospheric Dispersion," Rep. ORO-545. USAEC, Washington, D.C.

Holaday, D. A. (1959). The nature of wastes produced in the mining and milling of ores. *In* "Industrial Radioactive Waste Disposal." Hearings before Joint Committee on Atomic Energy.

Holaday, D. A. (1969). History of the exposure of miners to radon. *Health Phys.* 16, 547–552.

Holaday, D. A., Archer, V. E., and Lundin, F. (1968). A summary of United States exposure experiences in the uranium mining industry. *Proc. Symp. Diagnosis Treatment Deposited Radionuclides, 1968* p. 451. Excerpta Med. Found., Amsterdam.

Hollaender, A., ed. (1954). "Radiation Biology," Vol. 1. McGraw-Hill, New York.

Hollaender, A., ed. (1955). "Radiation Biology," Vol. 2. McGraw-Hill, New York.

Holland, J. Z. (1953). "A Meteorological Survey of the Oak Ridge Area," Rep. ORO-99. USAEC, Washington, D.C.

Holland, J. Z. (1959a). "Summary of New Data on Atmospheric Fallout," Rep. TID-5554. USAEC, Washington, D.C.

Holland, J. Z. (1959b). Stratospheric radioactivity data obtained by balloon sampling. *In* "Fallout from Nuclear Weapons Tests." Hearings before Congressional Joint Committee on Atomic Energy.

Holland, J. Z. (1963). Physical origin and dispersion of radioiodine. *Health Phys.* 9, 1095.

Holleman, D. F., Luick, J. R., and Whicker, F. W. (1971). Transfer of radiocesium from lichen to reindeer. *Health Phys.* 21, 657.

Holtzman, R. B. (1964). Lead-210 and polonium-210 in potable waters in Illinois. *In* "The Natural Radiation Environment" (J. A. S. Adams and W. M. Lowder, eds.), p. 227. Univ. of Chicago Press, Chicago, Illinois.

Holtzman, R. B., and Ilcewicz, F. H. (1966). Lead-210 and polonium-210 in tissues of cigarette smokers. *Science* 153, 1259–1260.

Horan, J. R., and Gammill, W. P. (1963). The health physics aspects of the SL-1 accident. *Health Phys.* 9, 177.

Howells, G. P., and Bath, D. (1969). Trace elements in the Hudson River. *In* "Development of a Biological Monitoring System and a Survey of Trace Metals, Radionuclides and Pesticide Residues in the Lower Hudson River," Final Report for the Period Jan. 1, 1966 to Dec. 31, 1968. New York University Institute of Environmental Medicine, New York.

Howells, G. P., and Lauer, G. J., eds. (1969). "Hudson River Ecology," Proc. 2nd Symp. Hudson River Ecol., N.Y. State Dept. of Environmental Conservation, New York.

Howells, H. (1966). Discharges of low-activity, radioactive effluent from the windscale works into the Irish Sea. *Proc. Disposal of Radioactive Wastes into Seas, Oceans and Surface Waters, 1966* p. 769. IAEA, Vienna.

Hubbert, M. K. (1971). Energy resources for power production. *Proc. Environ. Aspects Nucl. Power Sta., 1971.* IAEA, Vienna.

Hughes, B. C. (1969). Nuclear excavation design of a transisthmian sea-level canal. *Nucl. Appl. Technol.* **7**, 305–327.

Hultqvist, B. (1956). Studies on naturally occurring ionizing radiation. *Kgl. Sv. Vetenskapsakade., Handl.* [4] Supplement.

Hunter, H. F., and Ballou, N. E. (1951). Fission product decay rates. *Nucleonics* **9**, C2.

Hursh, J. B. (1953). "The Radium Content of Public Water Supplies," Rep. UR-257. University of Rochester, Rochester, New York.

Hursh, J. B., Lovaas, A., and Biltz, E. (1960). "Radium in Bone and Soft Tissues of Man," Rep. UR-581. University of Rochester, Rochester, New York.

Ichikawa, R. (1961). On the concentration factors of some important radionuclides in marine food organisms. *Bull. Jap. Soc. Sci. Fish.* **27**, 66–74.

International Atomic Energy Agency. (1960). *Proc. Disposal of Radioactive Wastes.* IAEA, Vienna.

International Atomic Energy Agency. (1966). "Manual on Environmental Monitoring in Normal Operation," Safety Ser. No. 16. IAEA, Vienna.

International Atomic Energy Agency. (1967a). Assessment of airborne radioactivity. *Proc. Instrum. & Techniques for the Assessment of Airborne Radioactivity in Nuclear Operations.* IAEA, Vienna.

International Atomic Energy Agency. (1967b). Disposal of radioactive wastes into the ground. *Proc. Disposal of Radioactive Wastes into the Ground, 1967.* IAEA, Vienna.

International Atomic Energy Agency. (1967c). "Risk Evaluation for Protection of the Public in Radiation Accidents," Safety Ser. No. 21. IAEA, Vienna.

International Atomic Energy Agency. (1967d). "Radiation Protection Standards for Radioluminous Timepieces," Recommendations of the European Nuclear Energy Agency and the International Atomic Energy Agency, Safety Ser. No. 23. IAEA, Vienna.

International Atomic Energy Agency. (1969). Handling of radiation accidents. *Proc. Handling of Radiation Accidents, 1969.* IAEA, Vienna.

International Atomic Energy Agency. (1970a). "Power and Research Reactors in Member States." IAEA, Vienna.

International Atomic Energy Agency. (1970b). "Directory of Whole Body Radioactivity Monitors." IAEA, Vienna.

International Atomic Energy Agency. (1970c). "Regulations for the Safe Transport of Radioactive Materials," 3rd rev. draft. IAEA, Vienna.

International Atomic Energy Agency. (1971a). "Disposal of Radioactive Wastes into Rivers, Lakes and Estuaries," Safety Ser. No. 36. IAEA, Vienna.

International Atomic Energy Agency. (1971b). "Intercalibration of Analytical Methods on Marine Environmental Samples," Progr. Rep. No. 4. IAEA, Monaco.

International Atomic Energy Agency. (1971c). *Proc. Environ. Aspects Nucl. Power Sta., 1971.* IAEA, Vienna.

International Commission on Radiation Protection. (1960). Report of Committee II on permissible dose for internal radiation. *Health Phys.* **3**, 1.

International Commission on Radiological Protection. (1965). "Principles of Environmental Monitoring Related to the Handling of Radioactive Materials," ICRP Publ. No. 7.

International Commission on Radiological Protection. (1966). Report of Committee I on the evaluation of risks from radiation. *Health Phys.* **12**, 240.

International Commission on Radiological Protection. (1969). "Radiosensitivity and Spatial Distribution of Dose," ICRP Publ. No. 14. Pergamon, Oxford.

International Commission on Radiation Units and Measurements. (1971). "Radiation Quantities and Units," ICRU Rep. No. 19. ICRU Publ., Washington, D.C.

Ishimaru, T., Hoshino, T., Ichimaru, M., Okada, H., Tomiyasu, T., Tsuchimoto, T., and Yamamoto, T. (1971). Leukemia in atomic bomb survivors, Hiroshima and Nagasaki, 1 October 1950–30 September 1966. *Radiat. Res.* **45**, 216–233.

Jaakkola, T., Puumala, H., and Miettinen, J. (1967). Microelement levels in environmental samples in Finland. *Radioecol. Concent. Processes, Proc. Int. Symp., 1966* p. 341.

Jablon, S., and Kato, H. (1970). Childhood cancer in relation to prenatal exposure to atomic-bomb radiation. *Lancet* pp. 1000–1003.

Jablon, S., Belsky, J. L., Tachikawa, K., and Steer, A. (1971). Cancer in Japanese exposed as children to atomic bombs. *Lancet* pp. 927–932.

Jacobi, W. (1964). The dose to the human respiratory tract by inhalation of short-lived ^{222}Rn- and ^{220}Rn-decay products. *Health Phys.* **10**, 1163.

Jacobs, D. G. (1968). "Sources of Tritium and its Behavior upon Release to the Environment," TID-24365. USAEC, Washington, D.C.

Jacobs, D. G., Barton, C. J., Kelly, M. J., Bowman, C. R., Hanna, S. R., Gifford, F. A., Jr., and Culkowski, W. M. (1972). "Theoretical Evaluation of Consumer Products from Project Gasbuggy," Final Report. Phase II. Hypothetical Population Exposures Outside the San Juan Basin, ORNL-4748. Oak Ridge Nat. Lab., Oak Ridge, Tennessee.

Japan Society for the Promotion of Science. (1956). "Research in the Effects and Influences of the Nuclear Bomb Test Explosions." Jap. Soc. Promotion Sci., Tokyo.

Jaworowski, Z. (1967). "Stable and Radioactive Lead in Environment and Human Body." Nuclear Energy Information Center, Warsaw.

Jaworowski, Z., Bilkiewicz, J., and Zylica, E. (1971). Ra^{226} in contemporary and fossil snow. *Health Phys.* **20**, 449.

Jinks, S. M., and Eisenbud, M. (1972). Concentration factors in the aquatic environment. *Radiation Data Rep.* **13**, 243.

Joint Committee on Atomic Energy. (1959a). "Fallout from Nuclear Weapons Tests."

Joint Committee on Atomic Energy. (1959b). "Biological and Environmental Effects of Nuclear War."

Joint Committee on Atomic Energy. (1959c). "Report of Fallout Prediction Panel," Hearings on Fallout from Nuclear Weapons Tests.

Joint Committee on Atomic Energy. (1962). "Space Nuclear Power Applications." Hearings before The Sub-Committee on Research, Development and Radiation.

Joint Committee on Atomic Energy. (1963). "Fallout, Radiation Standards, and Countermeasures." Hearings (Part 1) before the Subcommittee on Research, Development, and Radiation.

Joint Committee on Atomic Energy. (1965). "Federal Radiation Council Protective Action Guides." Hearings before the Subcommittee on Research, Development, and Radiation.

Joint Committee on Atomic Energy. (1968). "Commercial Plowshare Services." Hearings before the Subcommittee on Legislation, Joint Committee on Atomic Energy, Congress of the United States.

Joint Committee on Atomic Energy. (1969a). "Radiation Standards for Uranium Mining." Hearings before the Subcommittee on Research, Development, and Radiation.

Joint Committee on Atomic Energy. (1969b). "Selected Materials on Environmental Effects of Producing Electric Power."

Joint Comittee on Atomic Energy. (1969c). "Hearings on Environmental Effects of Producing Electric Power," Part 1.

Joint Committee on Atomic Energy (1970a). "Hearings on Environmental Effects of Producing Electric Power," Part 2.

Joint Committee on Atomic Energy. (1970b). "Naval Nuclear Propulsion Program." Hearings of Testimony of Vice Adm. H. G. Rickover.

Joint Committee on Atomic Energy. (1970c). "Hearings on AEC Authorizing Legislation for F.Y. 71." US Govt. Printing Office, Washington, D.C.

Joint Committee on Atomic Energy. (1971). "Use of Uranium Mill Tailings for Construction Purposes." Hearings before the Subcommittee on Raw Materials of the Joint Committee on Atomic Energy, Congress of the United States.

Kahn, B., Blanchard, R. L., Kolde, H. E., Krieger, H. L., Gold, S., Brinck, W. L., Averett, W. J., Smith, D. B., and Martin, A. (1971a). "Radiological Surveillance Studies at a Pressurized Water Nuclear Power Reactor." U.S. Environ. Protection Agency.

Kahn, B., Blanchard, R. L., Krieger, H. L., Kolde, H. E., Smith, D. B., Martin, A., Gold, S., Averett, W. J., Brinck, W. L., and Karches, G. J. (1971b). Radiological surveillance studies at a boiling water nuclear power reactor. Proc. Environ. Aspects Nucl. Power Sta., 1971. IAEA, Vienna.

Kaplan, S. I. (1964). Behavior of iodine released in containment vessels. Nucl. Safety 5, 254.

Kaplan, S. I. (1971). HTGR safety. Nucl. Safety 12, 438–447.

Karas, J. S., and Stanbury, J. B. (1965). Fatal radiation syndrome from an accidental nuclear excursion. N. Engl. J. Med. 272, 755–761.

Kathren, R. L., Day, W. C., Denham, D. H., and Brown, J. L. (1964). Health physics following a nuclear excursion: The LRL incident of 26 March 1963. Health Phys. 10, No. 3. 183.

Kato, H. (1971). Mortality in children exposed to the A-bombs while in utero, 1945–1969. Amer. J. Epidemiol. 93, 435–442.

Kauranen, P., and Miettinen, J. K. (1969). "^{210}Po and ^{210}Pb in the arctic food chain and the natural radiation exposure of Lapps. Health Phys. 16, 287–296.

Kaye, S. V., and Rohwer, P. S. (1970). "Dose-Estimation Studies Related to Proposed Construction of an Atlantic-Pacific Interoceanic Canal with Nuclear Explosives: Phase III," ORNL-4579. Oak Ridge Nat. Lab., Oak Ridge, Tennessee.

Keely, R. B., and Wenstrand, T. K. (1971). Impact of fuel reprocessing on the health physics profession. Health Phys., 20, No. 2, 143.

Keilholtz, G. W. (1971). Krypton-xenon removal systems. Nucl. Safety 12, 591–599.

Keilholtz, G. W., and Barton, C. J. (1965). "Behavior of Iodine in Reactor Containment Systems," Rep. No. 15. Nucl. Safety Inform. Cent., Oak Ridge Nat. Lab., Oak Ridge, Tennessee.

Keilholtz, G. W., and Battle, G. C., Jr. (1969). "Fission Product Release and Transport In Liquid Metal Fast Breeder Reactors," Rep. No. 37. Nuclear Safety Inform. Cent., Oak Ridge Nat. Lab., Oak Ridge, Tennessee.

Kelley, T. F. (1965). Polonium-210 content of mainstream cigarette smoke. Science 149, 537.

Kennedy, G. C. (1964). A proposal for a nuclear power program. In "Engineering with Nuclear Explosives," TID-7695, p. 305. USAEC, Washington, D.C.

Kent, C. E., Levey, S., and Smith, J. M. (1971). Effluent control for boiling water reactors. *Proc. Environ. Aspects Nucl. Power Sta., 1971.* IAEA, Vienna.

King, L. J., and McCarley, W. T. (1961). "Plutonium Release Incident of November 20," Rep. ORNL-2989. Oak Ridge Nat. Lab., Oak Ridge, Tennessee.

Kirk, W. P. (1972). "Krypton-85: A Review of the Literature and an Analysis of Radiation Hazards." Environmental Protection Agency, Office of Research and Monitoring, Washington, D.C.

Kleinman, M. T. (1971). "The Stratospheric Inventory of Pu-238," Rep. HASL-245. USAEC, Washington, D.C.

Klement, A. W. (1959). "A Review of Potential Radionuclides Produced in Weapons Detonations," *Rep.* WASH-1024. USAEC, Washington, D.C.

Klement, A. W. (1965). Radioactive fallout phenomena and mechanisms. *Health Phys.* **11,** 1265.

Klement, A. W., Jr. (1969). Radiological safety research for nuclear excavation projects—interoceanic canal studies. *U.S., Pub. Health Serv., Rep.* SWRHL-82.

Klevin, P. B., Weinstein, M. S., and Harris, W. B. (1956). Groundlevel contamination from stack effluents. *Amer. Ind. Hyg. Ass., Quart.* **17,** 189.

Knapp, A. H., (1961). "The Effect of Deposition Rate and Cumulative Soil Level on the Concentration of Strontium-90 in U.S. Milk and Food Supplies," Rep. TID-13945. USAEC, Washington, D.C.

Knapp, H. A. (1963). "Iodine-131 in Fresh Milk and Human Thyroids Following A Single Deposition of Nuclear Test Fallout," TID-19266. USAEC, Washington, D.C.

Knox, J. B. (1969). Nuclear excavation: Theory and applications. *Nucl. Appl. Technol.* **7,** 189–231.

Kobayashi, R., and Nagai, I. (1956). "Cooperation by the United States in the Radiochemical Analyses." Jap. Soc. Promotion Sci., Tokyo.

Koczy, F. F. (1958). Natural radioactivity as a tracer in the ocean. *Proc. 2nd U.N. Int. Conf. Peaceful Uses At. Energy, 1958* p. 351. United Nations, New York.

Koczy, F. F. (1960). The distribution of elements in the sea. *Proc., Disposal of Radioactive Wastes, 1959.* IAEA, Vienna.

Kohman, T. (1959). *In* "Radiation Hygiene Handbook" (H. Blatz, ed.), Sect. 6. McGraw-Hill, New York.

Kohman, T., and Saito, N. (1954). Radioactivity in geology and cosmology. *Annu. Rev. Nucl. Sci.* **4.**

Kolehmainen, S., Hasanen, E., and Miettinen, J. (1968). [137]Cs in the plants, plankton, and fish of the Finnish lakes and factors affecting its accumulation. *Proc. 1st Int. Congr. Radiat. Protect., 1968* p. 407. Pergamon, Oxford.

Kolehmainen, S., Takatalo, S., and Miettinen, J. (1969). A tracer experiment with I-131 in an oligotrophic lake. *Radioecol., Proc. Nat. Symp., 2nd. 1969* USAEC CONF-670503, p. 278.

Korff, S. A. (1971). Personal communication.

Krey, P. W. (1967). Atmospheric burnup of a plutonium-238 generator. *Science* **158,** 769.

Krey, P. W., and Hardy, E. P. (1970). "Plutonium in Soil Around the Rocky Flats Plant," HASL-235. USAEC, Washington, D.C.

Krey, P. W., Kleinman, M. T., and Krajewski, B. T. (1970). "Sr-90, Zr-95 and Pu-238 Stratospheric Inventories 1967–1969," HASL-227, pp. 39–69. USAEC, Washington, D.C.

Kulp, J. L., and Schulert, A. R. (1962). Strontium-90 in man. V. *Science* **136,** 619.

Kumatori, T., Ishihara, T., Ueda, T., and Miyoshi, K. (1965). "Medical Survey of Japanese Exposed to Fall-out Radiation in 1954, A Report after 10 Years."

Kuroda, P. K., Hodges, H. L., Fry, L. M., and Moore, H. E. (1962). Stratospheric residence time of strontium-90. *Science* **137,** 15.

Lacy, W. J., and Stangler, M. J. (1962). The postnuclear attack water contamination problem. *Health Phys.* **8,** 423.

Lamarsh, J. R. (1966). "Nuclear Reactor Theory." Addison-Wesley, Reading, Massachusetts.

Lamont, L. (1965). "Day of Trinity." Atheneum, New York.

Langham, W. H., ed. (1967). "Radiobiological Factors in Manned Space Flight," p. 8. Nat. Acad. Sci.—Nat. Res. Counc., Washington, D.C.

Langham, W. H., and Anderson, E. C. (1958a). Cesium-137 biospheric contamination from nuclear weapons tests. *Proc. Conf. Passage Fission Prod. Food Chains, 1958.* Harwell, England.

Langham, W. H., and Anderson, E. C. (1958b). Entry of radioactive fallout into the biosphere and man. *Bull. Swiss Acad. Med. Sci.* **14,** 434.

Lapple, C. E., and Sheppard, C. B. (1940). Calculation of particle trajectory. *Ind. Eng. Chem.* **32,** 605.

Larson, C. E., Lincoln, T. A., and Bahler, K. W. (1966). "Mortality Comparison," Rep. No. K-A-708. Union Carbide Corp., Oak Ridge, Tennessee.

Laurer, G. R., and Eisenbud, M. (1963). Low-level in vivo measurement of iodine-131 in humans. *Health Phys.* **9,** 4.

Laurer, G. R., and Eisenbud, M. (1968). In vivo measurements of nuclides emitting soft penetrating radiations. *Proc. Symp. Diagnosis Treatment of Deposited Radionuclides, 1968* p. 189. Excerpta Med. Found., Amsterdam.

Lederer, C. M., Hollander, J. M., and Perlman, I. (1967). "Table of Isotopes," 6th ed. Wiley, New York.

Leipunsky, O. I. (1957). Radioactive hazards from clean hydrogen bomb and fission atomic bomb explosions. *At. Energ.* **3,** 530.

Lengemann, F. W. (1966). Predicting the total projected intake of radioiodine from milk by man. I. The situation where no counter measures are taken. *Health Phys.* **12,** 825.

Lentsch, J. W., Wrenn, M. E., Kneip, T., and Eisenbud, M. (1970). Manmade radionuclides in the Hudson River estuary. *Proc. 5th Annu. Midyear Top. Symp., 1970* p. 499. Health Physics Society, Idaho Falls, Idaho.

Lentsch, J. W., Kneip, T. J., Wrenn, M. E., Howells, G. P., and Eisenbud, M. (1972). Stable manganese and Mn-54 distributions in the physical and biological components of the Hudson River estuary. *Radioecol., Proc. Nat. Symp. 3rd, 1971* (in press). Oak Ridge, Tennessee.

Lewis, E. B. (1959). Statement. *In* "Fallout from Nuclear Weapons Tests." Hearings before Joint Committee on Atomic Energy.

Lewis, E. B. (1963). Leukemia, multiple myeloma and aplastic anemia in American radiologists. *Science* **142,** 1492–1494.

Lewis, W. L. (1955). "Arthritis and Radioactivity." Christopher Publ. House, Boston, Massachusetts.

Libby, W. F. (1952). "Radiocarbon Dating." Univ. of Chicago Press, Chicago, Illinois.

Libby, W. F. (1958). Paper presented before Swiss Academy of Medicine, Lausanne, Switzerland.

Libby, W. F. (1959). Remarks at University of Washington, Mar. 13, 1959. *In* "Fallout from Nuclear Weapons Tests." Hearings before Joint Committee on Atomic Energy.

Lindeken, C. L. (1971). Natural terrestrial background variations between residences. *16th Annu. Meet., Health Phys. Soc., New York, 1971.*

Linderoth, C. E., and Pearce, D. W. (1959). "Operating Practices and Experiences at Hanford," Rep. TID-7621. USAEC, Washington, D.C.

Little, Arthur D., Inc. (1970). "An Assessment of the Economic Effects of Radiation Exposure Standards for Uranium Miners," Gen. Serv. Admin. Contr. GS-03S-33584.

Little J. B., and Radford, E. P., Jr. (1967). Polonium-210 in bronchial epithelium of cigarette smokers. *Science* 155, 606.

Lockhart, L. B. (1958). Atmospheric radioactivity studies at U.S. Naval Research Laboratory. *U.S. Nav. Res. Lab., Rep.* 5249.

Lockhart, L. B., Jr. (1964). Radioactivity of the radon-222 and radon-220 series in the air at ground level. *In* "The Natural Radiation Environment" (J. A. S. Adams and W. M. Lowder, eds.). Univ. of Chicago Press, Chicago, Illinois.

Lockhart, L. B., *et al.* (1964). Characteristics of air filter media used for monitoring airborne radioactivity. *U.S. Nav. Res. Lab., Rep.* 6054.

Lodge, J. P., Bien, G. S., and Suess, H. E. (1960). The carbon-14 content of urban airborne particulate matter. *Int. J. Air Pollut.* 2, 309.

Logsdon, J. E., and Chissler, R. I. (1970). "Radioactive Waste Discharges to the Environment from Nuclear Power Facilities," BRH/DER 70-2. Bur. Radiol. Health, Environ. Radiat. Div.

Logsdon, J. E., and Hickey, J. W. N. (1971). Radioactive waste discharges to the environment from a nuclear fuel reprocessing plant. *Radiol. Health Data Rep.* pp. 305–312. June, 1971.

Look Magazine. (1960). A sequel to atomic tragedy. *Look Mag.* April 12, 1960.

Lorenz, E. (1944). Radioactivity and lung cancer; a critical review in miners of Schneeberg and Joachimstahl. *J. Nat. Cancer Inst.* 5, 1.

Lough, S. A., Hamada, G. H., and Comar, C. L. (1960). Secretion of dietary strontium-90 and calcium in human milk. *Proc. Soc. Exp. Biol. Med.* 104, 194.

Lowder, W. M., and Condon, W. J. (1965). Measurement of the exposure of human populations to environmental radiation. *Nature (London)* 206, 658–662.

Lowder, W. M., and Solon, L. R. (1956). "Background Radiation," Rep. NYO-4712. USAEC, Washington, D.C.

Lowman, F. G. (1960). Marine biological investigations at the Eniwetok test site. *Proc. Disposal Radioactive Wastes, 1960.* IAEA, Vienna.

Lowman, F. G., Rice, T. R., and Richards, F. A. (1971). Accumulation and redistribution of radionuclides by marine organisms. *In* "Radioactivity in the Marine Environment." Nat. Acad. of Sci., Washington, D.C.

Loysen, P., Breslin, A. J., and DiGiovanni, H. J. (1956). "Experimental Collection, Efficiency of a Stratospheric Air Sampler," Rep. NYO-4708. USAEC, Washington, D.C.

Lucas, H. F., Edgington, D. N., and Markun, F. (1970). Natural thorium in human bone. *Health Phys.* 19, 739.

Lundin, F. E., Jr., Lloyd, J. W. Smith, E. M., Archer, V. E., and Holaday, D. A. (1969). Mortality of uranium miners in relation to radiation exposure, hard-rock mining and cigarette smoking—1950 through September 1967. *Health Phys.* 16, No. 5, 571.

Lundin, F. E., Jr., Wagoner, J. K., and Archer, V. E. (1971). "Radon Daughter Exposure and Respiratory Cancer Quantitative and Temporal Aspects," Joint Monogr. No. 1. Nat. Inst. for Occupational Safety & Health—Nat. Inst. Environ. Health Sci., Public Health Service.

McCluggage, W. C. (1971). The AEC accident record and recent changes in AEC manual. Chapter 0529. Proc. Int. Symp. Packaging Transport Radioactive Mater., 3rd, 1971. USAEC Rep. BNWL-SA-3906.

McEachern, P., Myers, W. G., and White, F. A. (1971). Uranium concentrations in surface air at rural and urban localities within New York state. Environ. Sci. Technol. 5, 700–703.

McElroy, J. L., Blasewitz, A. G., and Schneider, K. (1971). Status of the waste solidification demonstration program. Nucl. Technol. 12, 69–81.

Machta, L. (1959). "Hearings on Biological and Environmental Effects of Nuclear War." Joint Committee on Atomic Energy.

Machta, L., and List, R. J. (1959). Analysis of stratospheric strontium-90 measurements. In "Fallout from Nuclear Weapons Tests," Hearings before Joint Committee on Atomic Energy.

MacMahon, B. (1962). Prenatal x-ray exposure and childhood cancer. J. Nat. Cancer Inst. 28, 1173.

Madey, R. (1967). Space radiation dosimetry. Health Phys. 13, 345.

Magno, P. J., Kauffman, P. E., and Schleien, B. (1967). Plutonium in environmental and biological media. Health Phys. 13, 1325–1330.

Magno, P. J., Reavey, T., and Apidianakis, J. C. (1970a). "Liquid Waste Effluents from a Nuclear Fuel Reprocessing Plant," Rep. BRH-NERHL 70-2. Northeast. Radiol. Health Lab., Bur. Radiol. Health.

Magno, P. J., Groulx, P. R., and Apidianakis, J. C. (1970b). Lead-210 in air and total diets in U.S. Health Phys. 18, 383.

Mangeno, J. J., and Miles, M. E. (1970). Disposal of radioactive wastes from U.S. naval nuclear powered ships and their support facilities, 1969. Radiol. Health Data Rep. pp. 373–377.

Marley, W. G., and Fry, T. M. (1956). Radiological hazards from an escape of fission products and the implications in power reactor location. Proc. 1st Int. Conf. Peaceful Uses At. Energy, 1955. United Nations, New York.

Martin, J. E., Harward, E. D., Oakley, D. T., Smith, J. M., and Bedrosian, P. H. (1971). "Radioactivity from Fossil Fuel and Nuclear Power Plants," Rep. SM-146/19. Environmental Aspects of Nuclear Power Stations. IAEA, Vienna.

Martland, H. S. (1951). "Collection of Reprints on Radium Poisoning, 1925–1939." U.S. At. Energy Comm., Oak Ridge, Tennessee.

Mason, B. (1960). "Principles of Geochemistry." Wiley, New York.

Mauchline, J., and Templeton, W. L. (1964). Artificial and natural radioisotopes in the marine environment. In "Annual Review of Oceanography & Marine Biology" (H. Barnes, ed.), Vol. 2, pp. 229–279. Allen & Unwin, London.

Mawson, C. A. (1965). "Management of Radioactive Wastes." Van Nostrand-Reinhold, Princeton, New Jersey.

Mayneord, W. V., Radley, J. M., and Turner, R. C. (1958). The alpha-ray activity of humans and their environment. Proc. 2nd Int. Conf. Peaceful Uses At. Energy, 1958. United Nations, New York.

Mayneord, W. V., Turner, R. C., and Radley, J. M. (1960). Alpha activity of certain botanical materials. Nature (London) 187, 208.

Mays, C. W., Jee, W. S. S., Lloyd, R. D., Stover, B. J., Dougherty, J. H., and Taylor,

G. N. (1969). "Delayed Effects of Bone-Seeking Radionuclides." Univ. of Utah Press, Salt Lake City.

Mays, C. W., and Lloyd, R. D. (1972a). Bone sarcoma risk from ^{90}Sr. *In* "Biomedical Implications of Radiostrontium Exposure" (L. K. Bustad and M. Goldman, eds.), Div. Tech. Inform., USAEC, Washington, D.C.

Mays, C. W., and Lloyd, R. D. (1972b). Bone sarcoma incidence vs. alpha particle dose. *In* "The Radiobiology of Plutonium" (B. Stover and W. S. S. Jee, eds.). J. W. Press, University of Utah.

Meade, P. J. (1960). Meteorological aspects of the peaceful uses of atomic energy. *World Meteorol. Org., Tech. Note* **33**, Part 1, WMO-No. 97 TP 41.

Medical Research Council. (1956). "The Hazards to Man of Nuclear and Allied Radiations." HM Stationery Office, London.

Menczer, L. F. (1965). Radioactive ceramic glazes. *Radiol. Health Data* pp. 656–659.

Meneeley, G. R., and Linde, S. M., eds. (1965). "Second Symposium, Whole Body Counting and Effects of Internal Gamma Ray Emitting Radioisotopes." Thomas, Springfield, Illinois.

Menzel, R. G. (1960). Radioisotopes in soils: Effects of amendments on availability. *In* "Radioisotopes in the Biosphere" (R. S. Caldecott and L. A. Snyder, eds.), Chapter 3. Univ. of Minnesota Press, Minneapolis.

Menzel, R. G. (1964). Competitive uptake by plants of potassium, rubidium, cesium, calcium, strontium and barium from soils. *Soil Sci.* **77**, 419.

Menzel, R. G. (1965). Soil-plant relationships of radioactive elements. *Health Phys.* **11**, 1325.

Merlini, M., Girardi, F., and Pozzi, G. (1967). Activation analysis in studies of an aquatic ecosystem. *Proc. Nucl. Activation Techniques Life Sci., 1967.* IAEA, Vienna.

Merriam, G. R., and Focht, E. F. (1957). A clinical study of radiation cataracts and the relationship to dose. *Amer. J. Roentgenol., Radium Ther., Nucl. Med.* **77**, 759.

Meyer, W. (1971). An argument for a recoverable high-level waste container. *Nucl. News,* April, pp. 38–40.

Miettinen, J. K. (1969). Enrichment of radioactivity by arctic ecosystems in Finnish Lapland. *Radioecol., Proc. Nat. Symp. 2nd, 1969.* U.S. AEC CONF-670503.

Miller, E. C. (1966). "The Integrity of Reactor Pressure Vessels," Rep. No. 15. Nucl. Safety Inform., Oak Ridge Nat. Lab., Oak Ridge, Tennessee.

Miller, R. W. (1969). Delayed radiation effects in atomic-bomb survivors. *Science* **166**, 569–573.

Minx, R. P., Shleien, B., Klement, A. W., Jr., and Miller, C. R. (1971). "Estimates of Ionizing Radiation Doses in the United States, 1960–2000." Special Studies Group, Div. of Criteria & Standards, Office of Radiation Programs, Environmental Protection Agency.

Miskel, J. A. (1964). Characteristics of radioactivity produced by nuclear explosives. *Proc. 3rd Plowshare Symp. Eng. Nucl. Explosives, 1964.*

Mistry, K. B., Bharathan, K. G., and Gopal-Ayengar, A. R. (1970). Radioactivity in the diet of population of the Kerala coast including monazite bearing high radiation areas. *Health Phys.* **19**, 535.

Miyake, Y., and Saruhashi, K. (1960). Vertical and horizontal mixing rates of radioactive material in the ocean. *Proc. Disposal of Radioactive Wastes, 1959.* IAEA, Vienna.

Mochizuki, Y., Mowafy, R., and Pasternack, B. (1963). Weights of human thyroids in New York City. *Health Phys.* **9**, 219.

Moghissi, A. A., and Lieberman, R. (1970). Tritium body burden of children, 1967–1968. *Radiol. Health Data Rep.* pp. 227–231.

Mole, R. H. (1957). Shortening of life by chronic irradiation—the experimental facts. *Nature (London)* **180**, 456.

Morgan, J. M., Jr., Knapp, J. W., and Thompson, J. T. (1961). "A Study of the Possible Consequences and Costs of Accidents in the Transportation of High-Level Radioactive Materials," Rep. NYO-9772. USAEC, Washington, D.C.

Morgan, K. Z., and Turner, J. E. (1967). "Principles of Radiation Protection." Wiley, New York.

Morgan, K. Z., Snyder, W. S., and Ford, M. R. (1956). Maximum permissible concentration of radioisotopes in air and water for short period exposure. *Proc. 1st Int. Conf. Peaceful Uses At. Energy, 1955.* United Nations, New York.

Morris, P. L., and Klinsky, J. (1962). "Radiological Aspects of Water Supplies." State Hyg. Lab., University of Iowa, Iowa City.

Morse, J. G. (1963). Energy for remote areas. *Science* **139**, 1175.

Morse, R. S., and Welford, G. A. (1971). Dietary intake of ^{210}Pb. *Health Phys.* **21**, 53.

Mullins, L. J., and Leary, J. A. (1969–1970). ^{238}Pu for biomedical applications. *Isotop. Radiat. Technol.* **7**, 197–207.

Muth, H., Schraub, A., Aurand, K., and Hantke, H. H. (1957). Measurements of normal radium burdens. *Brit. J. Radiol., Suppl.* **7,**

Myrloi, M. G., and Wilson, J. G. (1951). On the proton component of the vertical cosmic-ray beam at sea level. *Proc. Phys. Soc., London, Sect. A* **64**, 404.

National Academy of Engineering. (1972). "Engineering for the Resolution of the Energy-Environment Dilemma." Committee on Power Plant Siting, National Academy of Sciences, Washington, D.C.

National Academy of Sciences—National Research Council. (1956). Pathologic effects of atomic radiation. *Nat. Acad. Sci.—Nat. Res. Counc., Publ.* **452.**

National Academy of Sciences—National Research Council. (1957a). Disposal of radioactive wastes on land. *Nat. Acad. Sci.—Nat. Res. Counc., Publ.* **519.**

National Academy of Sciences—National Research Council. (1957b). The effects of atomic radiation on oceanograph 7 and fisheries. *Nat. Acad. Sci.—Nat. Res. Counc., Publ.* **551.**

National Academy of Sciences—National Research Council. (1959). Radioactive waste disposal into Atlantic and Gulf coastal waters. *Nat. Acad. Sci.—Nat. Res. Counc., Publ.* **655.**

National Academy of Sciences—National Research Council. (1961a). Effects of inhaled radioactive particles. *Nat. Acad. Sci.—Nat. Res. Counc., Publ.* **848.**

National Academy of Sciences—National Research Council. (1961b). Long-term effects of ionizing radiations from external sources. *Nat. Acad. Sci.—Nat. Res. Counc., Publ.* **849.**

National Academy of Sciences—National Research Council. (1961c). Effects of ionizing radiation on the human hemapoietic system. *Nat. Acad. Sci.—Nat. Res. Counc., Publ.* **875.**

National Academy of Sciences—National Research Council. (1961d). Internal emitters. *Nat. Acad. Sci.—Nat. Res. Counc., Publ.* **833.**

National Academy of Sciences—National Research Council. (1962). Disposal of low-level radioactive waste into Pacific coastal waters. *Nat. Acad. Sci.—Nat. Res. Counc., Publ.* **985.**

National Academy of Sciences—National Research Council. (1963). Damage to livestock in event of nuclear war. *Nat. Acad. Sci.—Nat. Res. Counc., Publ.* **1078**.

National Academy of Sciences—National Research Council. (1964). Civil defense, project harbor summary report. *Nat. Acad. Sci.—Nat. Res. Counc., Publ.* **1237**.

National Academy of Sciences—National Research Council. (1969). "Civil Defense," Little Harbor Report. A Report to the Atomic Energy Commission by a Committee of the National Academy of Sciences, Washington, D.C.

National Academy of Sciences—National Research Council. (1970). "Disposal of Solid Radioactive Wastes in Bedded Salt Deposits." *Nat. Acad. Sci.—Nat. Res. Counc.*, Washington, D.C.

National Committee on Radiation Protection & Measurements. (1941). Safe handling of radioactive luminous compounds. NCRP Report No. 5. *Nat. Bur. Stand.* (*U. S.*), *Handb.* **27**.

National Committee on Radiation Protection & Measurements. (1957). Permissible dose from external sources of ionizing radiation. *Nat. Bur. Stand.* (*U. S.*), *Handb.* **59**, 1954; Addendum (1957).

National Committee on Radiation Protection & Measurements. (1959). Maximum permissible body burdens and maximum permissible concentrations of radionuclides in air and in water for occupational exposure. *Nat. Bur. Stand.* (*U. S.*), *Handb.* **69**.

National Council on Radiation Protection & Measurements. (1971). "Basic Radiation Protection Criteria," NCRP Rep. No. 39.

National Safety Council. (1971). "Accident Facts." Chicago, Illinois.

Nefzger, M. D., Miller, R. J., and Fujino, T. (1968). Eye findings in atomic bomb survivors of Hiroshima, and Nagasaki: 1963–1964. *Amer. J. Epidemiol.* **89**, 129.

Neill, R. H., Youmans, H. D., and Wyatt, J. L. (1971). Estimates of potential doses to various organs from X-radiation emissions from color television picture tubes. *Radiol. Health Data Rep.* **12**, 1–6.

Nelson, D. J. (1967). The prediction of ^{90}Sr uptake in fish using data on specific activities and biological half-lives. *Radioecol. Concentr. Processes, Proc. Int. Symp., 1966* p. 843. Pergamon, Oxford.

Nelson, D. J. (1969). Cesium, cesium-137, and potassium concentrations in the White crappie and some other clinch river fish. *Radioecol., Proc. Nat. Symp., 2nd, 1969.* U.S. AEC CONF-670503, p. 240.

New York State. (1959). "Protection from Radioactive Fallout," Special Task Force Report to Gov. Nelson A. Rockefeller.

Nichols, J. P., and Binford, F. T. (1971). "Status of Noble Gas Removal and Disposal," ORNL-TM-3515. Oak Ridge Nat. Lab., Oak Ridge, Tennessee.

Nishita, H., Romney, E. M., and Larson, K. H. (1961). Uptake of radioactive fission products by crop plants. *J. Agr. Food Chem.* **9**, 101.

Nordyke, M. D. (1969). Underground engineering applications. *Proc. Public Health Aspects of Peaceful Uses of Nuclear Explosives, 1969.* Pub. Health Serv. Rep. SWRHL-82.

Norton, H. T. (1957). "The Turbulent Diffusion of River Contaminants," *Rep.* HW-49195. USAEC, Washington, D.C.

Nuclear Safety Information Center. (1968). "Review of Methods of Mitigating Spread of Radioactivity from a Failed Containment System," NSIC Rep. No. 27. Oak Ridge Nat. Lab., Oak Ridge, Tennessee.

Nydal, R. (1968). Further investigation in the transfer of radiocarbon in nature. *J. Geophys. Res.* **73**, 3617–3635.

Oak Ridge National Laboratory. (1970). "Siting of Fuel Reprocessing Plants and Waste Management Facilities," Rep. ORNL-4451. USAEC, Washington, D.C.

Odum, E. P. (1963). "Ecology." Saunders, Philadelphia, Pennsylvania.

Okrent, D. (1965). A look at fast reactor safety. *Nucl. Safety* **6**, 317.

Ophel, I. L. (1963). The fate of radiostrontium in a freshwater community. *Radioecol., Proc. Nat. Symp., 1st, 1961* p. 213.

Ophel, I. L., and Judd, J. M. (1966). Effects of internally deposited radionuclides on the thermal tolerance of fish. *Proc. Disposal of Radioactive Wastes into Seas, Oceans and Surface Waters*, p. 825. IAEA, Vienna.

Ophel, I. L., and Judd, J. M. (1969). Strontium-Calcium relationships in aquatic food Chains. *Radioecol., Proc. Nat. Symp., 2nd, 1969.* U.S. AEC CONF-670503, p. 221.

Palmer, H. E., and Beasley, T. M. (1965). Iron-55 in humans and their foods. *Science* **149**, 431.

Parker, F. L., Schmidt, G. D., Cottrell, W. B., and Mann, L. A. (1961). Dispersion of radiocontaminants in an estuary. *Health Phys.* **6**, 66.

Parker, F. L., Churchill, M. A., Andrew, R. W., Frederick, B. J., Carrigan, P. H., Jr., Cragwall, J. S., Jr., Jones, S. L., Struxness, E. G., and Morton, R. J. (1966). Dilution, dispersion and mass transport of radionuclides in the Clinch and Tennessee Rivers. *Proc. Disposal of Radioactive Wastes into Seas, Oceans and Surface Waters, 1966* p. 33. IAEA, Vienna.

Parker, G. W., and Creek, G. E. (1958). Experiments on the release of fission products from molten reactor fuels. *Proc. 2nd U.N. Int. Conf. Peaceful Uses At. Energy, 1958* p. 1074. United Nations, New York.

Parker, H. M. (1969). The dilemma of lung dosimetry. *Health Phys.* **16**, No. 5, 553.

Parker, H. M., and Healy, J. W. (1956). Effects of an explosion of a nuclear reactor. *Proc. 1st Int. Conf. Peaceful Uses At. Energy, 1955* p. 482. United Nations, New York.

Parsly, L. F. (1971). "Removal lf Elemental Iodine from Reactor Containment Atmospheres by Spraying," Rep. ORNL-4623. Oak Ridge Nat. Lab., Oak Ridge, Tennessee.

Pasquill, F. (1961). The estimation of the dispersion of windborne material. *Meteorol. Mag.* **90**, 33.

Pasquill, F. (1962). "Atmospheric Diffusion." Van Nostrand-Reinhold, Princeton, New Jersey.

Patt, H. M., and Brues, A. M. (1954). The pathological physiology of radiation injury in the mammal. *In* "Radiation Biology" (A. Hollaender, ed.), Chapter 14. McGraw-Hill, New York.

Patterson, H. W., and Thomas, R. H. (1971). "Radiation and Risk—The Source Data," Rep. LBL-331. Lawrence Berkeley Lab., Univ. of California, Berkeley.

Pavlovskaya, N. A. (1960). Natural content and distribution of thorium in the human organism. *Med. Radiol.* **11**, 28.

Pearce, D. W., Linderoth, C. E., Nelson, J. L., and Ames, L. L. (1960). A review of radioactive waste disposal to the ground at Hanford. *Proc. Disposal of Radioactive Wastes, 1959.* IAEA, Vienna.

Pearson, J. E., and Jones, G. E. (1965). Emanation of radon-222 from soils and its use as a tracer. *J. Geophys. Res.* **70**, 5279.

Pendleton, R. C. (1962). Accumulation of Cs-137 through the aquatic food web. 3rd seminar, biological problems in water pollution. *U.S., Pub. Health Serv., Publ.* 999-WP-25.

Pendleton, R. C., and Hanson, W. C. (1958). Absorption of Cs-137 by components of an aquatic community. *Proc. 2nd U.N. Int. Conf. Peaceful Uses At. Energy, 1958.* United Nations, New York.

Pendleton, R. C., Lloyd, R. D., and Mays, C. W. (1963). Iodine-131 in Utah during July and August, 1962. *Science* **141,** 640.

Pendleton, R. C., Mays, C. W., Lloyd, R. D., and Brooks, A. L. (1964). Differential accumulation of ¹³¹I from local fallout in people and milk. *Proc. Hanford Symp. Biol. Radioiodine, 1964* p. 73. Pergamon, Oxford.

Penelle, G. (1968). "Description et analyse de l'accident de criticite survenu au reacteur Venus a Mol en date du 30 Decembre 1965." *Proc. Int. Congr. Radiat. Protect., 1st, 1968* p. 1223. Pergamon, Oxford.

Penna Franca, E., Fiszman, M., Lobao, N., Costa Ribeiro, C., Trindale, H., Dos Santos, P. L., and Batista, D. (1968). Radioactivity of Brazil nuts. *Health Phys.* **14,** 95–99.

Penna Franca, E., Fiszman, M., Lobao, N., Trindade, H., Costa Ribeiro, C., and Santos, P. L. (1970). Radioactivity in the diet in high background areas of Brazil. *Health Phys.* **19,** No. 5, 657.

Perkins, R. W., and Nielsen, J. M. (1965). Cosmic-ray produced radionuclides in the environment. *Health Phys.* **11,** 1297–1304.

Pertsov, L. A. (1964). "The Natural Radioactivity of the Biosphere." Atomizdat, Moscow. (Translated by Israel Program for Scientific Translations, Jerusalem, 1967.)

Petrow, H. G., and Strehlow, C. D. (1967). Spectrophotometric determination of thorium in bone ash using arsenazo. III. *Anal. Chem.* **39,** 265.

Petterssen, S. (1958). "Introduction to Meteorology," 2nd ed. McGraw-Hill, New York.

Pettigrew, G. L., Robinson, E. W., Schmidt, G. D. (1971). "State and Federal Control of Health Hazards from Radioactive Materials Other Than Materials Regulated Under the Atomic Energy Act of 1954." Food & Drug Admin.

Petukhova, E. V., and Knizhnikov, V. A. (1900). "Dietary Intake of Sr-90 and Cs-137," Publ. A/AC.82/G/L.1245. United Nations Scientific Committee, Sales Section, New York.

Pickering, R. J., Carrigan, P. H., Jr., Tamura, T., Abee, H. H., Beverage, J. W., and Andrew, R. W., Jr. (1966). Radioactivity in bottom sediment of the Clinch and Tennessee Rivers. *Proc. Disposal of Radioactive Wastes into Seas, Oceans and Surface Waters, 1966* p. 57. IAEA, Vienna.

Pifer, J. W., Toyooka, E. T., Murray, R. W., Ames, W. R., and Hempelmann, L. H. (1963). Neoplasms in children treated with x-rays for thymic enlargement. I. Neoplasms and mortality. *J. Nat. Cancer Inst.* **31,** 1333–1356.

Pneumo Dynamics Corporation. (1961). "Sea Disposal Container Test and Evaluation," Rep. TID-13226. USAEC, Washington, D.C.

Pohl-Ruling, J., and Scheminzky, F. (1954). Das Konzentrationsverhaltnis Blut/Luft bei der Radon-inhalation und die Radon-aufnahme in den Menschliehen Korper im Radioaktiven Thermalstollen von Badgastein/Bockstein. *Strahlentherapie* **95,** 267.

Polikarpov, G. G. (1966). "Radioecology of Aquatic Organisms." North-Holland Publ., Amsterdam (distributed by Van Nostrand-Reinhold, Princeton, New Jersey).

Polikarpov, G. G. (1967). General features of the concentration processes of radioactive substances by hydrobionts in different seas of the world ocean. *Radioecol. Concent. Processes, Proc. Int. Symp., 1966* p. 819.

Porter, C. R., Phillips, C. R., Carter, M. W., and Kahn, B. (1967). The cause of relatively high Cs-137 concentrations in Tampa, Florida, milk. *Radioecol. Concent. Processes, Proc. Int. Symp., 1966* p. 95. Pergamon, Oxford.

Preston, A., and Jeffries, D. F. (1969). The I.C.R.P. critical group concept in relation to the Windscale sea discharges. *Health Phys.* **16,** 33.

Pritchard, D. W. (1958). Factors affecting the disposal of fission products in estuarine and inshore environments. *Proc. 2nd U.N. Int. Conf. Peaceful Uses At. Energy, 1958.* United Nations, New York.

Pritchard, D. W. (1960). The application of existing oceanographic knowledge to the problem of radioactive waste disposal into the sea. *Proc. Disposal of Radioactive Wastes, 1959.* IAEA, Vienna.

Pritchard, D. W., Reid, R. O., Okubo, A., and Carter, H. H. (1971). Physical processes of water movement and mixing. *In* "Radioactivity in the Marine Environment." Nat. Acad. of Sci., Washington, D.C.

Prospero, J. M., and Carlson, T. N. (1970). Radon-222 in the N. Atlantic trade winds: Its relationship to dust transport from Africa. *Science* **167,** 974.

Prosser, D. L. (1970). SNAP-27 on the moon. *Isotop. Radiat. Technol.* **7,** 443.

Raabe, O. G. (1970). Estimation of the relative inhalation hazard of reactor inventory radionuclides. *Proc. Symp. Health Phys. Aspects of Nucl. Facility Siting, 1970* Vol. III, p. 619. Health Phys. Soc., New York.

Radford, E. P., Jr., and Hunt, V. R. (1964). Polonium-210: A volatile radioelement in cigarettes. *Science* **143,** 247.

Rajewsky, B., and Stahlhofen, W. (1966). ^{210}Po activity in the lungs of cigarette smokers. *Nature (London)* **209,** 1312–1313.

Ramsden, D. (1969). The measurement of plutonium-239 *in vivo. Health Phys.* **16,** 145–154.

Rand Corporation. (1953). "Worldwide Effects of Atomic Weapons," Rep. R-251-AEC. Santa Monica, California.

Rankama, K. (1954). "Isotope Geology." McGraw-Hill, New York.

Rankama, K. (1963). "Progress in Isotope Geology." Wiley, New York.

Rankama, K., and Sahama, T. G. (1950). "Geochemistry." Univ. of Chicago Press, Chicago, Illinois.

Reed, J. R., and Nelson, D. J. (1969). Radiostrontium uptake in blood and flesh in bluegills. *Radioecol. Proc. Nat. Symp., 2nd, 1969.* U.S. AEC CONF-670503, p. 226.

Reinig, W. C., ed. (1970). "Environmental Surveillance in the Vicinity of Nuclear Facilities." Thomas, Springfield, Illinois.

Revelle, R., and Schaefer, M. B. (1957). General considerations concerning the ocean as a receptacle for artificially radioactive materials. *Nat. Acad. Sci.—Nat. Res. Counc., Publ.* **551.**

Revelle, R., Folsom, T. R., Goldberg, E. D., and Isaacs, J. D. (1956). Nuclear science and oceanography. *Proc. 1st Int. Conf. Peaceful Uses At. Energy, 1955* p. 277. United Nations, New York.

Ritchie, J. C., Clebsch, E. E. C., and Rudolph, W. K. (1970). Distribution of fallout and natural gamma radionuclides in litter, humus and surface mineral soil layers under natural vegetation in the Great Smoky Mountains, North Carolina-Tennessee. *Health Phys.* **18,** 479–489.

Rivera, J. (1965). Radiation to bone from ^{90}Sr in New York City residents. *AEC Symp. Ser.* **5,** 737–742.

Roberts, H., Jr., and Menzel, R. G. (1961). Availability of exchangeable and non-exchangeable strontium-90 to plants. *J. Agr. Food Chem.* **9,** 95.

Robinson, E. W. (1968). The use of radium in consumer products. *U.S., Pub. Health Serv. Rep.* MORP 68-5.

Rogers, F. C. (1971). Underground nuclear power plants: Environmental and economic aspects. *Nucl. News* p. 36.

Roser, F. X., and Cullen, T. L. (1964). External radiation measurements in high background regions of Brazil. *In* "The Natural Radiation Environment" (J. A. S. Adams and W. M. Lowder, eds.), p. 825. Univ. of Chicago Press, Chicago, Illinois.

Roser, F. X., Kegel, G., and Cullen, T. L. (1964). Radiogeology of some high-background areas of Brazil. *In* "The Natural Radiation Environment" (J. A. S. Adams and W. M. Lowder, eds.), p. 855. Univ. of Chicago Press, Chicago, Illinois.

Runion, T. C. (1970). "Testimony before Joint Committee on Atomic Energy," Part 2, Vol. 1, p. 1704. Hearings on Environmental Effects of Producing Electric Power.

Russell, J. L., and Hahn, P. B. (1971). Public health aspects of iodine-129 from the nuclear power industry. *Radiol. Health Data Rep.* pp. 189–194.

Russell, R. S. (1960). The passage of fission products through food chains. *In* "Radio-isotopes in the Biosphere," Chapter 19. Univ. of Minnesota Press, Minneapolis.

Russell, R. S. (1965). Interception and retention of airborne material on plants. *Health Phys.* **11,** 1305–1315.

Russell, R. S. (1966). "Radioactivity and Human Diet." Pergamon, Oxford.

Russell, R. S., and Bruce, R. S. (1969). Environmental contamination with fallout from nuclear weapons. *Proc. Environ. Contam. by Radioactive Mater., 1969.* IAEA, Vienna.

Russell, W. L. (1965). Studies in mammalian radiation genetics. *Nucleonics* **23,** 1.

Russell, W. L. (1968). Recent studies on the genetic effects of radiation in mice. *Pediatrics* **41,** 223–230.

Saccomanno, G., Archer, V. E., Saunders, R. F., James, L. A., and Beckler, P. A. (1964). Lung cancer of uranium miners on the Colorado Plateau. *Health Phys.* **10,** 1195–1202.

Saenger, E. L., ed. (1963). "Medical Aspects of Radiation Accidents." USAEC, Washington, D.C.

Saenger, E. L., Thoma, G. E., and Tompkins, E. (1968). Incidence of leukemia following treatment of hyperthyroidism. *J. Amer. Med. Ass.* **205,** 147.

Salo, E. O., and Leet, W. L. (1969). The concentration of ^{65}Zn by oysters maintained in the discharge canal of a nuclear power plant. *Radioecol. Proc. Nat. Symp., 2nd, 1969.* U.S. AEC CONF-670503.

Samuels, L. D. (1964). A study of environmental exposure to radium in drinking water. *In* "The Natural Radiation Environment" (J. A. S. Adams and W. M. Lowder, eds.), p. 239. Univ. of Chicago Press, Chicago, Illinois.

Schaefer, H. J. (1971). Radiation exposure in air travel. *Science* **173,** 780–783.

Scheminzky, F. (1961). 25 Jahre Baderforschung in Gastein mit Verzeichnis der Wissenschaftlichen Veroffenlichungen bis 1960. *Sonderabdruck Badgasteiner Bladeblatt* Nos. 35 and 36.

Schneider, K. J., ed. (1969). "Waste Solidification Program," Vol. 1, Technology for Pot, Spray, and Phosphate Glass Solidification Processes. Rep. BNWL-1073. Battelle-Northwest, Richland, Washington.

Schneider, K. J. (1970). Solidification of radioactive wastes. *Chem. Eng. Progr.* **66,** No. 2.

Schneider, K. J., Bradshaw, R. L., Blasewitz, A. G., Blomeke, J. O., and McClain, W. C.

(1971). Status of solidification and disposal of highly radioactive liquid wastes from nuclear power in the U.S.A. *Proc. Environ. Aspects Nucl. Power Sta., 1971* p. 369. IAEA, Vienna.

Schrenk, H. H. *et al.* (1949). Air pollution in Donora, Pennsylvania. *U.S., Pub. Health Serv., Bull.* **306.**

Schuffelen, A. C. (1961). "On the Radioactive Contamination of Soils and Crops," Seminar on the Agricultural and Public Health Aspects of Radioactive Contamination. FAO—IAEA, World Health Organ., Geneva.

Schulz, R. K. (1965). Soil chemistry of radionuclides. *Health Phys.* **11,** 1317.

Scott, N. S. (1897). X-ray injuries. *Amer. X-Ray J.* **1,** 57.

Scott, R. L., Jr. (1971). Fuel-melting incident at the Fermi reactor on Oct. 5, 1966. *Nucl. Safety* **12,** 123.

Seaborg, G. T. (1958). "The Transuranic Elements." Addison-Wesley, Reading, Massachusetts.

Seaborg, G. T., and Bloom, J. (1970). Fast breeder reactors. *Sci. Amer.* **223,** 13.

Seelentag, W., and Schmier, H. (1963). Radiation exposure from luminous watch dials. *Radiol. Health Data* **4,** 209–213.

Seymour, A. M. (1966). Accumulation and loss of ^{65}Zn by oysters in a natural environment. *Proc. Disposal of Radioactive Wastes into Seas, Oceans & Surface Waters, 1966* p. 605. IAEA, Vienna.

Shafer, C. K. (1959). "Testimony before Joint Committee on Atomic Energy." Hearings on Biological and Environmental Effects of Nuclear War.

Shapiro, J. (1956). Radiation dosage from breathing radon and its daughter products. *AMA Arch. Ind. Health* **14,** 169.

Shapley, D. (1971). Rocky flats: Credibility gap widens on plutonium plant safety. *Science* **174,** 569–571.

Shearer, S. D., Jr., and Lee, G. F. (1964). Leachability of radium-226 from uranium mill solids and river sediments. *Health Phys.* **10,** 217.

Shearer, S. D., Jr., and Sill, C. W. (1969). Evaluation of atmospheric radon in the vicinity of uranium mill tailings. *Health Phys.* **17,** 77.

Shelton, F. H. (1959). Statement. *In* "Fallout from Nuclear Weapons Tests." Hearings before the Joint Committee on Atomic Energy.

Sherman, J. T. (1971). Uranium 1970. *Eng. Mining J.* **172,** 108–111.

Shipman, T. L., ed. (1961). Acute radiation death resulting from an accidental nuclear critical excursion. *J. Occup. Med.* **3,** No. 3, Spec. Suppl.

Shleien, B. (1969). Evaluation of radium-226 in total diet samples, 1964 to June, 1967. *Radiol. Health Data Rep.*

Shleien, B. (1970a). An evaluation of internal radiation exposure based on dose commitments from radionuclides in milk, food and air. *Health Phys.* **18,** 267–275.

Shleien, B. (1970b). "An Estimate of Radiation Doses Received by Individual Living in the Vicinity of a Nuclear Fuel Reprocessing Plant in 1968," BRH/NERHL 70-1. Dept. of Health, Education and Welfare.

Shleien, B., Glavin, T. P., and Friend, A. G. (1965). Particle size fractionation of airborne gamma-emitting radionuclides by graded filters. *Science* **147,** 290–292.

Shleien, B., Cochran, J. A., and Magno, P. J. (1970). Strontium-90, strontium-89, plutonium-239, and plutonium-238 concentrations in ground-level air, 1964–1969. *Environ. Sci. Technol.* **4,** 598–602.

Shor, R., Lafferty, R. H., and Baker, P. S. (1971). ^{90}Sr heat sources. *Isotop. Radiat. Technol.* **8,** 260.

Shuping, R. E., Phillips, C. R., and Moghissi, A. A. (1970). Krypton-85 levels in the environment determined from dated krypton gas samples. *Radiol. Health Data Rep.* **11**, 671–672.

Slade, D. H., ed. (1968). "Meteorology and Atomic Energy—1968," Rep. TID-24190. USAEC, Washington, D.C.

Slansky, C. M. (1971). Separation processes for noble gas fission products from the offgas of fuel reprocessing plants. *At. Energy Rev.* **9**, 423.

Smith, F. A., and Dzuiba, S. P. (1949). "Preliminary Observations of the Uranium Content of Photographic Materials." University of Rochester, Rochester, New York (unpublished memo.).

Smith, M. E. (1951). Meteorological factors in air pollution problems. *Amer. Ind. Hyg. Ass. Quart.* **12**, 151.

Smith, S. (1971). The need for plowshare gas. *Nucl. Technol.* **11**, 331–334.

Snelling, R. N. (1969). Environmental survey of uranium mill tailings pile, Tuba City, Ariz. *Radiol. Health Data Rep.* **10**, 475.

Snelling, R. N. (1971). Environmental survey of uranium mill tailings pile, Mexican Hat, Utah. *Radiol. Health Data Rep.* **12**, 17–28.

Soldano, B. A., and Ward, W. T. (1971). The utility of ice cubes as an absorbent for gaseous fission products. *Nucl. Technol.* **12**, 363–366.

Soldat, J. K. (1963). The relationship between I-131 concentrations in various environmental samples. *Health Phys.* **9**, 1167.

Soldat, J. K. (1965). Environmental evaluation of an acute release of ^{131}I to the atmosphere. *Health Phys.* **11**, 1009.

Solon, L. R. *et al.* (1958). External environmental radiation measurements in the United States. *Proc. 2nd U.N. Int. Conf. Peaceful Uses At. Energy, 1958* p. 740. United Nations, New York.

Spalding, R. F., and Sackett, W. (1972). Uranium in runoff from the Gulf of Mexico distributive province: Anomalous concentrations. *Science* **175**, 629.

Spiers, F. W. (1966). Dose to bone from strontium-90: Implications for the setting of maximum permissible body burden. *Radiat. Res.* **28**, 624–642.

Spiers, F. W. (1968). "Radioisotopes in the Human Body: Physical and Biological Aspects." Academic Press, New York.

Spiers, F. W., and Griffith, H. D. (1956). Measurements of local gamma-ray background in Leeds and Aberdeen. *Brit. J. Ratiol.* **29**, 175–176.

Spiess, H., and Mays, C. W. (1970). Bone cancers induced by ^{224}Ra (Th X) in children and adults. *Health Phys.* **19**, 713–729.

Spitsyn, V. I. *et al.* (1958). A study of the migration of radioelements in soils. *Proc. 2nd U.N. Int. Conf. Peaceful Uses At. Energy, 1958* p. 2207. United Nations, New York.

Spitsyn, V. I. *et al.* (1960). Sorption regularities in behavior of fission product elements during filtration of their solutions through ground. *Proc. Disposal of Radioactive Wastes, 1959.* IAEA, Vienna.

Starr, C. (1969). Social benefit versus technological risk. *Science* **165**, 1232.

Starr, C. (1971). Benefit-cost relationships in socio-technical systems. *Proc. Environ. Aspects Nucl. Power Sta., 1971.* IAEA, Vienna.

Stebbins, A. K. (1961). "Second Special Report on High Altitude Sampling Program," Defense At. Support Agency Rep. 539B.

Stebbins, A. K., and Minx, R. F. (1962). "The High Altitude Sampling Program."

Testimony before Joint Committee on Atomic Energy, Hearings on Radioactive Fallout.

Stehney, A. F. (1956). Studies of the radium content of humans arising from the natural radium of their environment. *Proc. 1st Int. Conf. Peaceful Uses At. Energy, 1955*. United Nations, New York.

Stein, J. L., Jaquish, R. E., and Sharpe, T. J. (1971). Evaluation of the sampling frequency of the pasteurized milk network. *Radiol. Health Data Rep.* **12**, 451–455.

Steinberg, E., and Glendenin, L. (1956). *Proc. 1st Int. Conf. Peaceful Uses At. Energy, 1955* p. 614. United Nations, New York.

Stevens, D. L., Jr. (1963). "A Brief History of Radium." Bur. Radiol. Health, U.S. Public Health Service.

Stewart, A., and Kneale, G. W. (1970). Radiation dose effects in relation to obstetric x-rays and childhood cancers. *Lancet* p. 1185.

Stewart, N. G. *et al.* (1957). World-wide deposition of long-lived fission products from nuclear test explosions. *U.K. At. Energy Auth., Res. Group*, Rep. MP/R 2354.

Stigall, G. E., Fowler, T. W., and Krieger, H. L. (1971). ^{131}I discharges from an operating boiling water reactor nuclear power station. *Health Phys.* **20**, 593.

Stoller, S. M., and Richards, R. B. (1961). "Reactor Handbook," Vol. II. Wiley (Interscience), New York.

Stonier, T. (1964). "Nuclear Disaster." World Publ. Co., Cleveland, Ohio.

Storer, J. B. (1965). Radiation resistance with age in normal and irradiated populations of mice. *Radiat. Res.* **25**, 435–459.

Storer, J. B. (1969). Late effects: Extrapolation to low dose rate exposure. *Health Phys.* **17**, 3.

Strom, G. (1968). Atompheric dispersion of stack effluents. *In* "Air Pollution" (A. C. Stern, ed.), Vol. I. Academic Press, New York.

Suess, H. E. (1958). The radioactivity of the atmosphere and hydrosphere. *Annu. Rev. Nucl. Sci.* **8**, 243.

Sutton, O. G. (1953). "Micrometeorology." McGraw-Hill, New York.

Sverdrup, H. V., Johnson, M. W., and Fleming, R. H. (1942). "The Oceans." Prentice-Hall, Englewood Cliffs, New Jersey.

Swindell, G. E. (1971). The 1970 review of the IAEA regulations for the safe transport of radioactive materials. *Proc. Int. Symp. Packaging & Transport. Radioactive Mater., 3rd, 1971*. U.S. AEC CONF-710801, Vol. 1.

Tabor, W. H. (1963). Operating experience of the Oak Ridge research reactor through 1962. *Nucl. Safety* **5**, 116.

Tajima, E. (1956). Airborne radioactivity. *Science* **123**, 211.

Tamplin, A. R., and Fisher, H. L. (1966). "Estimation of Dosage to Thyroids of Children in the U.S. from Nuclear Tests Conducted in Nevada during 1952 through 1955," Rep. UCRL-14707, Lawrence Radiat. Lab.

Taylor, L. S. (1971). "Radiation Protection Standards." Chem. Rubber Publ. Co., Cleveland, Ohio.

Templeton, W. L., and Brown, V. M. (1964). The relationship between the concentration of calcium and strontium and Sr-90 in wild brown trout and the concentration of stable elements in some waters of the U.K. and implications in radiological health studies. *Int. J. Air Water Pollut.* **8**, 49–75.

Templeton, W. L., Nakatani, R. E., and Held, E. E. (1971). Radiation effects. *In* "Radioactivity in the Marine Environment." National Academy of Sciences, Washington, D.C.

Teresi, J. D., and Newcombe, C. L. (1961). Calculations of maximum permissible concentrations of radioactive fallout in water and air based upon military exposure criteria. *Health Phys.* **4,** 275.

Thompson, S. E. (1965). "Effective Half-Life of Fallout Radionuclides on Plants with Special Emphasis on Iodine-131," Rep. UCRL-12388. Lawrence Radiat. Lab.

Thompson, T. J., and Beckerley, J. G. (1964). "The Technology of Nuclear Reactor Safety," Vol. 1. MIT Press, Cambridge, Massachusetts.

Totter, J. R., Zelle, M. R., and Hollister, H. (1958). Hazard to man of carbon-14. *Science* **128,** 1490.

Triffet, T. (1959). Basic properties and effects of fallout. *In* "Biological and Environmental Effects of Nuclear War." Hearings before the Joint Committee on Atomic Energy.

Tsivoglou, E. C. (1959). Radioactive waste disposal to surface waters. *In* "Industrial Radioactive Waste Disposal." Hearings before the Joint Committee on Atomic Energy.

Tsivoglou, E. C. et al. (1960a). Estimating human radiation exposure on the Animas River. *J. Amer. Water Works Ass.* **52,** 1271.

Tsivoglou, E. C., Stein, M., and Towne, W. S. (1960b). Control of radioactive pollution of the Animas River. *J. Water Pollut. Contr. Fed.* **32,** 262.

Tsuzuki, M. (1955). The experience concerning radioactive damage of Japanese fishermen by Biniki fallout. *Muench. Med. Wochenschr.* **97,** 988.

Unger, W. E., Browder, F. N., and Mann, S. (1971). Nuclear safety in American radiochemical processing plants. *Nucl. Safety* **12,** 234.

United Kingdom Agricultural Research Council. (1961). "Strontium-90 in Milk and Agricultural Materials in the United Kingdom 1959–1960," Rep. No. 4.

United Kingdom Atomic Energy Office. (1957). "Accident at Windscale No. 1 Pile on October 10, 1957." HM Stationery Office, London.

United Kingdom Atomic Energy Office. (1958). "Final Report on the Windscale Accident," HM Stationery Office, London.

United Nations. (1956). "Proceedings of the First International Conference on the Peaceful Uses of Atomic Energy, 1955." United Nations, New York.

United Nations. (1958). "Proceedings of the United Nations Second International Conference on the Peaceful Uses of Atomic Energy, 1958." United Nations, New York.

United Nations. (1965). Proceedings of The Third International Conference on the Peaceful Uses of Atomic Energy, 1965." United Nations, New York.

United Nations. (1972). Proceedings of The Fourth International Conference on the Peaceful Uses of Atomic Energy, 1971." United Nations, New York (in press).

United Nations Scientific Committee on the Effects of Atomic Radiation. (1958). 13th Session, Suppl. No. 17 (A/3838). United Nations, New York.

United Nations Scientific Committee on the Effects of Atomic Radiation. (1962). 17th Session, Suppl. No. 16 (A/5216). United Nations, New York.

United Nations Scientific Committee on the Effects of Atomic Radiation. (1964). 19th Session, Suppl. No. 14 (A/5814). United Nations, New York.

United Nations Scientific Committee on the Effects of Atomic Radiation. (1966). 21st Session, Suppl. No. 14 (A/6314). United Nations, New York.

United Nations Scientific Committee on the Effects of Atomic Radiation. (1969). 24th Session, Suppl. No. 13 (A/7613). United Nations, New York.

United States Atomic Energy Commission. (1957a). "Theoretical Possibilities and

Consequences of Major Accidents in Large Nuclear Power Plants," Rep. WASH-740. USAEC, Washington, D.C.

United States Atomic Energy Commission. (1957b). "Atomic Energy Facts." USAEC, Washington, D.C.

United States Atomic Energy Commission. (1960a). "Hazards Summary Report on Consolidated Edison Thorium Reactor," Docket No. 50-3, Exhibit K-5 (Revision 1). USAEC, Washington, D.C.

United States Atomic Energy Commission. (1960b). "Summary of Available Data on the Strontium-90 Content of Foods and of Total Diets in the United States," Rep. HASL-90. USAEC. Washington, D.C.

United States Atomic Energy Commission. (1961). "Investigation Board Report on the SL-1 Accident." USAEC, Washington, D.C.

U.S. Atomic Energy Commission. (1967). Luminescence dosimetry. *Proc. Int. Conf. on Luminescence Dosimetry, 1967*. U.S. AEC CONF-650637.

U.S. Atomic Energy Commission. (1968a). "Radioisotopes: Production & Development of Large-Scale Uses," Rep. WASH No. 1095. USAEC, Washington, D.C.

U.S. Atomic Energy Commission. (1968b). *In* "Meteorology and Atomic Energy, 1968." (D. H. Slade, ed.). Div. Tech. Inform., USAEC, Washington, D.C.

U.S. Atomic Energy Commission. (1969a). "Code of Federal Regulations, Title 10, Part 20." USAEC, Washington, D.C.

U.S. Atomic Energy Commission. (1969b). "Report on the May 11, 1969 Fire at the Rocky Flats Plant near Boulder, Colorado," USAEC Press Release No. M-257. USAEC, Washington, D.C.

U.S. Atomic Energy Commission. (1970a). "The Nuclear Industry—1970." USAEC, Washington, D.C.

U.S. Atomic Energy Commission. (1970b). "Survey Report on Structural Design of Piping Systems and Components," TID-25553. USAEC, Washington, D.C.

U.S. Atomic Energy Commission. (1971a). "Annual Report to Congress." US Govt. Printing Office, Washington, D.C.

U.S. Atomic Energy Commission. (1971b). "Statistical Data of the Uranium Industry." Grand Junction Operations Office, USAEC, Washington, D.C.

U.S. Atomic Energy Commission. (1971c). *Proc. Int. Symp. Packaging & Transport. Radioactive Mater., 1971* U.S. AEC CONF-710801, Vol. 1. USAEC, Washington, D.C.

U.S. Atomic Energy Commission. (1971d). "Operational Accidents and Radiation Exposure Experience Within the U.S. Atomic Energy Commission 1943–1970," Rep. WASH 1192. USAEC, Washington, D.C.

U.S. Atomic Energy Commission. (1971e). "Environmental Statement for Liquid Metal Fast Breeder Reactor Demonstration Plant" (Draft). USAEC, Washington, D.C.

U.S. Atomic Energy Commission. (1971f). "Division of Isotopes Development Research and Development Projects," TID-4066. USAEC, Washington, D.C.

U.S. Bureau of Census. (1971). "Statistical Abstract of the United States: 1971," 92nd ed.) Washington, D.C.

U.S. Department of Agriculture. (1957). "Soils." U.S. Dept. Agr., Washington, D.C.

U.S. Department of Defense. (1961). "Fallout Protection." U.S. Dept. Defence, Washington, D.C.

U.S. Department of State. (1946). "International Control of Atomic Energy," Publ. No. 2702. U.S. Dept. of State, Washington, D.C.

U.S. Government. (1968). "Considerations Affecting Steam Power Plant Site Selection," Office of Science & Technology, Washington, D.C.

U.S. Public Health Service. (1967). Radioassay procedures for environmental samples. *U.S., Pub. Health Serv.*, Publ. **999-RH-27**.

U.S. Public Health Service. (1969a). "Evaluation of Radon 222 Near Uranium Tailings Piles," DER 69-1. U.S. Pub. Health Serv., Washington, D.C.

U.S. Public Health Service. (1969b). Population Exposure Studies Section, Bureau of Radiological Health. Population dose from X-rays, U.S. 1964. (Estimates of gonad and genetically significant dose from the Public Health Service X-Ray Exposure Study.) *U.S., Pub. Health Serv.*, Publ. **2001**.

U.S. Public Health Service. (1970). "Survey of the Use of Radionuclides in Medicine," Rep. BRH/DMRE 70-1. Bureau of Radiological Health, Washington, D.C.

U.S. Public Health Service. (1971). "State and Federal Control of Health Hazards from Radioactive Materials Other Than Materials Regulated Under the Atomic Energy Act of 1954," Rep. BRH/DMRE 71-4. U.S. Pub. Health Serv., Washington, D.C.

U.S. Weather Bureau. (1955). "Meteorology and Atomic Energy." US Govt. Printing Office, Washington, D.C.

University of California, Davis; Lawrence Radiation Laboratory; American Society for Engineering Education; American Nuclear Engineering with Nuclear Explosives. *Proc. Plowshare Symp., 3rd, 1964.*

Upton, A. C. (1966). Radiobiological aspects of the supersonic transport. *Health Phys.* **12, 209.**

Van Cleave, Charl. D. (1966). "Late Somatic Effects of Ionizing Radiation," Rep. TD-24310. USAEC, Washington, D.C.

Van der Hoven, I. (1968). Section 5-3. *In* "Meteorology and Atomic Energy" (D. H. Slade, ed.), p. 000. USAEC, Washington, D.C.

Van Middlesworth, L. (1954). Radioactivity in animal thyroids from various areas. *Nucleonics* **12, 56.**

Villforth, J. C. (1964). Problems in radium control. *Pub. Health Rep.* **79, 337.**

Villforth, J. C., Robinson, E. W., and Wold, G. J. (1969). A review of radium incidents in the U.S.A. *Proc. Handling Radiat. Accidents, 1969.* IAEA, Vienna.

Vinogradov, A. P. (1953). "The Elementary Chemical Composition of Marine Organisms." Sears Foundation for Marine Research New Haven, Connecticut.

Volchok, H. L. (1970). "Worldwide Deposition of ^{90}Sr Through 1969," Rep. HASL-227. USAEC, Washington, D.C.

Volchok, H. L., and Kleinman, M. T. (1971). "Worldwide Deposition of Sr-90 Through 1970," Rep. HASL-243. USAEC, Washington, D.C.

Volchok, H. L., Knuth, R., and Kleinman, M. T. (1972). "Plutonium in the Neighborhood of Rocky Flats, Colorado: Airborne Respirable Particles," Rep. HASL-246. USAEC, Washington, D.C.

von Hevesey, G. (1966). Radioactive tracers and their application. *Isotop. Radiat. Technol.* **4, 9.**

Wadleigh, C. H. (1957). Growth of plants. *In* "Soils." US Dep. Agr., Washington, D.C.

Wald, N., Thoma, G. E., Jr., and Broun, G., Jr. (1962). Hematologic manifestations of radiation exposure in man. *Progr. Hematol.* **3, 1.**

Walker, D. H. (1970). Fission product release during a LOCA. *Proc. Health Phys. Aspects Nucl. Facility Siting, 1970* p. 113. Health Phys. Soc., New York.

Walton, G. N. (1961). Fission and fission products. *In* "Atomic Energy Waste, Its

Nature, Use, and Disposal" (E. Glueckauf, ed.), Chapter 1. Wiley (Interscience), New York.

Way, K., and Wigner, E. P. (1948). The rate of decay of fission products. *Phys. Rev.* **73**, 1318.

Webb, J. H. (1949). The fogging of photographic film by radioactive contaminants in cardboard packaging materials. *Phys. Rev.* **76**, 375.

Weems, S. J., Lyman, W. G., and Haga, P. B. (1970). The ice-condenser reactor containment system. *Nucl. Safety* **11**, 215.

Wegst, A. V., Pelletier, C. A., and Whipple, G. H. (1964). Detection and quantitation of fallout particles in a human lung. *Science* **143**, 957–959.

Weinberg, A. M., and Wigner, E. P. (1958). "The Physical Theory of Neutron Chain Reactors." Univ. of Chicago Press, Chicago, Illinois.

Weiss, E. S., Rallison, M. L., London, W. T., and Thompson, G. D. C. (1971). Thyroid nodularity in southwestern Utah school children exposed to fallout radiation. *Amer. J. Pub. Health, Nat. Health* **61**, 241–249.

Welander, A. (1969). Distribution of Radionuclides in the Environment of Eniwetok and Bikini Atolls. *Radioecol., Proc. Nat. Symp., 2nd, 1969.* U.S. AEC CONF-670503.

Welford, G. A. (1960). Urinary uranium levels in non-exposed individuals. *J. Ind. Hyg. Ass.* **21**, 68.

Welford, G. A., and Baird, R. (1967). Uranium levels in human diet and biological materials. *Health Phys.* **13**, 1321–1324.

Welford, G. A., and Collins, W. R. (1960). Fallout in New York City during 1958. *Science* **131**, 1711.

Welford, G. A., and Sutton, D. (1957). "Determination of the Uranium Content of the National Bureau of Standards Iron and Steel Chemical Standards," Rep. NYOO-4755. USAEC, Washington, D.C.

Werth, G. C. (1971). The Soviet program on nuclear explosives for the national economy. *Nucl. Technol.* **11**, 280–302.

Westinghouse Electric Corporation. (1971). "Systems Summary of a Westinghouse Pressurized Water Reactor Nuclear Power Plant." PWR Syst. Div., Pittsburgh, Pennsylvania.

Wilkening, M. H. (1952). Natural radioactivity as a tracer in the sorting of aerosols according to mobility. *Rev. Sci. Instrum.* **23**, 13.

Wilkening, M. H. (1964). Radon-daughter ions in the atmosphere. *In* "The Natural Radiation Environment" (J. A. S. Adams and W. M. Lowder, eds.), p. 359. Univ. of Chicago Press, Chicago, Illinois.

Williamson, M. M. (1969). Nuclear cratering applications. *Proc. Symp. Pub. Health Aspects Peaceful Uses Nucl. Explosives, 1969.* Pub. Health Serv. Rep. SWRHL-82.

Wilson, D. W., Ward, G. M., and Johnson, J. E. (1969). A quantitative model of the transport of Cs-137 from fallout to milk. *Proc. Environ. Contam. by Radioactive Mater., 1969.* IAEA, Vienna.

Wittels, M. C. (1966–1967). Stored energy in graphite and other reactor materials. *Nucl. Safety* **8**, 134.

Wollan, R. O., Staiger, J. W., and Boge, R. J. (1971). Disposal of low level radioactive waste at a large university incinerator. *Amer. Ind. Hyg. Ass., J.* **32**, 625.

Wong, K. M., and Noshkin, L. (1970). "^{239}Pu in Some Marine Organisms and Sediments," Rep. HASL-227. USAEC, Washington, D.C.

Wood, J. W., Tamagaki, H., Neriishi, S., Sato, T., Sheldon, W. F., Archer, P. G.,

Hamilton, H. B., and Johnson, K. G. (1969). Thyroid carcinoma in atomic bomb survivors—Hiroshima and Nagasaki. *Amer. J. Epidemiol.* **89,** 4.

Wooster, W. S., and Ketchum, B. H. (1957). "Transport and Dispersal of Radioactive Elements in the Sea," Rep. No. 551. Nat. Acad. of Sci., Washington, D.C.

World Health Organization. (1957). "Effects of Ionizing Radiation on Human Heredity." World Health Organ., Geneva.

World Health Organization. (1970). "Wholesomeness of Irradiated Food with Special Reference to Wheat, Potatoes, and Onions," Report of a Joint FAO/IAEA/WHO Expert Committee., World Health Organ., Geneva.

Wrenn, M. E. (1968). The dosimetry of ^{55}Fe. *Proc. Congr. Radiat. Protect., 1st, 1968* p. 843. Pergamon, Oxford.

Wrenn, M. E., and Cohen, N. (1967). Iron-55 from nuclear fallout in the blood of adults: Dosimetric implications and development of a model to predict levels in blood. *Health Phys.* **13,** 1075–1082.

Wrenn, M. E., Mowafy, R., and Laurer, G. R. (1964). ^{95}Zr-^{95}Nb in human lungs from fallout. *Health Phys.* **10,** 1051–1058.

Wrenn, M. E., Lentsch, J. W., Eisenbud, M., Lauer, J., and Howells, G. P. (1972a). Radiocesium distribution in water, sediment, and biota in the Hudson River Estuary from 1964 through 1970. *Radioecol., Proc. Nat. Symp., 3rd, 1971* (in press).

Wrenn, M. E., Rosen, J., and Cohen, N. (1972b). The *in vivo* measurement of Am-241 in man. *Proc. Symp. Assessment of Radioactive Organ and Body Burdens, 1971.* IAEA, Vienna (in press).

Wyerman, T. A., Farnsworth, R. K., and Stewart, G. L. (1970). Tritium in streams in the United States, 1961–1968. *Radiol. Health Data Rep.* **11,** 421–439.

Yamagata, N., and Yamagata, T. (1960). The concentration of cesium-137 in human tissues and organs. *Nat. Inst. Pub. Health* (Japan). **9,** 72.

Young, J. A., and Fairhall, A. W. (1968). Radiocarbon from nuclear weapons tests. *J. Geophys. Res.* **73,** 1185–1200.

Zapp, F. C. (1969). Testing of containment systems used with light-water-cooled power reactors. *Nucl. Safety* **10,** 308.

Zelle, M. R. (1960). Radioisotopes and the genetic mechanism: Mutagenic aspects. *In* "Radioisotopes in the Biosphere," Chapter 13. Univ. of Minnesota Press, Minneapolis.

Additional References

Attix, F. H., and Roesch, W. C., eds. (1968). "Radiation Dosimetry," Vol. 1, "Fundamentals." Academic Press, New York.

Beck, H. L., DeCampo, J. A., Gogolak, C. V., Lowder, W. M., McLaughlin, J. E., and Raft, P. D. (1972). New Perspectives on Low Level Environmental Radiation Monitoring Around Nuclear Facilities. *Nucl. Technol.* **14,** 232.

Bensen, D. W., and Sparrow, A. H., eds. (1971). Survival of Food Crops and Livestock in the Event of Nuclear War. *Proc. Symp. at Brookhaven Nat'l Lab. 1970.* USAEC CONF-700909.

Cummings, L. G. (1971). Third Party Liability Insurance and Government Indemnity Associated with the Transportation of Radioactive Materials. *Proc. 3rd Intl. Symp on Packaging and Transportation of Radioactive Materials.* USAEC CONF-710801 **2,** 516–531.

Dalla Valle, J. M. (1948). "Micromeritics, The Technology of Fine Particles," 2nd ed. Pitman, London.

Eisenbud, M. (1963). "Environmental Radioactivity." McGraw-Hill, New York.

Eisenbud, M., Laurer, G. R., Rosen, J. C., Cohen, N., and Thomas, J. (1969). *In Vivo* Measurement of Lead-210 as an Indicator of Cumulative Radon Daughter Exposure in Uranium Miners. *Health Phys.* **16,** 637.

Ergen, W. K. (1963). Site Criterion for Reactors with Multiple Containment. *Nucl. Safety* **4,** 8.

Feely, H. W., Biscaye, P. E., and Lagamarsino, R. J. (1965). Atmospheric Circulation Rates from Measurements of Nuclear Debris. Paper presented at the *CACR Symp. on Atmospheric Chemistry, Circulation and Aerosols, Visby, Sweden.* (Cited in UNSCEAR, 1966.)

Goldman, M., and Bustad, L. K., eds. (1972). "Biomedical Implications of Radiostrontium Exposure." USAEC Symp. Series 25.

Healy, J. W., and Baker, R. E. (1968). Radioactive Cloud-dose Calculations. *In* "Meteorology and Atomic Energy" (David H. Slade, ed.). USAEC Rept. No. TID-24190.

International Commission on Radiological Protection. (1968). Evaluation of Radiation Doses to Body Tissues from Internal Contamination Due to Occupational Exposure. ICRP Publ. No. 10. Pergamon, Oxford.

Marinelli, L. D. (1969). A Revised Estimate of Maximum Permissible Burden for [90]Sr. *In* "Delayed Effects of Bone-Seeking Radionuclides" (C. W. Mays, W.S.S. Jee, R. D. Lloyd, Betsy J. Stover, Jean H. Dougherty, and G. N. Taylor, eds.), p. 409. University of Utah Press, Salt Lake City, Utah.

Romney, E. M., and Davis, J. J. (1972). Ecological Aspects of Plutonium Dissemination in Terrestrial Environments. *Health Phys.* **22,** 551.

Smales, A., and Salmon, L. (1955). Determination by Radioactivation of Small Amounts of Rubidium and Cesium in Sea-Water & Related Materials of Geochemical Interest. *The Analyst* **80,** 37–50.

Wlodek, S. (1967). The Behavior of Cs-137 in Freshwater Reservoirs with Particular Reference to the Reactions of Plant Communities to Contamination. "Proc. Radioecological Concentration Processes" (B. Aberg and F. P. Hungate, eds.), p. 843. Pergamon, Oxford.

Subject Index

A

Accidents
deaths due to, in AEC program, 460–61
during transportation of radioactive materials, 288
environmental contamination from reactors, *see* Nuclear reactors
fallout in March, 1954, 401
Fermi reactor, meltdown, 428
Houston incident, 425
plutonium release at Oak Ridge, 417
record of AEC in regard to, 480
Rocky Flats, fire at, 430
SL-1 reactor explosion, 421
SNAP 9A, abortive reentry of, 428
Windscale reactor, 411
Actinium decay series, 162
Acute radiation effects, *see* Radiation, biological effects of
Advisory Committee on Reactor Safeguards, 265
Aerosols, *see* Particulates
Air-cooled reactors, 230, 244
^{41}Ar from, 244
Aquatic ecosystems, *see also* Oceans; Estuaries; Rivers; Aquatic organisms
behavior of suspended solids in, 139
concentration factors in, 152

diffusion of contaminants, 138
mixing characteristics, 141
role of sediments, 140
Aquatic organisms
effect of temperature on radiation effects, 138
radiation injury to, 138
Anthropecology, 10
Argon-41, from air-cooled reactors, 244
Argonaut, 231
Army Stationary Low Power Reactor, *see* SL-1
Artificial heart, isotopic power for, 285
Atmosphere, *see also* individual nuclides; Fallout; Diffusion
chemical and physical properties of, 88
coefficient of diffusivity, 89
concentrations and properties of natural radioactivity in, 161, 179, 180–184, 185, 187, 188
friction layer, 88
lapse rate, 90
residence time of dust, 117, 182
structure of, 90–91
temperature inversion, 92
transport mechanisms in, 90–112
turbulent diffusion, 92
vertical temperature gradient, 90
Atmospheric diffusion, *see* Diffusion

N